The G. H. Hardy Reader

CAMBRIDGE
UNIVERSITY PRESS

32 Avenue of the Americas, New York NY 10013-2473, USA

Cambridge University Press is part of the University of Cambridge.

It furthers the University's mission by disseminating knowledge in the pursuit of education, learning and research at the highest international levels of excellence.

www.cambridge.org
Information on this title: www.cambridge.org/9781107135550

About the Cover. The photo collage for the cover was designed by Do Good Design. The photo of Hardy as a schoolboy is reproduced with the permission of the Warden and Scholars of Winchester College, and that of Hardy as a Fellow of Trinity College with the permission of the Master and Fellows of Trinity College, Cambridge.

© 2015 by
The Mathematical Association of America (Incorporated)

This publication is in copyright. Subject to statutory exception and to the provisions of relevant collective licensing agreements, no reproduction of any part may take place without the written permission of Cambridge University Press.

First published 2016

A catalogue record for this publication is available from the British Library

ISBN 978-1-107-13555-0 Hardback

Cambridge University Press has no responsibility for the persistence or accuracy of URLs for external or third-party internet websites referred to in this publication, and does not guarantee that any content on such websites is, or will remain, accurate or appropriate.

The G. H. Hardy Reader

edited by

Donald J. Albers

Gerald L. Alexanderson

William Dunham

Council on Publications and Communications
Jennifer J. Quinn, Chair

Committee on Books
Fernando Gouvêa, Chair

Spectrum Editorial Board
Gerald L. Alexanderson, Co-Editor
James J. Tattersall, Co-Editor

Virginia M. Buchanan
Thomas L. Drucker
Richard K. Guy
Dominic Klyve
Shawnee L. McMurran
Daniel E. Otero
Jean J. Pedersen
Marvin Schaefer

SPECTRUM SERIES

The Spectrum Series of the Mathematical Association of America was so named to reflect its purpose: to publish a broad range of books including biographies, accessible expositions of old or new mathematical ideas, reprints and revisions of excellent out-of-print books, popular works, and other monographs of high interest that will appeal to a broad range of readers, including students and teachers of mathematics, mathematical amateurs, and researchers.

777 Mathematical Conversation Starters, by John de Pillis

99 Points of Intersection: Examples—Pictures—Proofs, by Hans Walser. Translated from the original German by Peter Hilton and Jean Pedersen

Aha Gotcha and Aha Insight, by Martin Gardner

All the Math That's Fit to Print, by Keith Devlin

Beautiful Mathematics, by Martin Erickson

Calculus and Its Origins, by David Perkins

Calculus Gems: Brief Lives and Memorable Mathematics, by George F. Simmons

Carl Friedrich Gauss: Titan of Science, by G. Waldo Dunnington, with additional material by Jeremy Gray and Fritz-Egbert Dohse

The Changing Space of Geometry, edited by Chris Pritchard

Circles: A Mathematical View, by Dan Pedoe

Complex Numbers and Geometry, by Liang-shin Hahn

Cryptology, by Albrecht Beutelspacher

The Early Mathematics of Leonhard Euler, by C. Edward Sandifer

The Edge of the Universe: Celebrating 10 Years of Math Horizons, edited by Deanna Haunsperger and Stephen Kennedy

Euler and Modern Science, edited by N. N. Bogolyubov, G. K. Mikhailov, and A. P. Yushkevich. Translated from Russian by Robert Burns

Euler at 300: An Appreciation, edited by Robert E. Bradley, Lawrence A. D'Antonio, and C. Edward Sandifer

Expeditions in Mathematics, edited by Tatiana Shubin, David F. Hayes, and Gerald L. Alexanderson

Five Hundred Mathematical Challenges, by Edward J. Barbeau, Murray S. Klamkin, and William O. J. Moser

The Genius of Euler: Reflections on his Life and Work, edited by William Dunham

The G. H. Hardy Reader, edited by Donald J. Albers, Gerald L. Alexanderson, and William Dunham

The Golden Section, by Hans Walser. Translated from the original German by Peter Hilton, with the assistance of Jean Pedersen.

The Harmony of the World: 75 Years of Mathematics Magazine, edited by Gerald L. Alexanderson with the assistance of Peter Ross

A Historian Looks Back: The Calculus as Algebra and Selected Writings, by Judith Grabiner

History of Mathematics: Highways and Byways, by Amy Dahan-Dalmédico and Jeanne Peiffer, translated by Sanford Segal

How Euler Did Even More, by C. Edward Sandifer

How Euler Did It, by C. Edward Sandifer

Illustrated Special Relativity Through Its Paradoxes: A Fusion of Linear Algebra, Graphics, and Reality, by John de Pillis and José Wudka

In the Dark on the Sunny Side: A Memoir of an Out-of-Sight Mathematician, by Larry Baggett

Is Mathematics Inevitable? A Miscellany, edited by Underwood Dudley

I Want to Be a Mathematician, by Paul R. Halmos

Journey into Geometries, by Marta Sved

JULIA: a life in mathematics, by Constance Reid

The Lighter Side of Mathematics: Proceedings of the Eugène Strens Memorial Conference on Recreational Mathematics & Its History, edited by Richard K. Guy and Robert E. Woodrow

Lure of the Integers, by Joe Roberts

Magic Numbers of the Professor, by Owen O'Shea and Underwood Dudley

Magic Tricks, Card Shuffling, and Dynamic Computer Memories: The Mathematics of the Perfect Shuffle, by S. Brent Morris

Martin Gardner's Mathematical Games: The entire collection of his Scientific American columns

The Math Chat Book, by Frank Morgan

Mathematical Adventures for Students and Amateurs, edited by David Hayes and Tatiana Shubin. With the assistance of Gerald L. Alexanderson and Peter Ross

Mathematical Apocrypha, by Steven G. Krantz

Mathematical Apocrypha Redux, by Steven G. Krantz

Mathematical Carnival, by Martin Gardner

Mathematical Circles Vol I: In Mathematical Circles Quadrants I, II, III, IV, by Howard W. Eves

Mathematical Circles Vol II: Mathematical Circles Revisited and Mathematical Circles Squared, by Howard W. Eves

Mathematical Circles Vol III: Mathematical Circles Adieu and Return to Mathematical Circles, by Howard W. Eves

Mathematical Circus, by Martin Gardner

Mathematical Cranks, by Underwood Dudley

Mathematical Evolutions, edited by Abe Shenitzer and John Stillwell

Mathematical Fallacies, Flaws, and Flimflam, by Edward J. Barbeau

Mathematical Magic Show, by Martin Gardner

Mathematical Reminiscences, by Howard Eves

Mathematical Treks: From Surreal Numbers to Magic Circles, by Ivars Peterson

A Mathematician Comes of Age, by Steven G. Krantz

Mathematics: Queen and Servant of Science, by E.T. Bell

Mathematics in Historical Context, by Jeff Suzuki

Memorabilia Mathematica, by Robert Edouard Moritz

More Fallacies, Flaws, and Flimflam, Edward J. Barbeau

Musings of the Masters: An Anthology of Mathematical Reflections, edited by Raymond G. Ayoub

New Mathematical Diversions, by Martin Gardner

Non-Euclidean Geometry, by H. S. M. Coxeter

Numerical Methods That Work, by Forman Acton

Numerology or What Pythagoras Wrought, by Underwood Dudley

Out of the Mouths of Mathematicians, by Rosemary Schmalz

Penrose Tiles to Trapdoor Ciphers . . . and the Return of Dr. Matrix, by Martin Gardner

Polyominoes, by George Martin

Power Play, by Edward J. Barbeau

Proof and Other Dilemmas: Mathematics and Philosophy, edited by Bonnie Gold and Roger Simons

The Random Walks of George Pólya, by Gerald L. Alexanderson

Remarkable Mathematicians, from Euler to von Neumann, by Ioan James

The Search for E.T. Bell, also known as John Taine, by Constance Reid

Shaping Space, edited by Marjorie Senechal and George Fleck

Sherlock Holmes in Babylon and Other Tales of Mathematical History, edited by Marlow Anderson, Victor Katz, and Robin Wilson

Six Sources of Collapse: A Mathematician's Perspective on How Things Can Fall Apart in the Blink of an Eye, by Charles R. Hadlock

Sophie's Diary, Second Edition, by Dora Musielak

Student Research Projects in Calculus, by Marcus Cohen, Arthur Knoebel, Edward D. Gaughan, Douglas S. Kurtz, and David Pengelley

Symmetry, by Hans Walser. Translated from the original German by Peter Hilton, with the assistance of Jean Pedersen.

The Trisectors, by Underwood Dudley

Twenty Years Before the Blackboard, by Michael Stueben with Diane Sandford

Who Gave You the Epsilon? and Other Tales of Mathematical History, edited by Marlow Anderson, Victor Katz, and Robin Wilson

The Words of Mathematics, by Steven Schwartzman

MAA Service Center
P.O. Box 91112
Washington, DC 20090-1112
800-331-1622   FAX: 301-206-9789

Contents

Overview   xiii

I   Biography

1. Hardy's Life   3

2. The Letter from Ramanujan to Hardy, 16 January 1913
   (*Collected Papers of Srinivasa Ramanujan*, Cambridge
   University Press, 1927, p. xxiii)   39

3. A Letter from Bertrand Russell to Lady Ottoline
   Morrell, 2 February 1913
   (*Ramanujan: Letters and Commentary*, Berndt and Rankin,
   Amer. Math. Society and London Math. Soc., 1995, pp. 44–45)   43

4. The Indian Mathematician Ramanujan
   (*Amer. Math. Monthly* 44 (1937), 137–155; from
   Ramanujan/Twelve Lectures, Cambridge University Press,
   1940, pp. 1–21)   47

5. "Epilogue" from *The Man Who Knew Infinity*,
   by Robert Kanigel
   (Scribner's, New York, 1991, pp. 361–373)   73

6. Posters of "Hardy's Years at Oxford," by R. J. Wilson   87

7. A Glimpse of J. E. Littlewood   101

8. A Letter from Freeman Dyson to C. P. Snow, 22 May
   1967, and Two Letters from Hardy to Dyson
   (*College Math. J.*, 25:1 (1994), 2–21)   109

9. Miss Gertrude Hardy   115

II Writings by and about G. H. Hardy

10 Hardy on Writing Books — 123

11 Selections from Hardy's Writings — 125

12 Selections from What Others Have Said about Hardy — 137

III Mathematics

13 An Introduction to the Theory of Numbers
(*Bull. Amer. Math. Soc.* 35:6 (1929), pp. 773–818) — 165

14 Prime Numbers
(*British Association Report* 10 (1915), pp. 350–354) — 199

15 The Theory of Numbers
(*British Association Report* 90 (1922), pp. 16–24) — 207

16 The Riemann Zeta-Function and Lattice Point Problems, by E. C. Titchmarsh
(*J. London Math. Soc.* 25 (1950), pp. 125–128) — 219

17 Four Hardy Gems — 225
  a. A Function — 226
  b. An Integral — 228
  c. An Inequality — 230
  d. An Application — 231

18 What Is Geometry?
(*Math. Gazette* 12 (1925), pp. 309–316) — 235

19 The Case against the Mathematical Tripos
(*Math. Gazette* 13 (1926), pp. 61–71) — 249

20 The Mathematician on Cricket, by C. P. Snow
(*Saturday Book, No. 8*, Hutchinson, London, 1948) — 267

21 Cricket for the Rest of Us, by John Stillwell — 277

22 A Mathematical Theorem about Golf
(*Math. Gazette* 29 (1945), pp. 226–227) — 285

23 Mathematics in War-Time
(*Eureka* 1–3 (1940), pp. 5–8) — 287

Contents   xi

24	**Mathematics** (*The Oxford Magazine* 48 (1930), pp. 819–821)	**291**
25	**Asymptotic Formulæ in Combinatory Analysis (excerpts) with S. Ramanujan** (*Proc. London Math. Soc.* (2) 17 (1918), pp. 75–115)	**295**
26	**A New Solution of Waring's Problem (excerpts), with J. E. Littlewood** (*Quart. J. Math.* 48 (1920), pp. 272–293)	**301**
27	**Some Notes on Certain Theorems in Higher Trigonometry** (*Math. Gazette* 3 (1906), pp. 284–288)	**305**
28	**The Integral $\int_0^\infty \frac{\sin x}{x} dx$ and Further Remarks on the Integral $\int_0^\infty \frac{\sin x}{x} dx$** (*Math. Gazette* 5 (1909), pp. 98–103; 8 (1916), pp. 301–303)	**311**

IV   Tributes

29	**Dr. Glaisher and the "Messenger of Mathematics"** (*Messenger of Mathematics* 58 (1929), pp. 159–160)	**325**
30	**David Hilbert** (*J. London Math. Soc.* 18 (1943), pp. 191–192)	**329**
31	**Edmund Landau (with H. Heilbronn)** (*J. London Math. Soc.* 13 (1938), pp. 302–310)	**333**
32	**Gösta Mittag-Leffler** (*J. London Math. Soc.* 3 (1928), pp. 156–160)	**343**

V   Book Reviews

33	**Osgood's *Calculus* and Johnson's *Calculus*** (*Math. Gazette* 4 (1907), pp. 307–309)	**351**
34	**Hadamard: *The Psychology of Invention in the Mathematical Field*** (*Math. Gazette* 30 (1946), pp. 111–115)	**355**
35	**Hulburt: *Differential and Integral Calculus*** (*Math. Gazette* 7 (1914), p. 337)	**363**

36	**Bôcher:** *An Introduction to the Study of Integral Equations* (*Math. Gazette* 5 (1910), pp. 208–209)	365
37	**Davison:** *Higher Algebra* (*Math. Gazette* 7 (1913), pp. 21–24)	367
38	**Zoretti:** *Leçons de Mathématiques Générales* (*Math. Gazette* 7 (1914), p. 338)	369

A Last Word — 371

Sources — 373
Acknowledgments — 381
Index — 385
About the Editors — 395

Overview

G. H. Hardy (1877–1947) ranks among the great mathematicians of the twentieth century. He was, as well, a colorful, mildly eccentric individual with expository skills of the first order.

Hardy was a product of the English educational system as it existed in the decades before 1900. He attended the prestigious Winchester College and then "went up" to Trinity College, Cambridge, where he distinguished himself as Fourth Wrangler and as winner of the Smith's Prize—honors that meant a lot in turn-of-the-century Britain, even if they leave modern readers somewhat puzzled. From there, his career took him to professorships at Cambridge and Oxford and to visiting positions at Princeton and Cal Tech.

Mathematically, Hardy is regarded as one of history's most accomplished analytic number theorists, although his papers ranged across other topics like divergent series, integration, and the theory of partitions. Throughout his career, he knew a Who's-Who of scholars, from Norbert Wiener (his student), to George Pólya (his co-author), to Bertrand Russell (his Cambridge colleague), to John Maynard Keynes (his friend). Of course, he is best remembered for his extraordinarily fruitful collaborations with J. E. Littlewood and Srinivasa Ramanujan. Clearly, G. H. Hardy kept good company.

His mathematical papers, in seven volumes, were published by Oxford University Press in the 1960s and 1970s. But there seemed to be the need for a book that gave a sense of Hardy, the man. For years, Don Albers of the Mathematical Association of America campaigned for just such a work. Everyone whom he approached agreed that it would be a worthy undertaking, but no one volunteered to do it.

Consequently, Don modified his vision from a full-blooded biography to a "reader" that would allow Hardy to speak for himself. Hardy was, after all, a gifted writer with a distinctive voice that could be inspirational, funny, or caustic—sometimes all at once. Accompanying his words would be material written by present-day authors to provide the necessary introductions and transitions. When Albers pitched his idea to the two of us—Alexanderson and Dunham—we thought we'd give it a try. This book is the result.

Our volume is divided into five main parts. We begin with our own Hardy biography, where we quote liberally from the man himself, especially from his classic, *A Mathematician's Apology*. In addition, we were fortunate to get the rights to reprint the Epilogue from Robert Kanigel's biography of Ramanujan, *The Man Who Knew Infinity* (Scribners, 1991).

Next comes a section called, 'Writings by and about G. H. Hardy." As the title suggests, we have selected polemics, quips, anecdotes, and other passages from Hardy and from those who knew him. In their subject matter, these roam far and wide, but we think they provide a better look at Hardy and his times.

Our third section focuses on Hardy's mathematics. Here we have selected pieces that he wrote for general mathematical audiences rather than for narrow specialists. For instance, we included his survey articles on the theory of numbers, on geometry, and even a tongue-in-cheek item titled, "A Mathematical Theorem on Golf." On a more technical level, we prepared our own accounts of four of Hardy's mathematical gems, ranging from his proof of a peculiar inequality to his introduction of what is now called the Hardy-Weinberg Law of genetics.

The fourth section contains some of Hardy's tributes to other mathematicians. These pieces—written as memorials to those he knew—have a personal dimension not found in the standard obituary.

Hardy was a regular reviewer of mathematics books, and our last section contains some of his reviews. When Hardy was enthusiastic about a book, his comments could soar; when he was not, he could be strikingly harsh. Either way, his reviews make for great reading.

Three words of warning. First, writings collected from so many different sources inevitably introduce repetition. Our authors did not coordinate with one another, so certain stories, phrases, and quotations—especially the famous ones—show up here more than once. We believed, however, that it was important to keep the original writings intact, so we shall risk the repetition.

Second, we are aware that certain statements in the book do not reflect current mathematical knowledge. Again, we chose to retain the original wording rather than try to update all such items (e.g., former conjectures that are now theorems).

Finally, Hardy regularly used what we today regard as sexist language. Many passages, taken literally, suggest that only males studied mathematics or that only males were in the professoriate. Of course, Hardy knew

Overview

better; after all, he mentored both Mary Cartwright and Olga Taussky. But such was the custom of the time.

As we assembled this book, the three of us were continually reminded that G. H. Hardy was a brilliant and unfailingly interesting character. This should be evident in the pages that follow.

Donald J. Albers
Gerald L. Alexanderson
William Dunham

I
Biography

1
Hardy's Life

Can you name a twentieth-century mathematician who has appeared as a character in a popular novel, an acclaimed play, a feature film, and a murder mystery? Better yet, can you name one who was described by a famous author as having written "... in his own clear and unadorned fashion, some of the most perfect English of his time?" [97, p. 151].

G. H. (Godfrey Harold) Hardy was that mathematician. He left deep footprints, both as a scholar and as a colorful, eccentric example of the English intellectual from a century ago.

(By the way, the novel is David Leavitt's *The Indian Clerk*, the play is Ira Hauptman's *Partition*, the film is *The Man Who Knew Infinity*, the mystery is Randall Collins's *The Case of the Philosophers' Ring*, and the famous author is C. P. Snow, whose admiration for Hardy's literary gifts knew no bounds.)

For all that has been written about Hardy, his own memoir, *A Mathematician's Apology*, in a sense tells it all. It is an eloquent essay reflecting upon his life as a creative mathematician. It is, as well, a poignant document, written at an age when waning powers deprived him of the ability to do first-class work in his beloved mathematics. The essay is preceded by a moving remembrance from the aforementioned novelist and physicist, C. P. Snow, who knew Hardy well when they were at Cambridge together. Their friendship continued up to the time of Hardy's death in 1947.

Although *A Mathematician's Apology* has its moments of melancholy—a word Hardy used to describe it on the opening page—it stands as an unrivaled manifesto for mathematics as a creative art. In it Hardy treated "pure" mathematics, which was, for him, the only kind that mattered. He of course recognized the practical side of the subject as it applied to real-world activities, but even here his praise was faint. He conceded that basic arithmetic has its uses; after all, "... the currency system of Europe conforms to it

3

approximately" [89, p. 382]. But his true feelings were clear. "It is undeniable," he wrote,

> ...that a good deal of elementary mathematics...has considerable practical utility. These parts of mathematics are, on the whole, rather dull; they are just the parts which have least aesthetic value. The "real" mathematics of the "real" mathematicians, the mathematics of Fermat and Euler and Gauss and Abel and Riemann, is almost wholly "useless"... [54, p. 119].

For someone of Hardy's sensibilities, a comparison between mathematics and the arts seemed obvious. His views were perhaps best expressed in this famous passage from the *Apology*:

> A mathematician, like a painter or a poet, is a maker of patterns. If his patterns are more permanent than theirs, it is because his are made with *ideas* [54, p. 84].

And he went on to say,

> The mathematician's patterns, like the painter's or the poet's, must be *beautiful*; the ideas, like the colours or the words, must fit together in a harmonious way. Beauty is the first test: there is no permanent place in the world for ugly mathematics [54, p. 85].

Hardy saw mathematical ideas as timeless, with an eternal appeal. He made a case for this view when he wrote:

> The Babylonian and Assyrian civilizations have perished; Hammurabi, Sargon, and Nebuchadnezzar are empty names; yet Babylonian mathematics is still interesting, and the Babylonian scale of 60 is still used in astronomy.

These ancient civilizations were followed by the Greeks, whose

> ...mathematics is the real thing. The Greeks first spoke a language which modern mathematicians can understand; as Littlewood said to me once, they are not clever schoolboys or 'scholarship candidates', but 'Fellows of another college'... Archimedes will be remembered when Aeschylus is forgotten, because languages die and mathematical ideas do not [54, pp. 80–81].

Can there be any remaining doubt about the provocative and engaging nature of Hardy's *Apology*? Atle Selberg, a Norwegian number theorist and early winner of the Fields Medal, called it "a great piece of literature"

[3, p. 266]. And the novelist Graham Greene, who reviewed the work, regarded it as the best account he had ever read of what it means to be a creative artist [27, p. 682].

If you are unfamiliar with *A Mathematician's Apology*, we suggest you put down this book, rush off to find a copy, and read it. You will not be disappointed.

For whatever reason, Britain has a tradition of producing scientists who are excellent writers. Charles Darwin, certainly the greatest British scientist this side of Newton, filled his *Origin of Species* with words of true beauty. Darwin's disciple, the comparative anatomist Thomas Huxley, was more than adept at turning a phrase. And Hardy's Cambridge colleague Bertrand Russell—mathematician, philosopher, and member of the Royal Society—won the Nobel Prize for *Literature* in 1950. Hardy, as a mathematician with a gift for writing, was in good company.

A Mathematician's Apology is a masterpiece. But there is more to be said about its author. Where he has described his work and given his views on important issues, we shall provide Hardy untouched and in his own words. That will constitute the bulk of this volume. In this introduction, however, we should sketch some of the milestones of his life and provide some recollections from those who knew him well.

Childhood and School

Who was G. H. Hardy? He was born in 1877 to a family of the professional class in Cranleigh, Surrey, England. His father held various positions in the preparatory section of Cranleigh School, and his mother was on the staff of another school nearby. The boy was brought up in a religious, Victorian household. His early fascination with mathematics was evident when he tried to factor the hymn numbers that appeared on the board in church. He had a sister Gertrude Edith, called "Gertie," who never married and remained close to her brother throughout his life. Indeed, Gertie may have been the only person who dared to call him "Harold."

Their mutual affection came in spite of a horrible childhood accident. We quote from Robert Kanigel:

> Gertrude lost her eye as a child, when Harold, playing carelessly with a cricket bat, struck her; she had to wear a glass eye for the rest of her life. The incident, however, did nothing to disrupt their sibling closeness, and may even have enhanced it. They were devoted to one another all their lives, kept in close touch almost as twins are said to, and for many years shared an apartment in London [97, p. 117].

Schoolboy Hardy before a soccer match at Winchester College, 1895.

At an appropriate age Hardy was enrolled in Cranleigh School, but eventually he was sent to boarding school at the famous and ancient Winchester College. This institution, with its harsh and forbidding reputation, was not to Hardy's tastes. Kanigel wrote that Hardy "... lived in a sort of intellectual ghetto within the college, a fortresslike complex of medieval, gray stone buildings, worlds apart from the sunny openness of Cranleigh" [97, p. 121].

Winchester had received its charter in 1382 when it was established by the local Bishop of Winchester, William of Wykeham. The good Bishop had been educated at New College, Oxford, so Winchester's graduates were expected to follow in his footsteps. Bucking tradition (and not for the last time), Hardy instead chose Trinity College, Cambridge, for it was stronger in mathematics than anything Oxford could boast. Trinity had been Newton's college, after all!

Upon leaving Winchester, Hardy never returned. He did, however, carry with him a lifelong distaste for mutton, a residue of Winchester's bizarre statute forcing the students to be served this for dinner five days a week [121, p. 452].

To this point, Hardy's mathematical education had not been an entirely satisfactory one. At Cranleigh and Winchester, his talents were such that he never took regular classes but instead was coached privately. Cambridge, of course, was famous beyond measure, but it had a drawback of its own: an

1. Hardy's Life

Scholar Hardy at Winchester College, 1895.

excessive emphasis on something called the "Tripos examination." We quote Kanigel's chilling description:

> The mathematical Tripos was impossibly arduous. You sat for four days of problems, often late into the evening, took a week's break, then came back for four more days... Here, sometimes on problem after problem, the brightest students, destined for distinguished mathematical careers, would not even know where to begin. It was a frightful ordeal... The Tripos, wrote one English-born mathematician years later, "became far and away the most difficult mathematical test that the world has ever known, one to which no university of the present day can show any parallel" [97, p. 129].

Worse, the mathematical skills necessary to succeed on this terrifying exam bore no resemblance to "modern" mathematics as practiced in the wider world. The research topics that were engaging mathematicians in France and Germany and the United States were nowhere to be found on the Cambridge Tripos. Yet, Kanigel observed, "... as [the Tripos] grew more demanding and more important, it also, in inimitable English fashion, took on the luster—and the deadweight—of Tradition" [97, p. 129].

So, when Hardy arrived at Cambridge, he found himself among students who were forced to expend far too much of their university careers in preparation for two dreaded weeks of testing during their final year. And the stakes were high. There was a ranking scheme that designated the best student as "Senior Wrangler," followed by Second Wrangler, Third Wrangler, and so on. Rewards for the top finish were significant. Shops in Cambridge actually sold postcards featuring the images of the Senior Wrangler (mathematicians as rock stars!).

Hardy wanted none of it. He raced through his studies, took the Tripos a year early, and perhaps as a consequence ended up as only Fourth Wrangler. He was later to campaign for the abolition of this famous, or perhaps infamous, examination. His plan of action was unequivocal: "I adhere to the view that the [Tripos] system is vicious in principle, and that the vice is too radical for what is usually called reform. I do not want to reform the Tripos, but to destroy it" [32, p. 71].

The Professor's Life

Hardy survived it all. He ended his student days after having received the prestigious Smith's Prize in mathematics —a far more telling indicator of mathematical ability than any wranglerhood—and was granted a Fellowship allowing him to stay at Trinity College. This would be his academic base for the next two decades.

In 1900, Hardy published an article, "On a class of definite integrals containing hyperbolic functions," in the *Messenger of Mathematics*. This was the first in his long line of 375 mathematical papers. In 1912 he coauthored his first paper with J. E. Littlewood. It was published in the *Proceedings of the Fifth International Congress of Mathematicians*, which was held that year in Cambridge and had brought to town such luminaries as Jacques Hadamard, E.H. Moore, Vito Volterra, and E. T. Whittaker [93, pp. 223–229].

That initial Hardy-Littlewood paper was titled, "Some Problems of Diophantine Approximation." Over the years these two great mathematicians collaborated on such topics as Fourier series, Waring's problem on the representation of integers as sums of kth powers, Goldbach's conjecture, and the Riemann zeta-function. Their 93 joint papers were of an extraordinary depth and significance. C. P. Snow claimed that the Hardy-Littlewood research team was the strongest mathematical partnership in history, and one can think of few others that even come close. Hardy's friend and colleague Harald Bohr, the mathematician brother of physicist Niels Bohr, was not far off the mark when

1. Hardy's Life

Hardy in 1900, the year in which he was elected a Fellow Of Trinity College.

he quipped that "nowadays there are only three great English mathematicians: Hardy, Littlewood, and Hardy-Littlewood" [9, pp xxvii–xxviii].

In many ways the Hardy-Littlewood collaboration was an odd one. The two men were scholars of roughly equal powers, but their personalities were quite different. Hardy was elegant and sophisticated, rather patrician in his English way. Littlewood had a rougher appearance and demeanor. Though they lived in the same college during much of their careers, their work was usually carried on via *letters* with the understanding, according to Bohr, that when one wrote to the other, the recipient "was under no obligation whatsoever to read it, let alone answer it." Further, it did not matter whether "one of them had not contributed the least bit to the contents of a paper under their common name; otherwise there would constantly arise quarrels and difficulties that now one, and now the other, would oppose being named co-author" [9, pp. xxvii–xxviii].

The collaboration was so peculiar that a theory developed to the effect that Hardy had created a fictional "Littlewood" in case something was published

Crisp & Co. Camb   Senior Wranglers, Cambridge, 1905.   Copyright.
Mr. J. E. Littlewood (bracketed with Mr. J. Mercer) Trinity College.

Senior Wrangler, J. E. Littlewood in 1905. Hardy and Littlewood met in 1906; their formal collaboration began in 1911. Together, they wrote more than one hundred papers.

that contained an error, which could then be blamed on his imaginary coauthor. Of course, there was no truth in this, and Hardy was more than willing to own up to a mistake. For instance, in the preface to the second edition to his Cambridge Tract, *The Integration of Functions of a Single Variable*, Hardy noted that he had earlier "reproduced a proof of Abel's which Mr. J. E. Littlewood afterwards discovered to be invalid. The correction of this error has led me to rewrite a few sections of the present edition completely." And in 1920 Hardy published an article on Waring's Problem in the *Transactions of the American Mathematical Society* that had to be corrected seven years later after "Miss G. K. Stanley, upon reading the 30-page manuscript, discovered a significant error in a formula that required a three-page set of corrections... to fix it up" [61, p. 845]. Miss Stanley was a Reader in Mathematics at Westfield College and had been a research student of Hardy's. One concludes from these episodes that (a) Littlewood was a real person, not a fiction upon whom to pin mathematical errors and (b) even great mathematicians can make mistakes. It is some comfort

1. Hardy's Life

for mere mortals to discover that the likes of Hardy (and Abel!) occasionally come to unwarranted conclusions.

During his years as a young scholar, Hardy was tapped by the Apostles, a secretive, highly select group of Cambridge intellectuals. Robert Kanigel described this shadowy organization:

> By the time Hardy joined, its ranks had included some of the most brilliant men Cambridge had ever produced. There was Tennyson, the poet; Whitehead, the philosopher; James Clerk Maxwell, the physicist; Bertrand Russell; and many others whose names are less recognizable today... A scientist, as one Apostle noted round this time, was elected only if "he was a very nice scientist." Hardy, apparently, was nice enough; his crystalline intellect, his sly charm, his aesthetic sensibilities, his good looks, his love of good conversation—all these would have endeared him to such men [97, p. 137].

Other Apostles during Hardy's era were E. M. Forster, Lytton Strachey, John Maynard Keynes, and Ludwig Wittgenstein —not to mention, alas, World War II spy Guy Burgess. It must have been a most extraordinary club.

If Hardy's membership in the Apostles was kept secret, everyone knew when he was elected a Fellow of the Royal Society in 1910. To become a member of this august scientific body at 33 was an achievement, although Isaac Newton had been elected at the age of 30 and Newton's precocious albeit obnoxious *protégé*, Fatio de Duillier, had been elected at 24. Hardy's admission to membership came on his second try, perhaps, as Edward Titchmarsh suggested in his obituary for the Royal Society, because the pure mathematician Hardy "had singularly little appreciation of science" [124, p. 450].

In 1920 Hardy left Cambridge to accept the venerable and prestigious Savilian Chair in Geometry at Oxford. The list of mathematicians who had previously held this post was long and distinguished and included Henry Briggs, John Wallis, Edmond Halley, and James Joseph Sylvester—and, after Hardy's vacating the chair, it went to Titchmarsh and Michael Atiyah, among others.

Although Hardy was happy at Oxford, he must have missed the intensity of mathematical discussion at Cambridge. In the *Oxford Magazine* from 1930, he gently scolded his new colleagues for allowing mathematics to be "overshadowed" by other academic pursuits [55, p. 819]. Perhaps it is no surprise that, in 1931, Hardy returned to Cambridge to assume the Sadleirian Chair of pure mathematics. This had previously been held by Arthur Cayley, Andrew Forsyth, and E. W. Hobson, and upon Hardy's retirement, by Louis J. Mordell, J.W.S. Cassels, and John Coates. That's good company indeed.

Although he stayed at Cambridge for the rest of his career, Hardy from time to time took the opportunity to visit at places like Caltech and Princeton

Hardy was elected a Fellow of the Royal Society in 1910.

and, as the years passed, to collect a growing number of honors and awards. Of particular note, in 1936 he received an honorary doctorate from Harvard University, where the citation recognized him as:

> A British mathematician who has led the advance to heights deemed inaccessible by previous generations [122, p. 221].

Hardy's Character

C. P. Snow was a close observer of G. H. Hardy. "His face was beautiful," recalled Snow from their first meeting,

> ... high cheek bones, thin nose, spiritual and austere but capable of dissolving into convulsions of internal gamin-like amusement... Cambridge at that time was full of unusual and distinguished faces—but even then, I thought that night, Hardy's stood out [54, pp. 9–10].

1. Hardy's Life

To Snow, Hardy was "unorthodox, eccentric, radical, ready to talk about anything" [54, p. 9]. But he was also "shy and self-conscious in all formal actions, and [he] had a dread of introductions" [54, p. 10]. Hardy revealed a competitive streak and was ever ready to judge people by their talents... or lack thereof. As Snow explained, Hardy, though born into the *haute bourgeoisie*,

> ... behaved much more like an aristocrat, or more exactly like one of the romantic projections of an aristocrat. Some of this attitude, perhaps, he had picked up from his friend Bertrand Russell. But most of it was innate. Underneath his shyness, he just didn't give a damn [54, p. 42].

Among Hardy's quips was, "It is never worth a first class man's time to express a majority opinion. By definition, there are plenty of others to do that" [54, p. 46].

Moreover, he was not impressed by those who had clawed their way to the heights of wealth and power. Hardy called such people "the large bottomed," a description, according to Snow, that was "... more psychological than anatomical" [54, p. 42]. Snow continued:

> To Hardy the large bottomed were the confident, booming, imperialist bourgeois English. The designation included most bishops, headmasters, judges, and all politicians, with the single exception of Lloyd George [54, p. 42].

Hardy seemed more tolerant of his professorial colleagues. And, even among the rarified heights of Cambridge dons, Hardy was recognized as someone truly remarkable. This was the case in spite of his esoteric mathematical specialty. As Snow observed in his foreword to *A Mathematician's Apology*, Hardy was "tolerant, loyal, extremely high-spirited, and in an undemonstrative way, fond of his friends." According to Snow, discussion in the combination room did not quite hit its stride until Hardy entered and made a comment or offered an observation that would get the conversation moving.

Titchmarsh once compiled a list of Hardy's likes and dislikes. Hardy liked English and French literature, detective stories, and *The Times* crossword puzzles, whereas he hated blood sports, politicians, mutton (of course), and mechanical gadgets of all kinds—watches, fountain pens, and telephones. And he liked cats while hating dogs.

The mathematician George Pólya, who knew him well, said that Hardy "... was strikingly good looking and very elegant when he put on a dinner jacket..." Yet there is the oft-told account of Hardy's not being able to look at himself in a mirror. Pólya reported:

Hardy in his rooms at Trinity College.

I heard the following story from Littlewood... Hardy travelled a lot, and when he went into a hotel room, he covered all the mirrors.

I never dared ask about it, but Littlewood did and Hardy said, "I cannot look at myself, I am so ugly" [108, p. 65].

For the same reason, apparently, he allowed very few pictures of himself to be taken. The most famous shows him slouched down in a wicker chair,

1. Hardy's Life

Hardy with Pólya in 1924. Hardy spent half of the year working with Pólya in Oxford and the other half with Littlewood in Cambridge.

peering over thin-rimmed glasses, looking aristocratic and scholarly. Pólya again:

> It was taken in his rooms at Trinity. He had, of course, been at public school (Winchester) as a boy, and when Leon Bowden saw this picture, he said, "To sit that way you have to have been educated in a public school" [108, p. 64].

Pólya in his photo album did have quite a few pictures of Hardy, enough so that Vladimir Drobot once remarked that Pólya had more pictures of Hardy than exist! [108, pp. 64, 65, 121]

Hardy's Mathematics

G. H. Hardy once described a mathematician as an observer of the "... astonishingly beautiful complex of logical relations ... " that constitute the discipline. Like a Himalayan adventurer, the mathematician must make sense of a new and challenging world and produce "maps" to guide others on their way. "Many of these maps," he wrote, "have been completed, while in others, and these, naturally, are the most interesting, there are vast uncharted regions" [89, p. 382].

Hardy conceded that much of mathematics seemed impenetrable to the non-expert. The one striking exception was number theory, which deals with concepts familiar to everyone: whole numbers, evens and odds, primes and composites. Yet, as he once observed, when considering these deceptively simple whole numbers, "There is no one so blind that he does not see them, and no one so sharp-sighted that his vision does not fail" [89, p. 385]. Has anyone better captured the charm and challenge of number theory?

In a survey issue of the *Journal of the London Mathematical Society* that E. C. Titchmarsh published in 1950, after Hardy's death, there were sections devoted to his work on divergent series, integral equations, inequalities, and Fourier series [116, pp. 125–128]. But his most important work was in analytic number theory. This involved the celebrated but unproved Riemann hypothesis, along with Waring's problem, the Goldbach conjecture, and partitions of integers. What he brought to these problems was ingenuity and enormous technical skill. George Pólya made this telling assessment:

> Hardy wrote very well and with great facility, but his papers, especially some of his joint papers with Littlewood, make no easy reading. The problems are very hard and the methods unavoidably very complex. He valued clarity, yet what he valued most in mathematics was not clarity but power, surmounting great obstacles that others abandoned in despair. He himself had very great power, and he was fascinated by the Riemann hypothesis [108, p. 751].

As this passage suggests, the Riemann hypothesis became a focus for Hardy. When he visited Harald Bohr in Copenhagen, they would often take walks. Pólya described their routine:

> First they sat down and talked, and then they went for a walk. As they sat down, they made up and wrote down an agenda. The first point of the agenda was always the same: 'Prove the Riemann hypothesis.'. . . This point was never carried out. Still, Hardy insisted that it should be written down each time" [109, p. 752].

Of course, the Riemann hypothesis has been the target of many mathematicians besides G. H. Hardy. Pólya, on his deathbed as he fell into and out of consciousness, asked to have his desk at home searched because he thought

1. Hardy's Life

Bohr carved in wood.

he had jotted down on a slip of paper a very promising idea for proving the Riemann hypothesis. No such slip was found.

The continuing fascination with this result comes from its profound, and profoundly intriguing, connection to the prime numbers. The hypothesis concerns the complex function $\zeta(s)$, which is defined to be the sum of the reciprocals of the sth powers of the positive whole numbers —that is, $\zeta(s) = \sum_{n=1}^{\infty} \frac{1}{n^s}$, where $s = a + bi$. The Riemann hypothesis claims that the points where this function is zero are all on the line in the complex plane whose real part is $1/2$ (with the exception of trivial zeros at negative integers on the real axis). In other words, if $\zeta(s) = 0$, then $s = \frac{1}{2} + bi$, for some real number b.

Georg Friedrich Bernhard Riemann had made this conjecture in a brilliant and groundbreaking 1859 paper. In Hardy's day, half a century later, no one had proved it, nor has anyone done so yet. However, one of the early advances in this direction came in 1914 when Hardy proved that there must be infinitely many solutions to $\zeta(s) = 0$ having the desired form. This fell far short of establishing that none exist elsewhere, but progress on this front is excruciatingly difficult; we must take what we can get. A broad sense of how Hardy proved his theorem can be found in E. C. Titchmarsh's "The Riemann Zeta-Function, and Lattice-Point Problems," reproduced later in this volume.

G. F. B. Riemann. Hardy made significant progress on the Riemann hypothesis in 1914, but as of this date no one has yet settled Riemann's conjecture that he framed in 1859.

A question comes immediately to mind: what does any of this have to do with the primes? Well, a lot. For a relatively elementary discussion of the link between Riemann's hypothesis and the prime numbers, we recommend John Derbyshire's 2003 book, *Prime Obsession*. There, we see how the connection arises from some ingenious formulas due to Euler in the 18th century and Riemann in the 19th.

Hardy, of course, did much more than obsess over the Riemann hypothesis. He was an accomplished analyst, a student of the history and philosophy of mathematics, and an occasional writer on set theory. In regard to the last of these, Hardy addressed the perplexing question of the axiom of choice—much debated in his day—by telling Bertrand Russell that he [Hardy] was "... disposed to assume it and hope for the best" [24].

In all it is not too much to say that Hardy transformed British mathematics. Pólya compared Hardy to the Hungarian mathematician, Lipót Fejér, who influenced so many young mathematicians in Hungary and, in that small country,

1. Hardy's Life

Pólya compared the influence of his teacher Leopold Fejér on mathematics in Hungary with that of Hardy in England.

led to a flowering of mathematics that survives to this day. There likewise arose a "Hardy School" of British mathematics. Pólya wrote that

> England had a great tradition in applied mathematics, starting with Newton, but did not contribute comparably to pure mathematics which was developed mainly in France and Germany. Hardy insisted on pure mathematics and his insistence changed the trend of mathematical work in England. (That he occasionally misjudged, and was unjust to, applied mathematics is of comparatively little importance) [109, p. 751].

Much has been made of Hardy's disdain for applied mathematics, as we have seen. It is thus ironic, if not a little perverse, that he is today remembered for a mathematical application to genetics: the Hardy-Weinberg Law. Titchmarsh recalled that, in a letter to *Science* in 1908,

> Hardy settled a debate about the proportions in which dominant and recessive Mendelian characters would be transmitted to a large mixed population. As it happens, the law is of central importance in the study of Rh-blood-groups and the treatment of haemolytic disease of the newborn. In the *Apology* Hardy wrote, 'I have never done anything "useful". No discovery of mine has made or is likely to make, directly or indirectly, for good or ill, the least difference to the amenity of the world.' It seems that there was at least one exception to this statement" [124, p. 449].

Later in this book, we shall consider the Hardy-Weinberg Law and show how it resolved an important question among geneticists.

We end our review of Hardy's mathematics with a passage from Norbert Wiener, who had studied under him at Cambridge. Wiener believed that:

> Hardy's approach to mathematics was that of a sportsman. To be a sportsman means to take a joy in difficulty, and to overcome difficulties according to a meticulously exact code of what is permissible and what is not permissible [127, p. 75].

Hardy's Religious Views

G. H. Hardy was a confirmed atheist. He seems to have had a conventional upbringing in the Church of England, although his factoring hymn numbers during the sermon was probably a bad omen for his theological future. He later observed:

> Religion is not a subject on which I can profess to have spent much thought. I have always taken what is, no doubt, a narrow view about religion. I have understood by religion a body of doctrine more or less resembling that which is preached by the Christian churches. To refute the arguments by which such doctrines have been supported is a dialectical exercise of the most elementary kind. This exercise I performed to my own satisfaction when at school, and there my interest in the matter has ended [58, p. 119].

Such an attitude makes it somewhat ironic that Hardy chose Trinity College, even with its high reputation in mathematics. It calls to mind the situation faced by that greatest of Trinity alumni, Isaac Newton. He had dismissed the

Christian doctrine of the Trinity as a papist plot, even though such a view was cause for dismissal from the College. Because of this Newton had to tread softly on the subject. Hardy was not so discreet. By his time, if you were talented enough, you did not have to adhere to any religious orthodoxy.

These issues followed Hardy to Oxford in 1920. There, he refused to attend the election of the Warden of New College because it was held in a religious building. His refusal to enter such venues—in spite of their architectural majesty—required a change in the rules. As Titchmarsh recalled, "The clause in the New College by-laws, enabling a fellow with a conscientious objection to being present in Chapel to send his vote to the scrutineers, was put in on his [*i.e.*, Hardy's] behalf" [124, p. 451].

At one point, Hardy characterized what it means, at minimum, to have a belief in God. He did so with a nod to William Shakespeare and Jane Austen:

> First, that there is at least one being who is not an animal or a man or any combination, finite or infinite, of animals or men. Secondly, that this being exists, in the sense in which the sun and moon and you and I exist, and Hamlet and Mr. Collins do not and never did exist... [58, p. 122].

It should come as no surprise that Hardy would put the Deity in the same category as Hamlet and Mr. Collins. "I do not profess to be unprejudiced about religion," he wrote, and he added:

> "... religious dogmas, as I have defined them, have an important characteristic, namely that of being false... It seems to me, for example, not less improbable that there is a God than that there is a tiger in the next room, or that Mr. Russell is a German spy [58, p. 123].

Hardy's atheism notwithstanding, he enjoyed describing God as his primary adversary. As Pólya put it, "Hardy's troubled relationship with God became his principal standing joke." When Pólya and Hardy, on one of their frequent strolls, came upon a church, Hardy would maneuver to keep Pólya between him and the building, because walking too near the church would perhaps tempt God to reach out and strike him dead with a lightning bolt.

The joke was on-going. Hardy loved sunshine—what Englishman doesn't?—so he would follow the sun to join George and Stella Pólya at their chalet high in the Alpine village of Engelberg. But on one occasion, as Pólya recalled,

> ... it rained all the time, and as there was nothing else to do, we played bridge: Hardy, who was quite a good bridge player, my wife, myself,

Stella Pólya took pictures of mathematicians visiting at home and in her travels with her husband George. Here she is with Hermann Weyl.

and a friend of mine, F. Gonseth, mathematician and philosopher. Yet after a while Gonseth had to leave, he had to catch a train. I was present as Hardy said to Gonseth: 'Please, when the train starts you open the window, you stick your head through the window, look up to the sky, and say in a loud voice: "I am Hardy."' ... You have to understand the underlying theory: When God thinks that Hardy has left, he will make good weather just to annoy Hardy [109, p. 753].

We recall that one of Hardy's friends was Harald Bohr. Again, in Pólya's words:

Hardy stayed in Denmark with Bohr until the very end of the summer vacation, and when he was obliged to return to England to start his lectures there was only a very small boat available (there was no airplane traffic at that time). The North Sea can be pretty rough and the probability that such a small boat would sink was not exactly zero. Still, Hardy took the boat, but sent a postcard to Bohr: "I proved the Riemann hypothesis. G. H. Hardy.". . . If the boat sinks and Hardy drowns, everybody must believe that he has proved the Riemann hypothesis. Yet God would not let Hardy have such a great honor and so he will not let the boat sink." A most ingenious form of insurance [109, pp. 752–753].

1. Hardy's Life

Pólya and Harald Bohr in Zurich, 1926. Harald Bohr, brother of Neils Bohr, was not only a mathematician, but a very good soccer player—a member of the Danish national team.

We end this section with an amusing example. It seems that, one evening at Oxford, Hardy (GHH) was tallying up the day's score in his on-going competition against God (G). Here's his scorecard, addressed to God [91, p. 215].

What a day.
Beat you at squash (GHH 67, G 33)
Good math. theorem (GHH 73, G 0)
Bishop dead [forgot his name] (GHH 33, G 0)
Ex-Chorister to be hanged (GHH 61, G 10).

This yields a total score of GHH 234, God 43. "What a day," indeed!

Poetry, Lists, and Cricket

More often than is common for a mathematician, Hardy wrote about poetry. For instance, he cited his friend and colleague A. E. Housman, who believed the importance of ideas in poetry is habitually exaggerated. Housman wrote "I cannot satisfy myself that there are any such things as poetical ideas... Poetry is not the thing said but a way of saying it." To support Housman's position, Hardy quoted from Shakespeare's *Richard II*:

> Not all the water in the rough rude sea
> Can wash the balm from an anointed King.

He then asked,

> Could lines be better, and could ideas be at once more trite and more false? The poverty of the ideas seems hardly to affect the beauty of the verbal pattern. A mathematician, on the other hand, has no material to work with but ideas, and so his patterns are likely to last longer, since ideas wear less with time than words [54, pp. 84–85].

Of course, Hardy here found a tie-in with mathematics.

One of the entertainments with which he amused himself was making up lists. As a case in point, he produced rankings of mathematicians. For natural mathematical talent, he placed Ramanujan at the top of one such list with 100 points. Below this were David Hilbert with a rank of 80 and his coauthor Littlewood with 30. He modestly assigned himself a score of 25 [97, p. 226].

But Hardy did not confine himself to mathematicians. In a letter to Pólya's wife Stella, he ranked the poets—with Hardy-esque precision—as follows:

> Shakespeare 100, Milton 73, Shelley 71, Tennyson 39, Browning 27, and Wilcox 2.

A modern reader might be unfamiliar with the last of these. It was Ella Wheeler Wilcox, who penned the lines, "Laugh, and the world laughs with you. Weep, and you weep alone." Hardy added that without this particularly graceful passage, Wilcox would have garnered only 0.07 points, probably as punishment for such lines as:

> So many gods, so many creeds,
> So many paths that wind and wind,
> While just the art of being kind
> Is all the sad world needs [6, p.70].

Regarding poetry, Hardy could be a stern critic. There is a tale of his visit to Harald Cramér and his wife in Stockholm. In the living room, "Hardy observed a vase where there was a single rose and a book of poems by Rabindranath Tagore. It was open to a page with a poem to a rose. Marta Cramér waited for

1. Hardy's Life   25

Hardy watching a rugby match on a chilly day in Cambridge, 1941.

his appraisal, but Hardy only said: Rabindranath Tagore, the bore of bores.' [3, p. 26]." Tagore, it must be noted, won the Nobel Prize for Literature in 1913.

One of Hardy's great passions was cricket (although, in the U.S., he became interested in baseball, initially perhaps out of a feeling of desperation). In looking back to Hardy's childhood, Kanigel described a world that "... resonated to the sound of cricket bats and the whirl of white flannel under the summer sun" [96, p. 122]. Hardy not only watched, but he also played the game up into his 60s.

Not surprisingly, Hardy combined his love of lists and cricket. He once imagined these opposing cricket teams of illustrious individuals from across the centuries [6, p. 66]:

P. G. H. Fender (capt)	Hobbs
D. Spinoza	Archimedes
A. Einstein	M. Angelo
A. de Rothschild	Shakespeare
Moses	Napoleon
David	H. Ford
H. Heine	Plato
B. Disraeli	Beethoven
God (F)	Christ
God (S)	Jack Johnson
God (H.G.)	Cleopatra

Two of Hardy's choices for cricket teams included Spinoza, Einstein, Archimedes, Christ, and several other famous figures. Note that God appears as Father, Son, and Holy Ghost—on the same team.

On the one team, P. G. H. Fender was a real cricketer. Heinrich Heine was a poet, and David is presumably the slayer of Goliath. The rest of the players are obvious, except for the probable confusion over Spinoza's first name (Baruch). Of course, we see there the Trinity: God (F) the Father, God (S) the Son, and God (H. G.) the Holy Ghost.

Atop the other team was the cricketer Sir John Berry Hobbs, whom the star-struck Hardy regarded with unabashed admiration. Indeed, if he found someone to be the very best in his or her field, Hardy said that that person was "in the Hobbs class." He would even insert this characterization into mathematical papers. Titchmarsh observed this would be utterly confusing to those unfamiliar with the game and who were often left wondering what Hardy meant or, worse, what the philosopher Thomas Hobbes had to do with anything.

In the second column, we deduce that M. Angelo was Michelangelo, that H. Ford was Henry Ford, and that Jack Johnson was the American boxing champion. Christ is also on the list, although Hardy did not explain how Christ and God (S) could play for competing teams [6, p. 66].

Hardy and Politics

Hardy's politics for the most part leaned leftward. This was especially evident with respect to the British involvement in World War I. In a short work titled *Bertrand Russell and Trinity/A College Controversy on the Last War*, Hardy described Russell's stance against the war as viewed from the perspective of the Cambridge dons. Russell's forthright opposition led to an expulsion from his Trinity lectureship in 1916 and eventually landed him in Brixton Prison. Years later, Hardy published, at his own expense, this account of the matter, one in which his sympathy for Russell's position is quite evident. The books were kept on a table outside Hardy's door, available to anyone who wanted to pick up a copy.

We note in passing that, for several years, Hardy and Russell had been together at Trinity College. They shared an interest in the foundations of mathematics, and Hardy wrote five papers on set theory and transfinite arithmetic between 1904 and 1910. He also contributed favorable reviews of Russell's *The Principles of Mathematics* and of Russell's work with A. N. Whitehead, the *Principia Mathematica*. Unfortunately, these two men with such literary talents never wrote anything together. Ivor Grattan-Guinness has discussed all of this in his article, "Russell and G. H. Hardy: A Study of their Relationship" [24, pp. 165–179].

While on the subject of Hardy and politics, we must mention his stance against anti-Semitism. It is well-known that in the 1930s there were

Bertrand Russell cast in bronze well after his political rehabilitation.

mathematicians who left Germany to escape the Nazi menace and others who stayed behind. Among the latter were Nazi apologists like Ludwig Bieberbach, a mathematician who had done excellent research in complex analysis.

Bieberbach's assessment of the social situation in Germany was of a wholly different order. He had formulated a theory that classified mathematicians into one of two categories: German or Nordic practitioners were vigorous, inventive, robust and productive; others, such as "Frenchmen and Jews," were overly fussy with too much concern for logical niceties. Gauss was cited as an example of the former. Cauchy was one of the latter. So was G. H. Hardy.

In a sarcastic note published in *Nature* in 1934, Hardy addressed Bieberbach's thesis with contempt. He conceded that such absurd theories might be spawned by various societal pressures, such as "Anxiety for one's own position" or "dread of falling behind the rising torrent of folly." But, wrote Hardy,

> Professor Bieberbach's reputation excludes such explanations of his utterances; and I find myself driven to the more uncharitable conclusion that he really believes them [50, p. 250].

In a similar vein, there is the story of Hardy's support for Norman Levinson, a brilliant American mathematician, a student of Norbert Wiener's, and a Jew. In

1. Hardy's Life

Norman Levinson was appointed to MIT with substantial assistance from Hardy. He was strongly influenced by Norbert Wiener and wrote several important papers on the Riemann hypothesis before his untimely death in 1975.

1936, Levinson was proposed for a position at MIT, which was, like so many U. S. institutions at the time, a bastion of anti-Semitism. As a consequence, Levinson's appointment seemed to be in trouble. Hardy knew of this situation from his friend Wiener. So, when he visited MIT and met its Provost, Vannevar Bush, Hardy was in fighting mode. We quote from the recollections of Zipporah Levinson, Norman's widow [116, pp. 4–5]:

> Vannevar Bush took him [Hardy] around, and every time he showed him a new thing, Hardy said, "What a marvelous theological institution this is!" Finally [Bush] said, "It's not a theological institution!" And Hardy said, "Then why don't you hire Levinson?"

A few months later, Levinson received his appointment.

Hardy as Teacher

As a lecturer Hardy was much in demand. He had important things to say, and he said them very well. Titchmarsh recalled his first-hand experience of a G. H. Hardy lecture:

> [Hardy] pursued it with an eager single-mindedness which the audience found irresistible. One felt, temporarily at any rate, that nothing else in the world but the proof of these theorems really mattered. There could have been no more inspiring director of the work of others [124, p. 451].

Dame Mary Cartwright was one of the most influential women in mathematics at Cambridge, where she was Mistress of Girton College. Cartwright had taken

E. C. Titchmarsh as a student at Oxford came under the influence of Hardy. They also shared a passion for cricket. He succeeded Hardy as the Savilian Professor of Geometry at Oxford.

1. Hardy's Life

Mary Cartwright was a regular in the Hardy-Littlewood Seminar. She became Mistress of Girton College, and, later, Dame Mary.

her D. Phil. at Oxford in 1930, with Hardy as her thesis advisor. After Hardy returned to Cambridge, Cartwright attended the so-called Hardy-Littlewood Seminar, which was to be conducted by these two great mathematicians. Littlewood took the first lecture, and Cartwright recalled,

> Hardy came in late, helped himself liberally to tea and began to ask questions. It seems as if he were trying to pin Littlewood on details, whereas Littlewood was trying to illustrate the main point while taking the details for granted. An irritated Littlewood told Hardy that he was not prepared to be heckled. I don't recall them ever being present together at any subsequent class. Thenceforth, Hardy and Littlewood alternated classes... Eventually, Littlewood ceased to participate, even though the class continued to be held in his rooms. Consequently

Hardy, listening attentively, in the Hardy-Littlewood Seminar, sometimes called the Conversation Class.

the class became known as 'the Hardy-Littlewood Seminar at which Littlewood was never present' [121, pp. 249].

Apparently the Hardy-Littlewood joint teaching model had some of the same peculiarities as their joint research activities.

Though Cartwright's mathematical focus veered in the direction of applied analysis, closer to Littlewood's interests than Hardy's, she spoke very favorably of Hardy's teaching. Typically his class, scheduled for early in the day, began about fifteen minutes late. Hardy would arrive with a beautifully crafted lecture

J. E. Littlewood in 1924. Curiously, Littlewood was an irregular at the Hardy-Littlewood Seminar.

Olga Taussky spent most of her career at Caltech. She impressed Hardy with her mathematics as well as her tea making ability.

and "... took immense trouble with his students whether they were good, bad, or indifferent" [121, pp. 247–249].

Olga Taussky and her husband John Todd spent the years before World War II in England. Todd is quoted as saying that "Olga was invited to talk at the famous Conversation Class conducted by G. H. Hardy ... This apparently went down well, for Hardy, after her talk, told her that she could use his name as a reference when applying for a position" [2, p. 10]. This shows a genuine regard for the young woman's talents and welfare.

Todd related another episode that seems to capture Hardy's eccentricities:

Once when Hardy's sister was visiting Cambridge, Miss Taussky, by now a Fellow, asked them both to tea at her college (Girton). Because Hardy had said to her on a previous occasion, 'No foreigner can make proper tea,' she took advice on this project. Miss Hardy arrived on time and they talked until Hardy arrived. He said he had lost his way (in Cambridge!). Miss Taussky suspected he was frightened of having to drink her tea. Anyway, Hardy got out of his sweaters and then tea was ready. Hardy said, 'I call that a good cup of tea.' Her formula is secret" [2, p. 10].

As mentioned earlier, Hardy was close to Mrs. Pólya (as they were often bridge partners) and, in addition to Mary Cartwright, took as a student another Oxford woman, G.K. Stanley, who happened to share his enthusiasm for cricket. Yet his relationships with women were not always so cordial. In a 1984 letter, Dame Mary recalled that Hardy's sister Gertrude

... adored him, but when she took a junior member of the staff of her school to have tea with Hardy, he ignored the young woman, and she found him difficult to talk to. "I might have been her shopping bag for all the notice he took of me."

Perhaps in the all-male world of Trinity, Hardy was shy around women unless he knew them as friends or students over a long period of time. Then again, he could well have treated men the same way if he didn't know them or couldn't share with them ideas about mathematics.

Hardy as Reviewer

Over Hardy's career he reviewed books for *The Times Literary Supplement, Nature, The Cambridge Review*, and *The Mathematical Gazette*. Many of his reviews were of texts used in secondary schools or colleges. It is perhaps surprising that someone of his enormous mathematical power should find it worthwhile to evaluate such texts, but so he did.

Sometimes his reviews were generous, even when the book was found to be wanting. Sometimes they were critical in a way that must have sent chills down the author's spine. We shall look at some excerpts here; a few complete reviews will be included in the body of this collection.

In 1930 Hardy reviewed David Eugene Smith's *A Source Book in Mathematics* and began with these words: "This is a very entertaining volume, a surprisingly successful attempt to do what nearly all good judges would have declared to be impossible" [82, pp. 197–198]. Twelve years later he was likewise enthusiastic about Richard Courant and Herbert Robbins's *What Is Mathematics?*, of which he wrote, "This is a thoroughly good book which deserves to run through many editions" [86, p. 673]. We note that both of these volumes are still in print and going strong, a fact perfectly in keeping with Hardy's assessment.

Sir Thomas Heath published an annotated translation of Euclid's *Elements* in 1908, which Hardy subsequently reviewed. Although he was not an uncritical fan of Euclid—Hardy characterized parts of the *Elements* as "positively bad"—he described it as "the greatest perhaps of all mathematics books" and gave Heath high marks for this scholarly edition. Not surprisingly, the sections where Euclid dealt with number theory, i.e., Books VII to IX, drew Hardy's highest praise. He believed that these are "... books to take away on a holiday and to read at breakfast and lunch and tea and dinner and in bed" [85, p. 421].

Some of Hardy's reviews were more measured. He liked *Principia Mathematica*, the mathematical/philosophical tome from Bertrand Russell and Alfred North Whitehead that is famous for its forbidding cascade of logical symbols. But Hardy offered this caveat: "The last page is a very natural place at which to open a book, and the appearance of the last page, with its crazy-looking symbolism, is appalling" [79, p. 321]. Initially suggesting that there might be "twenty or thirty" people in Britain who would read *Principia Mathematica*, Hardy later adjusted his estimate downward to "half a dozen" [*Ibid*].

In the review of a textbook, L. S. Hulburt's *Differential and Integral Calculus*, Hardy started off with:

> This book has many good points. It is readable and fairly accurate, and the examples are simple and interesting. The geometrical parts are the best, but even in the analytical parts, a good deal is done correctly which is bungled hopelessly in many books with wide circulation.

But he ended with this quirky observation:

> I may add that I cannot regard an elementary mathematical book as a very suitable place for instruction in the rudiments of French, and that in any case I have grave doubts whether 'Kō'sheé' is a very accurate phonetic rendering of the name of the great mathematician [70, p. 337].

And then there were those reviews that can only be described as "scathing." We offer just one example (whose unfortunate author will remain nameless):

> The publication of this book by the Cambridge Press can only be attributed to a reprehensible carelessness on the part of its expert advisers [68, p. 21].

A Preview of What's to Come

There is much more that could be said. But, for this, we shall for the most part use Hardy's own writings. Surely the most compelling of these is his account of the extraordinary mathematician, Srinivasa Ramanujan. Once, when

1. Hardy's Life

asked to give his greatest mathematical contribution, Hardy answered that it was his discovery of Ramanujan. We include here Hardy's description of this improbable discovery and of their subsequent collaboration. It is a story that brings to mind the old adage that truth is stranger than fiction. And it reinforces one of Hardy's most striking and poignant observations from late in his life:

> I still say to myself when I am depressed, and find myself forced to listen to pompous and tiresome people, "Well, I have done one thing you could never have done, and that is to have collaborated with both Littlewood and Ramanujan on something like equal terms" [54, p. 148].

Of course, we have many other examples of Hardy as writer. In what follows, we feature some of Hardy's essays on mathematics, his tributes to such figures as Gösta Mittag-Leffler and David Hilbert, and the aforementioned book reviews. We have included an article, "Mathematics in Wartime," that expands on Section 28 of the *Apology* where Hardy addressed the morality of mathematicians' working on armaments that could be used to kill vast numbers of people. Examples of his mathematical work appear in expository pieces under the title "Four Hardy Gems" and as other items in Section II (Mathematics). And we have a selection of stand-alone quotations by Hardy and by others (writing about him) that seem especially profound or funny or both.

From this introduction, it should be evident that Hardy was a brilliant individual and a powerful mathematician, a charming conversationalist with a refined sense of humor, an eccentric colleague and mentor—in short, a most interesting character.

George Pólya, collaborator and friend of Hardy.

Pólya observed that G. H. Hardy was the mathematician who impressed him most as a person. He said, "Hardy had quite a special personal charm. I cannot describe it, what the charm was, but everyone was attracted by him, men and women, mathematicians and non-mathematicians, intellectuals and very simple people."

We end with the words of Leonard Woolf, husband of Virginia, whose description of Hardy veered into the Shakespearean:

> Hardy was one of the most strange and charming of men. A "pure" mathematician of great brilliance, he became an F.R.S. and Savilian Professor of Geometry in Oxford. He had the eyes of a slightly startled fawn below the very beautiful and magnificent forehead of an infant prodigy. He gave one the feeling that he belonged more properly to Prospero's island than the Great Court of Trinity. [128, p. 110]

We hope you will find that reading Hardy can be both entertaining and enlightening. Please, see for yourself.

2

The Letter from Ramanujan to Hardy, 16 January 1913

(*Collected Papers of Srinivasa Ramanujan*, edited by G. H. Hardy,
R. V. Seshu Aiyar, and R. M. Wilson, Cambridge 1927, p. xxiii)*

Srinivasa Ramanujan, 1877–1920

---

* Reprinted with the permission of the Cambridge University Press.

From the Editors

This is the famous letter that Ramanujan wrote from India to introduce himself to Hardy. It started their important collaboration that was to have a significant impact on mathematics in the 20th century.

MADRAS, 16th *January* 1913

DEAR SIR,

I beg to introduce myself to you as a clerk in the Accounts Department of the Port Trust Office at Madras on a salary of only £20 per annum. I am now about 23 years of age. I have had no University education but I have undergone the ordinary school course. After leaving school I have been employing the spare time at my disposal to work at Mathematics. I have not trodden through the conventional regular course which is followed in a University course, but I am striking out a new path for myself. I have made a special investigation of divergent series in general and the results I get are termed by the local mathematicians as "startling".

Just as in elementary mathematics you give a meaning to a^n when n is negative and fractional to conform to the law which holds when n is a positive integer, similarly the whole of my investigations proceed on giving a meaning to Eulerian Second Integral for all values of n. My friends who have gone through the regular course of University education tell me that $\int_0^\infty x^{n-1} e^{-x} dx = \Gamma(n)$ is true only when n is positive. They say that this integral relation is not true when n is negative. Supposing this is true only for positive values of n and also supposing the definition $n\Gamma(n) = \Gamma(n+1)$ to be universally true, I have given meanings to these integrals and under the conditions I state the integral is true for all values of n negative and fractional. My whole investigations are based upon this and I have been developing this to a remarkable extent so much so that the local mathematicians are not able to understand me in my higher flights.

Very recently I came across a tract published by you styled *Orders of Infinity* in page 36 of which I find a statement that no definite expression has been as yet found for the number of prime numbers less than any given number. I have found an expression which very nearly approximates to the real result, the error being negligible. I would request you to go through the enclosed papers.

Being poor, if you are convinced that there is anything of value I would like to have my theorems published. I have not given the actual investigations nor the expressions that I get but I have indicated the lines on which I proceed. Being inexperienced I would very highly value any advice you give me. Requesting to be excused for the trouble I give you.

I remain, Dear Sir, Yours truly,

S. RAMANUJAN.

P.S. My address is S. Ramanujan, Clerk Accounts Department, Port Trust, Madras, India.

3

A Letter from Bertrand Russell to Lady Ottoline Morrell, 2 February 1913

(*Ramanujan: Letters and Commentary*, by Bruce C. Berndt and Robert A. Rankin, American Mathematical Society and London Mathematical Society, 1995, pp. 44–45)*

Bertrand Russell, ca. 1920.

---

* Reprinted with the kind permission of the publishers, the American Mathematical Society and the London Mathematical Society

From the Editors

This letter from Bertrand Russell to Lady Ottoline Morrell gives some idea of the level of excitement at Trinity over the discovery of the work of S. Ramanujan. Bertrand Russell is, of course, well-known, but in her day Lady Ottoline was widely recognized as part of the intellectual, literary and artistic community known as the Bloomsbury Group.

2 February 1913

Bertrand Russell to Lady Ottoline Morrell

<u>Sunday night</u>
My Darling,

No letter from you came this morning which was sad, but I expect it will come tomorrow by 1st post. I have had a very full day, Sanger and Norton in and out perpetually. I enjoyed Sanger very much indeed. But à propos of Wittgenstein I shocked him by saying I thought the present society a feeble lot. I finished reading poor Bosanquet, and will write my review of him tomorrow. Also I wrote a longish criticism of Karin's paper, which I sent her. In Hall I saw Arthur Dakyns and Halvey. Before dinner North appeared, much excited because he has decided to be an engineer, which is what he always wished, only it was thought his health wouldn't stand it; he is very happy about it. In Hall I found Hardy, and Littlewood in a state of wild excitement, because they believe they have discovered a second Newton, a Hindu clerk in Madras on £20 a year. He wrote to Hardy telling of some results he has got, which Hardy thinks quite wonderful, especially as the man has had only an ordinary school education. Hardy has written to the Indian Office and hopes to get the man here at once. It is private at present. I am quite excited hearing of it.

It is very hard to get time for Matter in term time. When I get back from London I shall have Royce, proofs, Preface to write, lectures, lunch and walk with Ward—besides Wittgenstein and the other people who drop in. The result is that I shall hardly have one spare moment till Friday evening, and then I shall be sleepy, so that brings it to Saturday. Of course Bosanquet has taken time. But practically I <u>can't</u> do <u>very</u> much except in vacations. In fact, I see I shan't get

3. A Letter from Bertrand Russell to Lady Ottoline Morrell, 2 February 1913

really seriously to work on it till after America. After that, I must do whatever is necessary in order to get on with it—leave here, probably. When one thinks of work, the years slip by at such a dreadful pace that one will be dead before one is really under way. It is so different from thinking of emotions, which makes life seem enormously long. One comes to so many ends and new beginnings, whereas in work one is always on the threshold. I find that in philosophy I grow every year more open-minded, less the slave of habitual opinions. It is a comfort, especially as I have taken great pains to have it so and avoid slavery to mental habits.

I am <u>very</u> happy, full of life and energy. I am longing to hear from you Dearest. All my love is with you every moment, my Heart.

Your B.

4
The Indian Mathematician Ramanujan

(*The American Mathematical Monthly* 44 (1937), 137–155. Reprinted in Hardy's *Ramanujan/Twelve Lectures on Subjects Suggested by His Life and Work*, Cambridge University Press, 1940, pp. 1–21)

From the Editors

This article is a transcript of a lecture delivered at Harvard University at the Tercentenary Conference of Arts and Sciences on August 31, 1936, recognizing the 300th year of Harvard. Hardy wrote two beautiful matching folio volumes of work by and about Ramanujan, albeit separated by roughly 20 years. We note that this article from the American Mathematical Monthly *provided the introductory essay for the collection of 1940.*

I have set myself a task in these lectures which is genuinely difficult and which, if I were determined to begin by making every excuse for failure, I might represent as almost impossible. I have to form myself, as I have never really formed before, and to try to help you to form, some sort of reasoned estimate of the most romantic figure in the recent history of mathematics; a man whose career seems full of paradoxes and contradictions, who defies almost all the canons by which we are accustomed to judge one another, and about whom all of us will probably agree in one judgment only, that he was in some sense a very great mathematician.

The difficulties in judging Ramanujan are obvious and formidable enough. Ramanujan was an Indian, and I suppose that it is always a little difficult for an Englishman and an Indian to understand one another properly. He was, at the best, a half-educated Indian; he never had the advantages, such as they are, of an orthodox Indian training; he never was able to pass the "First Arts

Examination" of an Indian university, and never could rise even to be a "Failed B.A." He worked, for most of his life, in practically complete ignorance of modern European mathematics, and died when he was a little over thirty and when his mathematical education had in some ways hardly begun. He published abundantly—his published papers make a volume of nearly 400 pages—but he also left a mass of unpublished work which had never been analysed properly until the last few years. This work includes a great deal that is new, but much more that is rediscovery, and often imperfect rediscovery; and it is sometimes still impossible to distinguish between what he must have rediscovered and what he may somehow have learnt. I cannot imagine anybody saying with any confidence, even now, just how great a mathematician he was and still less how great a mathematician he might have been.

These are genuine difficulties, but I think that we shall find some of them less formidable than they look, and the difficulty which is the greatest for me has nothing to do with the obvious paradoxes of Ramanujan's career. The real difficulty for me is that Ramanujan was, in a way, my discovery. I did not invent him—like other great men, he invented himself—but I was the first really competent person who had the chance to see some of his work, and I can still remember with satisfaction that I could recognise at once what a treasure I had found. And I suppose that I still know more of Ramanujan than any one else, and am still the first authority on this particular subject. There are other people in England, Professor Watson in particular, and Professor Mordell, who know parts of his work very much better than I do, but neither Watson nor Mordell knew Ramanujan himself as I did. I saw him and talked with him almost every day for several years, and above all I actually collaborated with him. I owe more to him than to anyone else in the world with one exception, and my association with him is the one romantic incident in my life. The difficulty for me then is not that I do not know enough about him, but that I know and feel too much and that I simply cannot be impartial.

I rely, for the facts of Ramanujan's life, on Seshu Aiyar and Ramachaundra Rao, whose memoir of Ramanujan is printed, along with my own, in his *Collected Papers*. He was born in 1887 in a Brahmin family at Erode near Kumbakonam, a fair-sized town in the Tanjore district of the Presidency of Madras. His father was a clerk in a cloth-merchant's office in Kumbakonam, and all his relatives, though of high caste, were very poor.

He was sent at seven to the High School of Kumbakonam, and remained there nine years. His exceptional abilities had begun to show themselves before he was ten, and by the time that he was twelve or thirteen he was recognised as a quite abnormal boy. His biographers tell some curious stories of his early years. They say, for example, that soon after he had begun the study of trigonometry,

4. The Indian Mathematician Ramanujan

Srinivasa Ramanujan—Bronze bust by Paul Granlund, 1984.

he discovered for himself "Euler's theorems for the sine and cosine" (by which I understand the relations between the circular and exponential functions), and was very disappointed when he found later, apparently from the second volume of Loney's *Trigonometry*, that they were known already. Until he was sixteen he had never seen a mathematical book of any higher class. Whittaker's *Modern analysis* had not yet spread so far, and Bromwich's *Infinite series* did not exist. There can be no doubt that either of these books would have made a tremendous difference to him if they could have come his way. It was a book of a very different kind, Carr's *Synopsis*, which first aroused Ramanujan's full powers.

Carr's book (*A synopsis of elementary results in pure and applied mathematics*, by George Shoobridge Carr, formerly Scholar of Gonville and Caius College, Cambridge, published in two volumes in 1880 and 1886) is almost unprocurable now. There is a copy in the Cambridge University Library, and there happened to be one in the library of the Government College of Kumbakonam, which was borrowed for Ramanujan by a friend. The book is not in any sense a great one, but Ramanujan has made it famous, and there is no doubt that it influenced him profoundly and that his acquaintance with it marked the real starting-point of his career. Such a book must have had its qualities, and Carr's, if not a book of any high distinction, is no mere third-rate textbook, but a book written with some real scholarship and enthusiasm and with a style and individuality of its own. Carr himself was a private coach in London, who came to Cambridge as an undergraduate when he was nearly forty, and was 12th Senior Optime in the Mathematical Tripos of 1880 (the same year in which he published the first volume of his book). He is now completely forgotten, even in his own college, except in so far as Ramanujan has kept his name alive; but he must have been in some ways rather a remarkable man.

I suppose that the book is substantially a summary of Carr's coaching notes. If you were a pupil of Carr, you worked through the appropriate sections of the *Synopsis*. It covers roughly the subjects of Schedule A of the present Tripos (as these subjects were understood in Cambridge in 1880), and is effectively the "synopsis" it professes to be. It contains the enunciations of 6165 theorems, systematically and quite scientifically arranged, with proofs which are often little more than cross-references and are decidedly the least interesting part of the book. All this is exaggerated in Ramanujan's famous notebooks (which contain practically no proofs at all), and any student of the notebooks can see that Ramanujan's ideal of presentation had been copied from Carr's.

Carr has sections on the obvious subjects, algebra, trigonometry, calculus and analytical geometry, but some sections are developed disproportionally, and particularly the formal side of the integral calculus. This seems to have been Carr's pet subject, and the treatment of it is very full and in its way definitely good. There is no theory of functions; and I very much doubt whether Ramanujan, to the end of his life, ever understood at all clearly what an analytic function is. What is more surprising, in view of Carr's own tastes and Ramanujan's later work, is that there is nothing about elliptic functions. However Ramanujan may have acquired his very peculiar knowledge of this theory, it was not from Carr.

On the whole, considered as an inspiration for a boy of such abnormal gifts, Carr was not too bad, and Ramanujan responded amazingly.

Through the new world thus opened to him (say his Indian biographers),[1] Ramanujan went ranging with delight. It was this book which awakened his genius. He set himself to establish the formulae given therein. As he was without the aid of other books, each solution was a piece of research so far as he was concerned.... Ramanujan used to say that the goddess of Namakkal inspired him with the formulae in dreams. It is a remarkable fact that frequently, on

---

[1] Quotations (except those from my own memoir of Ramanujan) are from Seshu Aiyar and Ramachaundra Rao.

4. The Indian Mathematician Ramanujan

rising from bed, he would note down results and rapidly verify them, though he was not always able to supply a rigorous proof....

I have quoted the last sentences deliberately, not because I attach any importance to them—I am no more interested in the goddess of Namakkal than you are—but because we are now approaching the difficult and tragic part of Ramanujan's career, and we must try to understand what we can of his psychology and of the atmosphere surrounding him in his early years.

I am sure that Ramanujan was no mystic and that religion, except in a strictly material sense, played no important part in his life. He was an orthodox high-caste Hindu, and always adhered (indeed with a severity most unusual in Indians resident in England) to all the observances of his caste. He had promised his parents to do so, and he kept his promises to the letter. He was a vegetarian in the strictest sense—this proved a terrible difficulty later when he fell ill—and all the time he was in Cambridge he cooked all his food himself, and never cooked it without first changing into pyjamas.

Now the two memoirs of Ramanujan printed in the *Papers* (and both written by men who, in their different ways, knew him very well) contradict one another flatly about his religion. Seshu Aiyar and Ramachaundra Rao say

> Ramanujan had definite religious views. He had a special veneration for the Namakkal goddess.... He believed in the existence of a Supreme Being and in the attainment of Godhead by men.... He had settled convictions about the problem of life and after...;

while I say

> ...his religion was a matter of observance and not of intellectual conviction, and I remember well his telling me (much to my surprise) that all religions seemed to him more or less equally true....

Which of us is right? For my part I have no doubt at all; I am quite certain that I am.

Classical scholars have, I believe, a general principle, *difficilior lectio potior*—the more difficult reading is to be preferred—in textual criticism. If the Archbishop of Canterbury tells one man that he[2] believes in God, and another that he does not, then it is probably the second assertion which is true, since otherwise it is very difficult to understand why he should have made it, while there are many excellent reasons for his making the first whether it be true or false. Similarly, if a strict Brahmin like Ramanujan told me, as he certainly

---

[2] The Archbishop.

did, that he had no definite beliefs, then it is 100 to 1 that he meant what he said.

This was no sufficient reason why Ramanujan should outrage the feelings of his parents or his Indian friends. He was not a reasoned infidel, but an "agnostic" in its strict sense, who saw no particular good, and no particular harm, in Hinduism or in any other religion. Hinduism is, far more, for example, than Christianity, a religion of observance, in which belief counts for extremely little in any case, and, if Ramanujan's friends assumed that he accepted the conventional doctrines of such a religion, and he did not disillusion them, he was practising a quite harmless, and probably necessary, economy of truth.

The Indian stamp issued in 1962 to commemorate the 75th anniversary of Ramanujan's birth.

This question of Ramanujan's religion is not itself important, but it is not altogether irrelevant, because there is one thing which I am really anxious to insist upon as strongly as I can. There is quite enough about Ramanujan that is difficult to understand, and we have no need to go out of our way to manufacture mystery. For myself, I liked and admired him enough to wish to be a rationalist about him; and I want to make it quite clear to you that Ramanujan, when he was living in Cambridge in good health and comfortable surroundings, was, in spite of his oddities, as reasonable, as sane, and in his way as shrewd a person as anyone here. The last thing which I want you to do is to throw up your hands and exclaim "here is something unintelligible, some mysterious manifestation of the immemorial wisdom of the East!" I do not believe in the immemorial wisdom of the East, and the picture which I want to present to you is that of a man who had his peculiarities like other distinguished men, but a man in whose society one could take pleasure, with whom one could drink tea and discuss politics or mathematics; the picture in short, not of a wonder from the East, or

4. The Indian Mathematician Ramanujan

an inspired idiot, or a psychological freak, but of a rational human being who happened to be a great mathematician.

Until he was about seventeen, all went well with Ramanujan.

> In December 1903 he passed the Matriculation Examination of the University of Madras, and in the January of the succeeding year he joined the Junior First in Arts class of the Government College, Kumbakonam, and won the Subrahmanyam scholarship, which is generally awarded for proficiency in English and Mathematics...,

but after this there came a series of tragic checks.

> By this time, he was so absorbed in the study of Mathematics that in all lecture hours—whether devoted to English, History, or Physiology—he used to engage himself in some mathematical investigation, unmindful of what was happening in the class. This excessive devotion to mathematics and his consequent neglect of the other subjects resulted in his failure to secure promotion to the senior class and in the consequent discontinuance of the scholarship. Partly owing to disappointment and partly owing to the influence of a friend, he ran away northward into the Telugu country, but returned to Kumbakonam after some wandering and rejoined the college. As owing to his absence he failed to make sufficient attendances to obtain his term certificate in 1905, he entered Pachaiyappa's College, Madras, in 1906, but falling ill returned to Kumbakonam. He appeared as a private student for the F.A. examination of December 1907 and failed....

Ramanujan does not seem to have had any definite occupation, except mathematics, until 1912. In 1909 he married, and it became necessary for him to have some regular employment, but he had great difficulty in finding any because of his unfortunate college career. About 1910 he began to find more influential Indian friends, Ramaswami Aiyar and his two biographers, but all their efforts to find a tolerable position for him failed, and in 1912 he became a clerk in the office of the Port Trust of Madras, at a salary of about £30 a year. He was then nearly twenty-five. The years between eighteen and twenty-five are the critical years in a mathematician's career, and the damage had been done. Ramanujan's genius never had again its chance of full development.

There is not much to say about the rest of Ramanujan's life. His first substantial paper had been published in 1911, and in 1912 his exceptional powers began to be understood. It is significant that, though Indians could befriend him, it was only the English who could get anything effective done. Sir Francis Spring and Sir Gilbert Walter obtained a special scholarship for him, £60 a year, sufficient for a married Indian to live in tolerable comfort. At the beginning of 1913 he wrote to me, and Professor Neville and I, after many difficulties, got him to England in 1914. Here he had three years of uninterrupted activity, the results of which you can read in the *Papers*. He fell ill in the summer of 1917, and never really recovered, though he continued to work, rather spasmodically, but with no real sign of degeneration, until his death in 1920. He became a Fellow of the Royal Society early in 1918, and a Fellow of Trinity College, Cambridge, later in the same year (and was the first Indian elected to either society). His last mathematical letter on "Mock-Theta functions", the subject of Professor Watson's presidential address to the London Mathematical Society last year, was written about two months before he died.

The real tragedy about Ramanujan was not his early death. It is of course a disaster that any great man should die young, but a mathematician is often comparatively old at thirty, and his death may be less of a catastrophe than it seems. Abel died at twenty-six and, although he would no doubt have added a great deal more to mathematics, he could hardly have become a greater man. The tragedy of Ramanujan was not that he died young, but that, during his five unfortunate years, his genius was misdirected, sidetracked, and to a certain extent distorted.

I have been looking again through what I wrote about Ramanujan sixteen years ago, and, although I know his work a good deal better now than I did then, and can think about him more dispassionately, I do not find a great deal which I should particularly want to alter. But there is just one sentence which now seems to me indefensible. I wrote

> Opinions may differ about the importance of Ramanujan's work, the kind of standard by which it should be judged, and the influence which it is likely to have on the mathematics of the future. It has not the simplicity and the inevitableness of the very greatest work; it would be greater if it were less strange. One gift it shows which no one can deny, profound and invincible originality. He would probably have been a greater mathematician if he could have been caught and tamed a little in his youth; he would have discovered more that was new, and that, no doubt, of greater importance. On the other hand he would have been less of a Ramanujan, and more of a European professor, and the loss might have been greater than the gain...

and I stand by that except for the last sentence, which is quite ridiculous sentimentalism. There was no gain at all when the College at Kumbakonam rejected the one great man they had ever possessed, and the loss was irreparable; it is the worst instance that I know of the damage that can be done by an

4. The Indian Mathematician Ramanujan

inefficient and inelastic educational system. So little was wanted, £60 a year for five years, occasional contact with almost anyone who had real knowledge and a little imagination, for the world to have gained another of its greatest mathematicians.

Ramanujan's letters to me, which are reprinted in full in the *Papers*, contain the bare statements of about 120 theorems, mostly formal identities extracted from his notebooks. I quote fifteen which are fairly representative. They include two theorems, (1.14) and (1.15), which are as interesting as any but of which one is false and the other, as stated, misleading. The rest have all been verified since by somebody; in particular Rogers and Watson found the proofs of the extremely difficult theorems (1.10)–(1.12).

(1.1) $\quad 1 - \dfrac{3!}{(1!2!)^3}x^2 + \dfrac{6!}{(2!4!)^3}x^4 - \cdots$

$\qquad = \left(1 + \dfrac{x}{(1!)^3} + \dfrac{x^2}{(2!)^3} + \cdots\right)\left(1 - \dfrac{x}{(1!)^3} + \dfrac{x^2}{(2!)^3} - \cdots\right).$

(1.2) $\quad 1 - 5\left(\dfrac{1}{2}\right)^3 + 9\left(\dfrac{1.3}{2.4}\right)^3 - 13\left(\dfrac{1.3.5}{2.4.6}\right)^3 + \cdots = \dfrac{2}{\pi}.$

(1.3) $\quad 1 + 9\left(\dfrac{1}{4}\right)^4 + 17\left(\dfrac{1.5}{4.8}\right)^4 + 25\left(\dfrac{1.5.9}{4.8.12}\right)^4 + \cdots = \dfrac{2^{\frac{3}{2}}}{\pi^{\frac{1}{2}}\{\Gamma(\frac{3}{4})\}^2}.$

(1.4) $\quad 1 - 5\left(\dfrac{1}{2}\right)^5 + 9\left(\dfrac{1.3}{2.4}\right)^5 - 13\left(\dfrac{1.3.5}{2.4.6}\right)^5 + \cdots = \dfrac{2}{\{\Gamma(\frac{3}{4})\}^4}.$

(1.5) $\quad \displaystyle\int_0^\infty \dfrac{1 + \left(\frac{x}{b+1}\right)^2}{1 + \left(\frac{x}{a}\right)^2} \cdot \dfrac{1 + \left(\frac{x}{b+2}\right)^2}{1 + \left(\frac{x}{a+1}\right)^2} \cdots dx$

$\qquad = \dfrac{1}{2}\pi^{\frac{1}{2}} \dfrac{\Gamma(a + \frac{1}{2})\Gamma(b+1)\Gamma(b - a + \frac{1}{2})}{\Gamma(a)\Gamma(b + \frac{1}{2})\Gamma(b - a + 1)}.$

(1.6) $\quad \displaystyle\int_0^\infty \dfrac{dx}{(1+x^2)(1+r^2x^2)(1+r^4x^2)\cdots}$

$\qquad = \dfrac{\pi}{2(1 + r + r^3 + r^6 + r^{10} + \cdots)}.$

(1.7) If $\alpha\beta = \pi^2$, then

$\qquad \alpha^{-\frac{1}{4}}\left(1 + 4\alpha \displaystyle\int_0^\infty \dfrac{xe^{-\alpha x^2}}{e^{2\pi x} - 1}dx\right) = \beta^{-\frac{1}{4}}\left(1 + 4\beta \displaystyle\int_0^\infty \dfrac{xe^{-\beta x^2}}{e^{2\pi x} - 1}dx\right).$

(1.8) $\quad \displaystyle\int_0^a e^{-x^2}dx = \dfrac{1}{2}\pi^{\frac{1}{2}} - \dfrac{e^{-a^2}}{2a+} \dfrac{1}{a+} \dfrac{2}{2a+} \dfrac{3}{a+} \dfrac{4}{2a + \cdots}.$

(1.9) $$4\int_0^\infty \frac{xe^{-x\sqrt{5}}}{\cosh x}dx = \frac{1}{1+}\frac{1^2}{1+}\frac{1^2}{1+}\frac{2^2}{1+}\frac{2^2}{1+}\frac{3^2}{1+}\frac{3^2}{1+\cdots}.$$

(1.10) If $u = \dfrac{x}{1+}\dfrac{x^5}{1+}\dfrac{x^{10}}{1+}\dfrac{x^{15}}{1+\cdots}$, $v = \dfrac{x^{\frac{1}{5}}}{1+}\dfrac{x}{1+}\dfrac{x^2}{1+}\dfrac{x^3}{1+\cdots}$, then

$$v^5 = u\frac{1-2u+4u^2-3u^3+u^4}{1+3u+4u^2+2u^3+u^4}.$$

(1.11) $$\frac{1}{1+}\frac{e^{-2\pi}}{1+}\frac{e^{-4\pi}}{1+\cdots} = \left\{\sqrt{\left(\frac{5+\sqrt{5}}{2}\right)} - \frac{\sqrt{5}+1}{2}\right\}e^{\frac{2}{5}\pi}.$$

(1.12) $$\frac{1}{1+}\frac{e^{-2\pi\sqrt{5}}}{1+}\frac{e^{-4\pi\sqrt{5}}}{1+\cdots}$$

$$= \left[\frac{\sqrt{5}}{1+\sqrt[5]{\left\{5^{\frac{3}{4}}\left(\frac{\sqrt{5}-1}{2}\right)^{\frac{5}{2}}-1\right\}}} - \frac{\sqrt{5}+1}{2}\right]e^{2\pi/\sqrt{5}}.$$

(1.13) If $F(k) = 1 + \left(\frac{1}{2}\right)^2 k + \left(\frac{1.3}{2.4}\right)^2 k^2 + \cdots$ and $F(1-k) = \sqrt{(210)}F(k)$, then

$$k = (\sqrt{2}-1)^4(2-\sqrt{3})^2(\sqrt{7}-\sqrt{6})^4(8-3\sqrt{7})^2(\sqrt{10}-3)^4$$
$$\times (4-\sqrt{15})^4(\sqrt{15}-\sqrt{14})^2(6-\sqrt{35})^2.$$

(1.14) The coefficient of x^n in $(1-2x+2x^4-2x^9+\cdots)^{-1}$ is the integer nearest to

$$\frac{1}{4n}\left(\cosh \pi\sqrt{n} - \frac{\sinh \pi\sqrt{n}}{\pi\sqrt{n}}\right).$$

(1.15) The number of numbers between A and x which are either squares or sums of two squares is

$$K\int_A^x \frac{dt}{\sqrt{(\log t)}} + \theta(x),$$

where $K = 0.764\ldots$ and $\theta(x)$ is very small compared with the previous integral.

I should like you to begin by trying to reconstruct the immediate reactions of an ordinary professional mathematician who receives a letter like this from an unknown Hindu clerk.

The first question was whether I could recognise anything. I had proved things rather like (1.7) myself, and seemed vaguely familiar with (1.8). Actually (1.8) is classical; it is a formula of Laplace first proved properly by Jacobi; and (1.9) occurs in a paper published by Rogers in 1907. I thought that, as an expert in definite integrals, I could probably prove (1.5) and (1.6), and did so, though with a good deal more trouble than I had expected. On the whole the integral formulae seemed the least impressive.

The series formulae (1.1)–(1.4) I found much more intriguing, and it soon became obvious that Ramanujan must possess much more general theorems and was keeping a great deal up his sleeve. The second is a formula of Bauer well known in the theory of Legendre series, but the others are much harder than they look. The theorems required in proving them can all be found now in Bailey's Cambridge Tract on hypergeometric functions.

The formulae (1.10)–(1.13) are on a different level and obviously both difficult and deep. An expert in elliptic functions can see at once that (1.13) is derived somehow from the theory of "complex multiplication", but (1.10)–(1.12) defeated me completely; I had never seen anything in the least like them before. A single look at them is enough to show that they could only be written down by a mathematician of the highest class. They must be true because, if they were not true, no one would have had the imagination to invent them. Finally (you must remember that I knew nothing whatever about Ramanujan, and had to think of every possibility), the writer must be completely honest, because great mathematicians are commoner than thieves or humbugs of such incredible skill.

The last two formulae stand apart because they are not right and show Ramanujan's limitations, but that does not prevent them from being additional evidence of his extraordinary powers. The function in (1.14) is a genuine approximation to the coefficient, though not at all so close as Ramanujan imagined, and Ramanujan's false statement was one of the most fruitful he ever made, since it ended by leading us to all our joint work on partitions. Finally (1.15), though literally "true", is definitely misleading (and Ramanujan was under a real misapprehension). The integral has no advantage, as an approximation, over the simpler function

$$(1.16) \qquad \frac{Kx}{\sqrt{(\log x)}},$$

found in 1908 by Landau. Ramanujan was deceived by a false analogy with the problem of the distribution of primes. I must postpone till later what I have to say about Ramanujan's work on this side of the theory of numbers.

It was inevitable that a very large part of Ramanujan's work should prove on examination to have been anticipated. He had been carrying an impossible handicap, a poor and solitary Hindu pitting his brains against the accumulated wisdom of Europe. He had had no real teaching at all; there was no one in India from whom he had anything to learn. He can have seen at the outside three or four books of good quality, all of them English. There had been periods in his life when he had access to the library in Madras, but it was not a very good one; it contained very few French or German books; and in any case Ramanujan did not know a word of either language. I should estimate that about two-thirds of Ramanujan's best Indian work was rediscovery, and comparatively little of it was published in his lifetime, though Watson, who has worked systematically through his notebooks, has since disinterred a good deal more.

The great bulk of Ramanujan's published work was done in England. His mind had hardened to some extent, and he never became at all an "orthodox" mathematician, but he could still learn to do new things, and do them extremely well. It was impossible to teach him systematically, but he gradually absorbed new points of view. In particular he learnt what was meant by proof, and his later papers, while in some ways as odd and individual as ever, read like the works of a well-informed mathematician. His methods and his weapons, however, remained essentially the same. One would have thought that such a formalist as Ramanujan would have revelled in Cauchy's Theorem, but he practically never used it,[3] and the most astonishing testimony to his formal genius is that he never seemed to feel the want of it in the least.

It is easy to compile an imposing list of theorems which Ramanujan rediscovered. Such a list naturally cannot be quite sharp, since sometimes he found a part only of a theorem, and sometimes, though he found the whole theorem, he was without the proof which is essential if the theorem is to be properly understood. For example, in the analytic theory of numbers he had, in a sense, discovered a great deal, but he was a very long way from understanding the real difficulties of the subject. And there is some of his work, mostly in the theory of elliptic functions, about which some mystery still remains; it is not possible, after all the work of Watson and Mordell, to draw the line between what he may have picked up somehow and what he must have found for himself. I will take only cases in which the evidence seems to me tolerably clear.

Here I must admit that I am to blame, since there is a good deal which we should like to know now and which I could have discovered quite easily. I saw Ramanujan almost every day, and could have cleared up most of the obscurity

---

[3] Perhaps never. There is a reference to "the theory of residues" on p. 129 of the *Papers*, but I believe that I supplied this myself.

4. The Indian Mathematician Ramanujan

by a little cross-examination. Ramanujan was quite able and willing to give a straight answer to a question, and not in the least disposed to make a mystery of his achievements. I hardly asked him a single question of this kind; I never even asked him whether (as I think he must have done) he had seen Cayley's or Greenhill's *Elliptic functions*.

I am sorry about this now, but it does not really matter very much, and it was entirely natural. In the first place, I did not know that Ramanujan was going to die. He was not particularly interested in his own history or psychology; he was a mathematician anxious to get on with the job. And after all I too was a mathematician, and a mathematician meeting Ramanujan had more interesting things to think about than historical research. It seemed ridiculous to worry him about how he had found this or that known theorem, when he was showing me half a dozen new ones almost every day.

I do not think that Ramanujan discovered much in the classical theory of numbers, or indeed that he ever knew a great deal. He had no knowledge at all, at any time, of the general theory of arithmetical forms. I doubt whether he knew the law of quadratic reciprocity before he came here. Diophantine equations should have suited him, but he did comparatively little with them, and what he did do was not his best. Thus he gave solutions of Euler's equation

$$(1.17) \qquad x^3 + y^3 + z^3 = w^3,$$

such as

$$(1.18) \qquad \begin{cases} x = 3a^2 + 5ab - 5b^2, & y = 4a^2 - 4ab + 6b^2, \\ z = 5a^2 - 5ab - 3b^2, & w = 6a^2 - 4ab + 4b^2; \end{cases}$$

and

$$(1.19) \qquad \begin{cases} x = m^7 - 3m^4(1+p) + m(2 + 6p + 3p^2), \\ y = 2m^6 - 3m^3(1+2p) + 1 + 3p + 3p^2, \\ z = m^6 - 1 - 3p - 3p^2, \quad w = m^7 - 3m^4 p + m(3p^2 - 1); \end{cases}$$

but neither of these is the general solution.

He rediscovered the famous theorem of von Staudt about the Bernoullian numbers:

$$(1.20) \qquad (-1)^n B_n = G_n + \frac{1}{2} + \frac{1}{p} + \frac{1}{q} + \cdots + \frac{1}{r},$$

where $p, q, \ldots$ are those odd primes such that $p-1, q-1, \ldots$ are divisors of $2n$, and G_n is an integer. In what sense he had proved it it is difficult to say, since he found it at a time of his life when he had hardly formed any definite concept of proof. As Littlewood says, "the clear-cut idea of what is *meant* by a proof, nowadays so familiar as to be taken for granted, he perhaps did not possess at all; if a significant piece of reasoning occurred somewhere, and the total mixture of evidence and intuition gave him certainty, he looked no further". I shall have something to say later about this question of proof, but I postpone it to another context in which it is much more important. In this case there is nothing in the proof that was not obviously within Ramanujan's powers.

There is a considerable chapter of the theory of numbers, in particular the theory of the representation of integers by sums of squares, which is closely bound up with the theory of elliptic functions. Thus the number of representations of n by two squares is

(1.21) $$r(n) = 4\{d_1(n) - d_3(n)\},$$

where $d_1(n)$ is the number of divisors of n of the form $4k + 1$ and $d_3(n)$ the number of divisors of the form $4k + 3$. Jacobi gave similar formulae for 4, 6 and 8 squares. Ramanujan found all these, and much more of the same kind.

He also found Legendre's theorem that n is the sum of 3 squares except when it is of the form

(1.22) $$4^a(8k + 7),$$

but I do not attach much importance to this. The theorem is quite easy to guess and difficult to prove. All known proofs depend upon the general theory of ternary forms, of which Ramanujan knew nothing, and I agree with Professor Dickson in thinking it very unlikely that he possessed one. In any case he knew nothing about the number of representations.

Ramanujan, then, before he came to England, had added comparatively little to the theory of numbers; but no one can understand him who does not understand his passion for numbers in themselves. I wrote before

> He could remember the idiosyncrasies of numbers in an almost uncanny way. It was Littlewood who said that every positive integer was one of Ramanujan's personal friends. I remember going to see him once when he was lying ill in Putney. I had ridden in taxi-cab No. 1729, and remarked that the number seemed to me rather a dull one, and that I hoped that it was not an unfavourable omen. "No," he replied, "it is a very interesting number; it is the smallest number expressible as a sum of two cubes in two different ways."[4] I asked him, naturally, whether he could tell me the solution of the corresponding problem

---

[4] $1729 = 12^3 + 1^3 = 10^3 + 9^3$.

4. The Indian Mathematician Ramanujan

for fourth powers; and he replied, after a moment's thought, that he knew no obvious example, and supposed that the first such number must be very large.

In algebra, Ramanujan's main work was concerned with hypergeometric series and continued fractions (I use the word algebra, of course, in its old-fashioned sense). These subjects suited him exactly, and here he was unquestionably one of the great masters. There are three now famous identities, the "Dougall-Ramanujan identity"

$$(1.23) \quad \sum_{n=0}^{\infty} (-1)^n (s+2n) \frac{s^{(n)}}{1^{(n)}} \frac{(x+y+z+u+2s+1)^{(n)}}{(x+y+z+u+s)_{(n)}} \prod_{x,y,z,u} \frac{x_{(n)}}{(x+s+1)^{(n)}}$$

$$= \frac{s}{\Gamma(s+1)\Gamma(x+y+z+u+s+1)}$$

$$\times \prod_{x,y,z,u} \frac{\Gamma(x+s+1)\Gamma(y+z+u+s+1)}{\Gamma(z+u+s+1)},$$

where

$$a^{(n)} = a(a+1)\cdots(a+n-1), \quad a_{(n)} = a(a-1)\cdots(a-n+1),$$

and the "Rogers-Ramanujan identities"

$$(1.24) \quad \begin{cases} 1 + \dfrac{q}{1-q} + \dfrac{q^4}{(1-q)(1-q^2)} + \dfrac{q^9}{(1-q)(1-q^2)(1-q^3)} + \cdots \\ \quad = \dfrac{1}{(1-q)(1-q^6)\cdots(1-q^4)(1-q^9)\cdots}, \\ 1 + \dfrac{q^2}{1-q} + \dfrac{q^6}{(1-q)(1-q^2)} + \dfrac{q^{12}}{(1-q)(1-q^2)(1-q^3)} + \cdots \\ \quad = \dfrac{1}{(1-q^2)(1-q^7)\cdots(1-q^3)(1-q^8)\cdots}, \end{cases}$$

in which he had been anticipated by British mathematicians, and about which I shall speak in other lectures.[5] As regards hypergeometric series one may say, roughly, that he rediscovered the formal theory, set out in Bailey's tract, as it was known up to 1920. There is something about it in Carr, and more in Chrystal's *Algebra*, and no doubt he got his start from that. The four formulae (1.1)–(1.4) are highly specialised examples of this work.

---

[5] See Lectures VI and VII.

His masterpiece in continued fractions was his work on

$$\text{(1.25)} \qquad \cfrac{1}{1+} \cfrac{x}{1+} \cfrac{x^2}{1+\cdots},$$

which includes the theorems (1.10)–(1.12). The theory of this fraction depends upon the Rogers-Ramanujan identities, in which he had been anticipated by Rogers, but he had gone beyond Rogers in other ways and the theorems which I have quoted are his own. He had many other very general and very beautiful formulae, of which formulae like Laguerre's

$$\text{(1.26)} \qquad \frac{(x+1)^n - (x-1)^n}{(x+1)^n + (x-1)^n} = \cfrac{n}{x+} \cfrac{n^2-1}{3x+} \cfrac{n^2-2^2}{5x+\cdots}$$

are extremely special cases. Watson has recently published a proof of the most imposing of them.

It is perhaps in his work in these fields that Ramanujan shows at his very best. I wrote

> It was his insight into algebraical formulae, transformation of infinite series, and so forth, that was most amazing. On this side most certainly I have never met his equal, and I can compare him only with Euler or Jacobi. He worked, far more than the majority of modern mathematicians, by induction from numerical examples; all his congruence properties of partitions, for example, were discovered in this way. But with his memory, his patience, and his power of calculation he combined a power of generalisation, a feeling for form, and a capacity for rapid modification of his hypotheses, that were often really startling, and made him, in his own peculiar field, without a rival in his day.

I do not think now that this extremely strong language is extravagant. It is possible that the great days of formulae are finished, and that Ramanujan ought to have been born 100 years ago; but he was by far the greatest formalist of his time. There have been a good many more important, and I suppose one must say greater, mathematicians than Ramanujan during the last fifty years, but not one who could stand up to him on his own ground. Playing the game of which he knew the rules, he could give any mathematician in the world fifteen.

In analysis proper Ramanujan's work is inevitably less impressive, since he knew no theory of functions, and you cannot do real analysis without it, and since the formal side of the integral calculus, which was all that he could learn from Carr or any other book, has been worked over so repeatedly and so intensively. Still, Ramanujan rediscovered an astonishing number of the most beautiful analytic identities. Thus the functional equation for the Riemann

4. The Indian Mathematician Ramanujan

Zeta-function

$$\zeta(s) = \sum_{n=1}^{\infty} \frac{1}{n^s},$$

namely

(1.27) $$\zeta(1-s) = 2(2\pi)^{-s} \cos\frac{1}{2}s\pi\, \Gamma(s)\zeta(s),$$

stands (in an almost unrecognisable notation) in the notebooks. So does Poisson's summation formula

(1.28) $$\alpha^{\frac{1}{2}}\left\{\frac{1}{2}\phi(0) + \phi(\alpha) + \phi(2\alpha) + \cdots\right\}$$
$$= \beta^{\frac{1}{2}}\left\{\frac{1}{2}\psi(0) + \psi(\beta) + \psi(2\beta) + \cdots\right\},$$

where

$$\psi(x) = \sqrt{\left(\frac{2}{\pi}\right)}\int_0^\infty \phi(t)\cos xt\, dt$$

and $\alpha\beta = 2\pi$; and so also does Abel's functional equation

(1.29) $$L(x) + L(y) + L(xy) + L\left\{\frac{x(1-y)}{1-xy}\right\} + L\left\{\frac{x(1-x)}{1-xy}\right\} = 3L(1)$$

for

$$L(x) = \frac{x}{1^2} + \frac{x^2}{2^2} + \frac{x^3}{3^2} + \cdots.$$

He had most of the formal ideas which underlie the recent work of Watson and of Titchmarsh and myself on "Fourier kernels" and "reciprocal functions"; and he could of course evaluate any evaluable definite integral. There is one particularly interesting formula, viz.

(1.30) $$\int_0^\infty x^{s-1}\{\phi(0) - x\phi(1) + x^2\phi(2) - \cdots\}dx = \frac{\pi\phi(-s)}{\sin s\pi},$$

of which he was especially fond and made continual use. This is really an "interpolation formula", which enables us to say, for example, that, under certain conditions, a function which vanishes for all positive integral values of its argument must vanish identically. I have never seen this formula stated explicitly by anyone else, though it is closely connected with the work of Mellin and others.

I have left till last the two most intriguing sides of Ramanujan's early work, his work on elliptic functions and in the analytic theory of numbers.

The first is probably too specialised and intricate for anyone but an expert to understand, and I shall say nothing about it now.[6] The second subject is still more difficult (as anyone who has read Landau's book on primes or Ingham's tract will know), but anyone can understand roughly what the problems of the subject are, and any decent mathematician can understand roughly why they defeated Ramanujan. For this was Ramanujan's one real failure; he showed, as always, astonishing imaginative power, but he proved next to nothing, and a great deal even of what he imagined was false.

Here I am obliged to interpolate some remarks on a very difficult subject: *proof* and its importance in mathematics. All physicists, and a good many quite respectable mathematicians, are contemptuous about proof. I have heard Professor Eddington, for example, maintain that proof, as pure mathematicians understand it, is really quite uninteresting and unimportant, and that no one who is really certain that he has found something good should waste his time looking for a proof. It is true that Eddington is inconsistent, and has sometimes even descended to proof himself. It is not enough for him to have direct knowledge that there are exactly

$$136.2^{256}$$

protons in the universe; he cannot resist the temptation of proving it; and I cannot help thinking that the proof, whatever it may be worth, gives him a certain amount of intellectual satisfaction. His apology would no doubt be that "proof" means something quite different for him from what it means for a pure mathematician, and in any case we need not take him too literally. But the opinion which I have attributed to him, and with which I am sure that almost all physicists agree at the bottom of their hearts, is one to which a mathematician ought to have some reply.

I am not going to get entangled in the analysis of a particularly prickly concept, but I think that there are a few points about proof where nearly all mathematicians are agreed. In the first place, even if we do not understand exactly what proof is, we can, in ordinary analysis at any rate, recognise a proof when we see one. Secondly, there are two different motives in any presentation of a proof. The first motive is simply to secure conviction. The second is to exhibit the conclusion as the climax of a conventional pattern of propositions, a sequence of propositions whose truth is admitted and which are arranged in accordance with rules. These are the two ideals, and experience shows that, except in the simplest mathematics, we can hardly ever satisfy the first ideal

---

[6] See Lecture XII.

4. The Indian Mathematician Ramanujan

without also satisfying the second. We may be able to recognise directly that 5, or even 17, is prime, but nobody can convince himself that

$$2^{127} - 1$$

is prime except by studying a proof. No one has ever had an imagination so vivid and comprehensive as that.

A mathematician usually discovers a theorem by an effort of intuition; the conclusion strikes him as plausible, and he sets to work to manufacture a proof. Sometimes this is a matter of routine, and any well-trained professional could supply what is wanted, but more often imagination is a very unreliable guide. In particular this is so in the analytic theory of numbers, where even Ramanujan's imagination led him very seriously astray.

There is a striking example, which I have very often quoted, of a false conjecture which seems to have been endorsed even by Gauss and which took about 100 years to refute. The central problem of the analytic theory of numbers is that of the distribution of the primes. The number $\pi(x)$ of primes less than a large number x is approximately

$$(1.31) \qquad \frac{x}{\log x};$$

this is the "Prime Number Theorem", which had been conjectured for a very long time, but was never established properly until Hadamard and de la Vallée-Poussin proved it in 1896. The approximation errs by defect, and a much better one is

$$(1.32) \qquad \operatorname{li} x = \int_0^x \frac{dt}{\log t}.\ ^7$$

In some ways a still better one is

$$(1.33) \qquad \operatorname{li} x - \tfrac{1}{2}\operatorname{li} x^{\frac{1}{2}} - \tfrac{1}{3}\operatorname{li} x^{\frac{1}{3}} - \tfrac{1}{5}\operatorname{li} x^{\frac{1}{5}} + \tfrac{1}{6}\operatorname{li} x^{\frac{1}{6}} - \tfrac{1}{7}\operatorname{li} x^{\frac{1}{7}} + \cdots$$

(we need not trouble now about the law of formation of the series). It is extremely natural to infer that

$$(1.34) \qquad \pi(x) < \operatorname{li} x,$$

at any rate for large x, and Gauss and other mathematicians commented on the high probability of this conjecture. The conjecture is not only plausible but is supported by *all* the evidence of the facts. The primes are known up to

---

[7] The integral is a 'principal value'. See §2.2.

10,000,000, and their number at intervals up to 1,000,000,000, and (1.34) is true for every value of x for which data exist.

In 1912 Littlewood proved that the conjecture is false, and that there are an infinity of values of x for which the sign of inequality in (1.34) must be reversed. In particular, there is a number X such that (1.34) is false for some x less than X. Littlewood proved the existence of X, but his method did not give any particular value, and it is only very recently that an admissible value, viz.

$$X = 10^{10^{10^{34}}}$$

was found by Skewes. I think that this is the largest number which has ever served any definite purpose in mathematics.

The number of protons in the universe is about

$$10^{80}.$$

The number of possible games of chess is much larger, perhaps

$$10^{10^{50}}$$

(in any case a second-order exponential). If the universe were the chess-board, the protons the chessmen, and any interchange in the position of two protons a move, then the number of possible games would be something like the Skewes number. However much the number may be reduced by refinements on Skewes's argument, it does not seem at all likely that we shall ever know a single instance of the truth of Littlewood's theorem.

This is an example in which the truth has defeated not only all the evidence of the facts and of common sense but even a mathematical imagination so powerful and profound as that of Gauss; but of course it is taken from the most difficult parts of the theory. No part of the theory of primes is really easy, but up to a point simple arguments, although they will prove very little, do not actually mislead us. For example, there are simple arguments which might lead any good mathematician to the conclusion

(1.35) $$\pi(x) \sim \frac{x}{\log x}$$

of the Prime Number Theorem,[8] or, what is the same thing, to the conclusion that

(1.36) $$p_n \sim n \log n,$$

where p_n is the n-th prime number.

---

[8] $f(x) \sim g(x)$ means that the ratio f/g tends to unity.

4. The Indian Mathematician Ramanujan

In the first place, we may start from Euler's identity

(1.37) $$\prod_p \frac{1}{1-p^{-s}} = \frac{1}{(1-2^{-s})(1-3^{-s})(1-5^{-s})\cdots}$$

$$= \frac{1}{1^s} + \frac{1}{2^s} + \frac{1}{3^s} + \cdots = \sum_n \frac{1}{n^s}.$$

This is true for $s > 1$, but both series and product become infinite for $s = 1$. It is natural to argue that, when $s = 1$, the series and the product should diverge in the same sort of way. Also

(1.38) $$\log \prod \frac{1}{1-p^{-s}} = \sum \log \frac{1}{1-p^{-s}}$$

$$= \sum \frac{1}{p^s} + \sum \left(\frac{1}{2p^{2s}} + \frac{1}{3p^{3s}} + \cdots \right),$$

and the last series remains finite for $s = 1$. It is natural to infer that

$$\sum \frac{1}{p}$$

diverges like

$$\log \left(\sum \frac{1}{n} \right),$$

or, more precisely, that

(1.39) $$\sum_{p \leq x} \frac{1}{p} \sim \log \left(\sum_{n \leq x} \frac{1}{n} \right) \sim \log \log x$$

for large x. Since also

$$\sum_{n \leq x} \frac{1}{n \log n} \sim \log \log x,$$

formula (1.39) indicates that p_n is about $n \log n$.

There is a slightly more sophisticated argument which is really simpler. It is easy to see that the highest power of a prime p which divides $x!$ is

$$\left[\frac{x}{p} \right] + \left[\frac{x}{p^2} \right] + \left[\frac{x}{p^3} \right] + \cdots,$$

where $[y]$ denotes the integral part of y. Hence

$$x! = \prod_{p \leq x} p^{[x/p]+[x/p]^2+\cdots},$$

(1.40) $$\log x! = \sum_{p \leq x} \left(\left[\frac{x}{p}\right] + \left[\frac{x}{p^2}\right] + \cdots \right) \log p.$$

The left-hand side of (1.40) is practically $x \log x$, by Stirling's Theorem. As regards the right-hand side, one may argue: squares, cubes, ... of primes are comparatively rare, and the terms involving them should be unimportant, and it should also make comparatively little difference if we replace $[x/p]$ by x/p. We thus infer that

$$x \sum_{p \leq x} \frac{\log p}{p} \sim x \log x, \quad \sum_{p \leq x} \frac{\log p}{p} \sim \log x,$$

and this again just fits the view that p_n is approximately $n \log n$.

This is broadly the argument used, naturally in a less naïve form, by Tchebychef, who was the first to make substantial progress in the theory of primes, and I imagine that Ramanujan began by arguing in the same sort of way, though there is nothing in the notebooks to show. All that is plain is that Ramanujan found the form of the Prime Number Theorem for himself. This was a considerable achievement; for the men who had found the form of the theorem before him, like Legendre, Gauss, and Dirichlet, had all been very great mathematicians; and Ramanujan found other formulae which lie still further below the surface. Perhaps the best instance is (1.15). The integral is better replaced by the simpler function (1.16), but what Ramanujan says is correct as it stands and was proved by Landau in 1909; and there is nothing obvious to suggest its truth.

The fact remains that hardly any of Ramanujan's work in this field had any permanent value. The analytic theory of numbers is one of those exceptional branches of mathematics in which proof really is everything and nothing short of absolute rigour counts. The achievement of the mathematicians who found the Prime Number Theorem was quite a small thing compared with that of those who found the proof. It is not merely that in this theory (as Littlewood's theorem shows) you can never be sure of the facts without the proof, though this is important enough. The whole history of the Prime Number Theorem, and the other big theorems of the subject, shows that you cannot reach any real understanding of the structure and meaning of the theory, or have any sound instincts to guide you in further research, until you have mastered the proofs. It

4. The Indian Mathematician Ramanujan

is comparatively easy to make clever guesses; indeed there are theorems, like "Goldbach's Theorem",[9] which have never been proved and which any fool could have guessed.

The theory of primes depends upon the properties of Riemann's function $\zeta(s)$, considered as an analytic function of the complex variable s, and in particular on the distribution of its zeros; and Ramanujan knew nothing at all about the theory of analytic functions. I wrote before

> Ramanujan's theory of primes was vitiated by his ignorance of the theory of functions of a complex variable. It was (so to say) what the theory might be if the Zeta-function had no complex zeros. His method depended upon a wholesale use of divergent series.... That his proofs should have been invalid was only to be expected. But the mistakes went deeper than that, and many of the actual results were false. He had obtained the dominant terms of the classical formulae, although by invalid methods; but none of them are such close approximations as he supposed.
>
> This may be said to have been Ramanujan's one great failure...

and if I had stopped there I should have had nothing to add, but I allowed myself again to be led away by sentimentalism. I went on to argue that "his failure was more wonderful than any of his triumphs", and that is an absurd exaggeration. It is no use trying to pretend that failure is something else. This much perhaps we may say, that his failure is one which, on the balance, should increase and not diminish our admiration for his gifts, since it gives us additional, and surprising, evidence of his imagination and versatility.

But the reputation of a mathematician cannot be made by failures or by rediscoveries; it must rest primarily, and rightly, on actual and original achievement. I have to justify Ramanujan on this ground, and that I hope to do in my later lectures.

NOTES ON LECTURE I

p. 1. This lecture is a reprint of one delivered at the Harvard Tercentenary Conference of Arts and Sciences on August 31, 1936, and published in the *American Math. Monthly*, 44 (1937), 137–155.

pp. 7–8. For (1.1) see Preece (2); for (1.2) and (1.3), Hardy (4, 8); for (1.4), Hardy (4), Whipple (1) and Watson (7). All these formulae are special

---

[9] "Any even, number greater than 2 is the sum of two primes".

cases of much more general formulae discussed in Bailey's Cambridge Tract *Generalised hypergeometric series* (no. 32, 1935). See also Lecture VII.

For (1.5) and (1.6), see. no. 11 of the *Papers*, and Hardy (5, 6).

There are formulae of the same type as (1.7) in a paper by Hardy in the *Quarterly Journal of Math*. 35 (1904), 193–207. The formula (1.7) itself is proved by Preece (2). See also no. 11 of the *Papers*.

For (1.8), see Watson (2); for (1.9), Preece (4); for (1.10)–(1.12), Watson (4, 6); and for (1.13), Watson (8).

As regards (1.15), see Landau, *Archiv der Math. und Physik* (3), 13 (1908), 305–315, and Stanley (1), Miss Stanley shows just how Ramanujan's statement is misleading. See also Lecture IV. (B).

p. 10. The Librarian of the University of Madras has very kindly sent me a copy of the catalogue published in 1914, which makes it plain that the library was better equipped than I had supposed. For example it possessed two standard French treatises on elliptic functions (Appell and Lacour, Tannery and Molk) as well as the books of Cayley and Greenhill. It seems plain from other evidence that Ramanujan knew something of the English books but nothing of the French ones.

p. 11. Euler found the general *rational* solution of (1.17), and his solution was afterwards simplified by Binet and other writers. See, for example, Hardy and Wright, 198–202. A number of special solutions similar to (1.18) will be found in Dickson's *History*, ii, 500 *et seq*. The solution (1.19) is substantially the same as one found by J. R. Young (Dickson, *History*, ii, 554).

The simplest known solution of

$$x^4 + y^4 = z^4 + t^4$$

is Euler's

$$158^4 + 59^4 = 134^4 + 133^4 = 635318657.$$

See Dickson, *Introduction*, 60–62, and *History*, ii, 644–647. Euler gave a solution involving two parameters, but no 'general' solution is known.

There is a proof of von Staudt's theorem, due to R. Rado, in Hardy and Wright, 89–92.

The quotation from Littlewood is from his review of the *Papers*.

p. 12. The general theory of the representation of numbers as sums of an even number of squares is discussed in Lecture IX. For Legendre's 'three square' theorem see Landau, *Vorlesungen*, i, 114–122.

p. 13. Laguerre's formula (1.26) is formula (18) of Ch. XII of the 'second edition' of Ramanujan's notebooks, See Watson (10), 146.

For 'the most imposing' formula see Watson (14).

4. The Indian Mathematician Ramanujan

p. 14. The equation (1.29) was rediscovered by Rogers, *Proc. London Math. Soc.* (2), 4 (1907), 169–189, and is attributed to him in the *Papers*, 337; but it is to be found in a posthumous fragment of Abel (*Œuvres*, ii, 193).

p. 15. For (1.30) see Lecture XI.

p. 17. Skewes, *Journal London Math. Soc.* 8 (1933), 277–283. Skewes assumes the truth of the Riemann Hypothesis, but he has since found a (much larger) value for x independent of the hypothesis. This work is still unpublished.

p. 18. For (1.40) see, for example, Hardy and Wright, 342; Ingham, 20; Landau, *Handbuch* 75–76.

Hardy in his rooms at New College, Oxford.

5

"Epilogue" from The Man Who Knew Infinity

Robert Kanigel

(Scribner's, New York, 1991, pp. 361–373)*

From the Editors

This biography of Ramanujan is a remarkable piece of story-telling, charmingly written and well-researched. It is, of course, a biography of Ramanujan, not Hardy, but the latter plays a significant role. The author graciously allowed us to reprint here the final chapter that describes the collaboration between the two men during their years together at Cambridge.

By the time he learned of Ramanujan's death, Hardy had already left Trinity.

In December 1919, about when Ramanujan, at his doctor's insistence, prepared to leave Kumbakonam for Madras, Hardy was writing J. J. Thomson, master of Trinity College, with the news that he had accepted the Savilian Professorship at Oxford. "The post carries with it a Fellowship at New College, the acceptance of which will vacate my Fellowship here automatically," he wrote. One reason for the move was the increasing load of administrative responsibility he bore at Cambridge. "If I wish to preserve full opportunities for the researches which are the principal permanent happiness of my life," he had decided, he would need a position offering "more leisure and less responsibility." At Oxford, he'd been assured, he would get that.

---

* Copyright 1991 by Robert Kanigel

Unmentioned in the letter but probably weighing more on him than administrative chores were the hard feelings left over from the war, the infighting that surrounded the Russell affair, and the departure of Ramanujan. "If it had not been for the Ramanujan collaboration, the 1914–1918 war would have been darker for Hardy," wrote Snow. "It was the work of Ramanujan which was Hardy's solace during the bitter college quarrels." Now Ramanujan was gone. Trinity, his home for thirty years, had grown ugly to him. He scarcely spoke with some of his colleagues. Earlier, he had urged W. H. Young, an older mathematician who had spent much of his professional life abroad, to apply for the Savilian chair, only to ask him to withdraw his name, which Young did. "Hardy," recalled Young's son, Laurence, also a mathematician, "felt he must get away."

Oxford was the other great English university, less than a hundred miles away. In *Camford Observed*, Jasper Rose and John Ziman tried to bring it and Cambridge within the compass of a single account: "Oxford is a city of wide and noble thoroughfares, the High, the Broad, St. Giles; in Cambridge all the streets straggle. The great buildings of Oxford face the streets, tall and imposing, and form a series of breathtaking vistas. The great buildings of Cambridge are more isolated, less emphatic, more secretive, giving on to college courts and gardens. Oxford is more coherent; Cambridge more diffuse. Oxford overwhelms—Cambridge beguiles." Academically, Cambridge tipped slightly more toward the sciences, Oxford toward the classics.

New College was one of Oxford's two dozen or so distinct colleges. The place was like a walled city, with medieval battlements, pierced by tall, narrow slots through which archers could fire their bows, still enclosing two corners of it and forming a backdrop to the shrubs, trees, and bushes of the College Garden. In moving to New College, Hardy was coming full circle. The college was founded by William of Wykeham in 1379, eight years before he founded Winchester as a feeder to it. It had been the destination of many of Hardy's abler Winchester classmates twenty-five years before and would probably have been his as well had he not, his head turned by that St. Aubyn book, opted for Trinity instead.

Hardy had been in Oxford just a few months when he received the news from Madras:

By direction of the [University] Syndicate, I write to communicate to you, with feelings of deep regret, the sad news of the death of Mr. S. Ramanujan, F.R.S., which took place on the morning of the 26th April.

5. "Epilogue" from The Man Who Knew Infinity

"It was a great shock and surprise to me to hear of Mr. Ramanujan's death," Hardy replied in a letter to Dewsbury. But was there, in what he wrote next, the barest breath of defensiveness?

> When he left England the general opinion was that, while still very ill, he had turned the corner towards recovery; he had even gained over a stone in weight (at one period he had wasted away almost to nothing). And the last letter I had from him (about two months ago) was quite cheerful and full of mathematics.

Ramanujan *had*, after all, been entrusted to Hardy's care. Was Hardy—in a not uncommon sort of response to a loved one's death—now trying to assure himself that Ramanujan's final decline had come only *after* he had been placed safely aboard the ship to India?

About the impact of Ramanujan's death on Hardy there can be no doubt:

> For my part, it is difficult for me to say what I owe to Ramanujan—his originality has been a constant source of suggestion to me ever since I knew him, and his death is one of the worst blows I have ever had.

At Oxford, the specter of the Great War still hung heavily over Hardy, as it did all across Europe. Feeling against Germany ran deep. "Let us trust," one English scientist had written *Nature* in the closing months of the war, "that for the next twenty years at least all Germans will be relegated to the category of persons with whom honest men will decline to have any dealings." Mathematicians were not immune to the bitterness; many in England and France felt that Central European mathematicians should be banned from international mathematical congresses.

Hardy had been revolted by the war's stupid savagery, hated the whole idea of old men sending boys off to die, and felt cruelly cut off from his mathematical friends on the Continent. Now, the war over, he tried to heal the wounds. He wrote the London *Times* protesting some of the vengeful imbecilities being bruited about. He cooperated in the peacemaking efforts of Gösta Mittag-Leffler, long-time editor of *Acta Mathematica*, a Swedish mathematical journal founded in 1882 midst similar tensions among mathematicians following the Franco-Prussian War. He wrote with his views of the war to the great German mathematician Edmund Landau; his own views, Landau wrote back with a mathematician's touch, had been the same—except "with trivial changes of sign."

While visiting Germany in 1921, Hardy wrote Mittag-Leffler: "For my part, I have in no respect modified my former views, and am in no circumstances prepared to take part in, subscribe to, or assist in any manner directly or indirectly, any Congress from which, for good reasons or for bad, mathematicians of particular countries are excluded." He had boycotted one such congress in Strasbourg in 1920 from which Germans, Austrians, and Hungarians had been kept out, and he would boycott another in Toronto four years later.

The armistice, the departure of Ramanujan, his own move to Oxford, and Ramanujan's death had all come within eighteen months. But by all accounts, Hardy fell in easily at New College, felt at home there in a way he never had in Cambridge. It was, Snow tells us, "the happiest time of his life." He was accepted. His new Oxford friends made a fuss over him. His conversational flamboyance found new and appreciative ears. Sometimes, it seemed, everyone in the Common Room—the Oxford term for what back at Cambridge was the Combination Room—waited to hear what Hardy was going to talk about.

Meanwhile, his collaboration with Littlewood, conducted largely by mail, continued. He was at the height of his mathematical powers, the zenith of his fame. Mary Cartwright would recall how her bare mention of "Professor Hardy's class" to a college porter drew a response revealing "a far greater respect for Hardy than the customary deference of those days of any college porter to any don."

For the academic year 1928–1929, Hardy exchanged places with Princeton's Oswald Veblen and spent the year in the United States, mostly at Princeton University. While in America, he kept up a busy lecture schedule; he spoke at Lehigh University, for example, on January eleventh, at Ohio State on the eighteenth, at the University of Chicago on the twenty-first. During February and March, he was in residence at California Institute of Technology. At the end of the year, Princeton's president asked him to stay a bit longer. Hardy wrote back that though he'd had "a delightful time," his duties at Oxford demanded his return.

On this or one of his other trips to America, he developed a taste for baseball. Babe Ruth, it was said of him, "became a name as familiar in his mouth as that of the cricketer Hobbs." One time, he was sent a book, inscribed by "Iron Man" Coombs, stuffed with problems in baseball tactics. "It is a *wonderful* book," he wrote a postcard saying. "I try to solve one problem a day (e.g. 1 out, runners at 1st and 2nd, batsman [the cricket term] hits a moderate paced grounder rather wide to 2nd baseman's left hand—he being right handed. Should he try for a double play 1st to second or 2nd to first? I *think* the former.)"

His love of cricket and tennis, of course, continued unabated. In tennis, he steadily improved his game. In one snapshot taken while Hardy was at Oxford, it is a bright, sunny day in late spring or early fall and Hardy, looking absolutely

5. *"Epilogue" from* The Man Who Knew Infinity

smashing in his white tennis gear, stands midst a group of about a dozen other players. They are Beautiful People, ca. 1925 or so, and Hardy, clutching his racket, wearing long-legged tennis togs and a heavy shawl sweater under a jacket, is one of them.

Though happy at Oxford, by 1931 Hardy was back in Cambridge, as Sadleirian Professor, following the death of E. W. Hobson; Cambridge was, after all, still the center, far more than Oxford, of English mathematics, and he was now being offered its senior mathematical chair. Another reason, according to Snow, was that the two universities had different rules about retirement; whereas Oxford would turn him out of his rooms at sixty-five, at Trinity he could occupy them until he died.

For a time, Hardy would periodically return to Oxford for a few weeks at a time to captain the New College Senior Common Room cricket team. And, of course, he was always there at Lord's for the annual match between Cambridge and Oxford. "There he was at his most sparkling, year after year," wrote Snow. "Surrounded by friends, men and women, he was quite released from shyness; he was the centre of all our attention, which he by no means disliked; and one could often hear the party's laughter from a quarter of the way round the ground."

Hardy's formerly unpopular antiwar views were now, when not forgotten, actually applauded. The young Cambridge mathematicians, Snow records, "were delighted to have him back: he was a *real* mathematician, they said; not like those Diracs and Bohrs the physicists were always talking about: he was the purest of the pure." It was, as Laurence Young portrayed it later, a golden age of Cambridge mathematics. "Spiritually and intellectually, Cambridge was suddenly at least the equal of Paris, Copenhagen, Princeton, Harvard, and of Warsaw, Leningrad, Moscow." A sprinkle of foreign visitors to Cambridge had now, as Jews and others sought escape from Hitler's Germany, become a torrent.

Beginning around 1933, Hardy, in cooperation with the Society for the Protection of Science and Learning, used his influence to get Jews and others driven from their jobs to England and other safe havens. Mathematicians of the stature of Riesz, Bohr, and Landau were among those who got out. "Hardy, in many ways, was other-worldly," A. V. Hill wrote, "but in his deep solicitude for the dangers and difficulties of his colleagues he showed not only a broad humanity but a fine and resolute loyalty to the universal integrity and brotherhood of learning."

Hardy resigned from at least one German organization of which he had been a member—not because it was German, but because of what it *did*. "My attitude towards German connections of this kind," he wrote Mordell in the early Nazi period, "is that I do nothing unless I am positively forced to; but if anti-Semitism becomes an ostensible part of the programme of any periodical or institution, then I cannot remain in it."

In 1934, Hardy wrote to *Nature* responding to a University of Berlin professor who purported to show the influence of blood and race upon creative style in mathematics. There were, it seems, "J-type" and "S-type" mathematicians, the former of good Aryan stock, the latter Frenchmen and Jews. Hardy icily surveyed Professor Bieberbach's assertions, made a show of seeking ground on which to excuse them, finally found himself "driven to the more uncharitable conclusion that he really believes them true."

Hardy's sympathies lay invariably with the underdog, and his political views were decidedly left-wing. Until about 1927, he was active in the National Union of Scientific Workers, even made recruiting speeches on its behalf. In one, as J. B. S. Haldane paraphrased it later, he said to his audience of scientists "that although our jobs were very different from a coalminer's, we were much closer to coalminers than capitalists. At least we and the miners were both skilled workers, not exploiters of other people's work, and if there was going to be a line-up he was with the miners." Visitors to Hardy's rooms often noted that on his mantelpiece stood photographs of Einstein, the cricketer Jack Hobbs—and Lenin.

But within the mathematical community, he, Littlewood, and those in their camp stood squarely in the Establishment. English mathematicians, Hardy wrote in 1934, no longer labored under "the superstition that it is impossible to be 'rigorous' without being dull, and that there is some mysterious terror to exact thought." The revolution he had helped usher in a quarter century before had won the day. Indeed, some would grumble later that it had actually impeded progress in such fields as algebra, topology, functional analysis, and other topics within pure mathematics. By the 1930s, in any case, Hardy was seen as part of the older generation.

During these years, the honors, large and small, rolled in. On March 6, 1929, the one-hundredth, anniversary of the death of the great Norwegian mathematician Abel, Hardy, in the presence of the king of Norway, received an honorary degree from the University of Oslo.

On December 27, 1932, he got the Chauvenet Prize, awarded every three years for a mathematical paper published in English, for his "An Introduction to the Theory of Numbers."

On February 29, 1934, he received a letter, on the hammer-and-sickle embossed stationery of the Soviet Union, from J. Maisky, the Soviet ambassador to Britain, congratulating him on his election as an honorary member of the Academy of Sciences in Leningrad.

5. "Epilogue" from The Man Who Knew Infinity

The universities of Athens, Harvard, Manchester, Sofia, Birmingham, and Edinburgh awarded him degrees. He received the Royal Medal of the Royal Society in 1920, its Sylvester Medal in 1940. He was made an honorary member of many of the leading foreign scientific academies. Without a doubt, he was the most distinguished mathematician in Britain.

To this period, his prime, much Hardy lore is owed. One year, Hardy's New Year's resolutions were to:

1. Prove the Riemann hypothesis.
2. Make 211 no out in the fourth innings of the last test match at the Oval [which was something like hitting a grand slam home run while behind by three runs in the ninth inning of the World Series' final game].
3. Find an argument for the nonexistence of God which shall convince the general public.
4. Be the first man at the top of Mt. Everest.
5. Be proclaimed the first president of the U.S.S.R. of Great Britain and Germany.
6. Murder Mussolini.

Another story neatly combined his love of cricket, his pleasure in the sun, his warfare with God, and his madcap bent. One of his collaborators, Marcel Riesz, was staying at the place Hardy shared with his sister in London. Hardy ordered him to step outside, open umbrella clearly in view, and yell up to God, "I am Hardy, and I am going to the British Museum." This, of course, would draw a lovely day from God, who had nothing better to do than thwart Hardy. Hardy would then scurry off for an afternoon's cricket, fine weather presumably assured.

In long talks with Hardy beginning in 1931 and extending over the next fifteen or so years, C. P. Snow came away steeped in Hardy's "old brandy" sensibilities. By old brandy Hardy meant any "taste that was eccentric, esoteric, but just within the confines of reason." For example, he once wrote Snow that "the half-mile from St. George's Square to the Oval [in London] is my old brandy nomination for the most distinguished walk in the world." Old brandy was a sort of studied eccentricity—youthful foolishness transformed into a "mature" form, made a little self-conscious, ossified...

And that, indeed, is what had happened to Hardy. Somehow, he had become an old man. Even by the fall of 1931, when he was fifty-four, you could see signs of it. Back in Cambridge for the year, Norbert Wiener noticed that "by now, Hardy had become an aged and shriveled replica of the young man whom I had met in Russell's rooms" twenty years before. Hardy knew it, too. On his return to Cambridge, he was distressed by all the new, young faces he saw among the mathematicians. "There is," he wrote, "something very intimidating to an older man in such youthful quickness and power."

One day in 1939, while dusting his bookcase, Hardy had his first heart attack. He was sixty-two at the time. In its wake, he could no longer play tennis, or squash, or cricket. His creativity waned. One listing of his most important papers (it included every one of those on which he had worked with Ramanujan) included none beyond 1935. Now, his output declined by even a crude quantitative yardstick—from half a dozen or so per year in the late 1930s to one or two per year.

His waning mathematical powers depressed him. So did the new war with Germany. But around 1941, when young Freeman Dyson came up to Cambridge from Hardy's old school, Winchester, and for two years attended his lectures, Dyson couldn't see it. To him Hardy was still a god. He and three other advanced students all sat around a table in a small room in the old Arts School, listened, and watched Hardy from a few feet away:

> He lectured like Wanda Landowska playing Bach, precise and totally lucid, but displaying his passionate pleasure to all who could see beneath the surface.... Each lecture was carefully prepared, like a work of art, with the intellectual denouement appearing as if spontaneously in the last five minutes of the hour. For me these lectures were an intoxicating joy, and I used to feel sometimes an impulse to hug that little old man in the white cricket-sweater two feet away from me, to show him somehow how desperately grateful we were for his willingness to go on talking.

Hardy retired from the Sadleirian Chair in 1942.

The year before, a photographer for the British magazine *Picture Post* snapped his picture at a rugby match. There he was on a chilly winter day, cigarette in hand, all rolled up in flannels, watching Cambridge defeat Oxford, 9 to 6. The photograph later appeared in one of the volumes of his collected papers. His sister didn't like it. "It makes him look so old," she said.

But he *was* old.

By 1946, he was virtually an invalid. Snow pictured him as "physically failing, short of breath after a few yards' walk." His sister came to nurse him (though the Trinity rules were so strict that she had to leave his rooms at night).

In early 1947, he tried to kill himself by swallowing barbiturates. But he took so many that he vomited them up, hit his head on the lavatory basin, and wound up with an ugly black eye for his trouble.

5. "Epilogue" from The Man Who Knew Infinity

Later that year, the Royal Society notified him that he was to receive its highest honor, the Copley Medal. "Now I know that I must be pretty near the end," he told Snow. "When people hurry up to give you honorific things there is exactly one conclusion to be drawn."

On November 24, Snow wrote his brother Philip: "Hardy is now dying (how long it will take no one knows, but he hopes it will be soon) and I have to spend most of my spare time at his bedside."

It *was* soon. Hardy died on December 1, 1947, the day he was to be presented the Copley Medal. He left his substantial savings and the royalties of his books, once having provided for his sister, to the London Mathematical Society. "His loss," wrote Norbert Wiener, "brought us the sense of the passing of a great age."

The 1939 heart attack began the long physical and emotional slide that led to his suicide attempt. And it was in its wake, about a month after France fell to the Nazis, that he put the finishing touches to *A Mathematician's Apology*, his paean to mathematics. Snow saw the *Apology* "as a book of haunting, sadness," the work of a man long past his creative prime—and knowing it. "It is a melancholy experience for a professional mathematician to find himself writing about mathematics," wrote Hardy. Painters despised art critics? Well, the same went for any creative worker, a mathematician included. But writing *about* mathematics, rather than doing it, was all that was left him.

And yet, the sadness is at the prospect of a rich, full life nearing its end, not bitterness at a life ill-spent. Pride runs through the *Apology*, too, and pleasure, and deep satisfaction.

> I still say to myself when I am depressed, and find myself forced to listen to pompous and tiresome people, "Well, I have done one thing *you* could never have done, and that is to have collaborated with both Littlewood and Ramanujan on something like equal terms."

Ramanujan. All these twenty years later, Ramanujan remained part of him, a bright beacon, luminous in his memory.

"Hardy," said Mary Cartwright, his student during the 1920s and whom Hardy would describe as the best woman mathematician in England, "practically never spoke of things about which he felt strongly." Yet at one remove from his listener, on the printed page, he became a little freer. And there he revealed Ramanujan's hold on him: "I owe more to him," he wrote, "than to any one else in the world with one exception [Littlewood?] and my association with him is the one romantic incident in my life."

In the years after his death, Hardy began rummaging through Ramanujan's papers and notebooks. That, as many other mathematicians were to learn, could be tough going. After arriving in Oxford, Hardy wrote Mittag-Leffler that he had prepared a short paper from Ramanujan's manuscripts, "but it was hardly substantial enough for the *Acta* [the journal Mittag-Leffler edited]. I am now trying to make a more important one. But it is not possible to do it very rapidly, as all of Ramanujan's work requires most careful editing."

By 1921, he had culled from Ramanujan's papers enough to prepare a sequel to Ramanujan's work on congruence properties of partitions. The manuscript from which he was working, Hardy wrote in a note appended to the paper, which appeared in *Mathematische Zeitschrift*, "is very incomplete, and will require very careful editing before it can be published in full. I have taken from it the three simplest and most striking results, as a short but characteristic example of the work of a man who was beyond question one of the most remarkable mathematicians of his time."

Hardy's own papers over the years were fairly littered with Ramanujan's name: "Note on Ramanujan's Trigonometrical Function $c_q(n)$ and Certain Series of Arithmetical Functions" appeared in 1921; "A Chapter From Ramanujan's Notebook" in 1923; "Some Formulae of Ramanujan" in 1924. Then more in the mid-1930s: "A Formula of Ramanujan in the Theory of Primes"; "A Further Note on Ramanujan's Arithmetical function $\tau(n)$." Hardy appreciated the debt he owed Ramanujan and Littlewood: "All my best work," he wrote, "has been bound up with theirs, and it is obvious that my association with them was the decisive event of my life."

Hardy was thirty-seven when he met Ramanujan, living out his boyhood dream as a Fellow of Trinity, already an F.R.S. But his collaboration with Littlewood had only just begun, and he would come to view his early contribution to mathematics, though formidable by standards other than his own, as unspectacular.

Then, abruptly, Ramanujan entered his life.

Ramanujan was, if nothing else, a living, breathing reproach to the Tripos system Hardy despised. Sheer intuitive brilliance coupled to long, hard hours on his slate made up for most of his educational lacks. And he was so devoted to mathematics that he couldn't bother to study the other subjects he needed to earn a college degree. This "poor and solitary Hindu pitting his brains against the accumulated wisdom of Europe," as Hardy called him, had rediscovered a century of mathematics and made new discoveries that would captivate mathematicians for the next century. (And all without a Tripos coach.)

Is it any wonder Hardy was beguiled?

From then on, over the next thirty-five years, Hardy did all he could to champion Ramanujan and advance his mathematical legacy. He encouraged

5. "Epilogue" from The Man Who Knew Infinity

Ramanujan. He acknowledged his genius. He brought him to England. He trained him in modern analysis. *And*, during Ramanujan's life and afterward, he placed his formidable literary skills at his service.

"Hardy wrote exquisite English," the *Manchester Guardian* would say of him, citing especially his obituary notice of Ramanujan as "among the most remarkable in the literature about mathematics." To mathematician W. N. Bailey, it was "one of the most fascinating obituary notices that I have ever read." And it was Hardy's book on Ramanujan, more than anything he knew about him otherwise, that convinced Ashis Nandy to make Ramanujan a prime subject of his own book. Hardy's pen fired the imagination, shaping Ramanujan's reception by the mathematical world.

It began in 1916, when Hardy reported to the university authorities in Madras on Ramanujan's work in England; one look at it and they asked that it be prepared for publication. In it, Hardy wrote of the "curious and interesting formulae" Ramanujan had in his possession, of how Ramanujan possessed "powers as remarkable in their way as those of any living mathematician," that his gifts were "so unlike those of a European mathematician trained in the orthodox school," that his work displayed "astonishing individuality and power," that in Ramanujan "India now possesses a pure mathematician of the first order." This was scarcely the sort of language apt to pass unnoticed among mathematicians, Indian or British, accustomed to the sort of flat, gray prose normally appearing in their journals. Someone once said of Hardy that "conceivably he could have been an advertising genius or a public relations officer." Here was the evidence for it.

Hardy's long obituary of Ramanujan appeared first in the *Proceedings of the London Mathematical Society* in 1921, a little later in the *Proceedings of the Royal Society*, then again in Ramanujan's *Collected Papers* in 1927. In it, he told Ramanujan's *story*. He invested it with feeling. His language lingered in memory. "One gift [Ramanujan's work] has which no one can deny," he concluded'—"profound and invincible originality."

> He would probably have been a greater mathematician if he had been caught and tamed a little in his youth; he would have discovered more that was new, and that, no doubt, of greater importance. On the other hand he would have been less of a Ramanujan, and more of a European professor, and the loss might have been greater than the gain.

Snow, who first met Hardy in 1931, revealed that for all Hardy's shyness, "about his discovery of Ramanujan, he showed no secrecy at all." Mary Cartwright recalled that "Hardy was terribly proud, and rightly, of having discovered Ramanujan." Ramanujan had enriched his life. He didn't *want* to forget Ramanujan, and he didn't.

On February 19, 1936, Hardy wrote S. Chandrasekhar from Cambridge: "I am going to give some lectures (here and at Harvard) on Ramanujan during the summer." They would become the basis for *Ramanujan: Twelve Lectures on Subjects Suggested By His Life and Work*. "A labor of love," one reviewer called it.

The Harvard lecture was part of the great university's celebration of the three-hundredth anniversary of its founding. The bash culminated in three grand Tercentenary Days, from September 16 to 18. Harvard Yard, now a great outdoor theater with seventeen thousand seats, was awash with silk hats, crimson bunting, and colorful academic costumes. On the second evening, at 9:00 P.M., upward of half a million people lined both banks of the Charles River for two hours of fireworks.

The following morning, under a steady drizzle and dark, brooding clouds (the leading edge of a hurricane moving up the Atlantic Coast), sixty-two of the world's most distinguished biologists, chemists, anthropologists, and other scholars received honorary degrees. They marched in a procession from Widener Library and took their places on stands erected in front of the pillars of Memorial Church at the yard's north end. Psychoanalyst Carl Jung was among them. So was Jean Piaget, the pioneer student of child development. So was English astrophysicist Sir Arthur Eddington. So was Hardy. The citation honoring him, slipped within the red leather presentation book stamped with Harvard's *Veritas* seal, called him "a British mathematician who has led the advance to heights deemed inaccessible by previous generations."

During his stay at Harvard, Hardy was put up at the house of a prominent lawyer, who later became a United States senator. Both he and his host, according to one account, were worried: whatever would they talk about? "The lawyer was no better prepared to discuss Zeta functions than the mathematician to comment upon the rule in Shelley's case." So they seized on their common enthusiasm—baseball. The Red Sox were in town, and Hardy was at Fenway often to watch them.

In the weeks prior to the grand finale, a Tercentenary Conference of Arts and Sciences brought to Harvard more than twenty-five hundred scholars for lectures under broad rubrics of knowledge like "The Place and Functions of Authority," and "The Application of Physical Chemistry to Biology." Einstein's wife was ill, so he sent word that he couldn't come. Nor could Werner Heisenberg, author of the uncertainty principle, who was advised at the last minute that he was needed for eight weeks' service in Hitler's army. Their absence notwithstanding, it was an august group, including no fewer than eleven Nobel Prize winners. "Highbrows at Harvard," *Time* headed its account. The *New York Times* covered some of the public lectures, including Hardy's.

5. *"Epilogue" from* The Man Who Knew Infinity

At about nine in the evening of the conference's first day, Hardy—wizened, gray, and almost sixty now—rose before his audience in New Lecture Hall. "I have set myself a task in these lectures which is genuinely difficult," he told them, in the measured cadences that were the mark of his speech and of his prose.

> and which, if I were determined to begin by making every excuse for failure I might represent as almost impossible. I have to form myself, as I have never really formed before, and to try to help you to form, some sort of reasoned estimate of the most romantic figure in the recent history of mathematics; a man whose career seems full of paradoxes and contradictions, who defies almost all the canons by which we are accustomed to judge one another, and about whom all of us will probably agree in one judgment only, that he was in some sense a very great mathematician.

And then Hardy, the memory still fresh of the day a quarter century before when an envelope stuffed with formulas arrived in the mail from India, began to tell about his friend, Ramanujan.

6

Posters of "Hardy's Years at Oxford"

R. J. Wilson

Emeritus Professor of Pure Mathematics, The Open University;
Emeritus Professor of Geometry, Gresham College, London;
Former Fellow of Keble College, Oxford

On 3 October 2013 the Mathematics Department of the University of Oxford moved into its magnificent new Institute, the Andrew Wiles Building.

As part of its increased outreach activities I proposed a regular series of large historical posters that could be displayed around the building featuring Oxford mathematicians. The Department was keen on my proposal and has been funding the whole project. My collaborator is Raymond Flood, a predecessor of mine as President of the British Society for the History of Mathematics, a successor as Gresham Professor of Geometry in London, and a co-editor (with John Fauvel) of *Oxford Figures: Eight Centuries of the Mathematical Sciences* (Oxford University Press, 2000, 2013).

Although mainly remembered as a Cambridge mathematician, G. H. Hardy spent eleven of his most productive years (1920–1931) as Oxford's Savilian Professor of Geometry. For some years I have given lectures on *G. H. Hardy's Oxford Years*, and this seemed the obvious starting point for the series; I thus already had much material (both textual and photographic) for my six posters, which I prepared with comments by Raymond Flood, and the Institute's External Relations Manager Dyrol Lumbard arranging a professional designer and generally making the arrangements.

Future series of posters will feature *John Wallis* (already under preparation by Raymond Flood), to be followed by *Lewis Carroll—Oxford Mathematician, Early Mathematics in Oxford* (including the Merton School), and *Three Victorian Savilian Professors* (Baden Powell, Henry Smith, and James Joseph Sylvester). These will be available on the Institute's website (www.maths.ox.ac.uk) as they appear.

G H Hardy's Oxford years

Godfrey Harold Hardy (1877–1947) was the most important British pure mathematician of the first half of the 20th century. Although he is usually thought of as a Cambridge man, his years from 1920 to 1931 as Savilian Professor of Geometry at Oxford University were actually his happiest and most productive. At Oxford he was at the prime of his creative life, and wrote over 100 papers there, including many of his most important investigations with his long-term Cambridge collaborator J E Littlewood.

"I was at my best at a little past forty, when I was a professor at Oxford."
> G H Hardy
> *A Mathematician's Apology*

"…the happiest time of his life…"
> C P Snow
> Introduction to
> *A Mathematician's Apology*

" He preferred the Oxford atmosphere and said that they took him seriously, unlike Cambridge."
> J E Littlewood
> *Littlewood's Miscellany*

The G H Hardy years posters were conceived by Robin Wilson with the assistance of Raymond Flood and Dyrol Lumbard, and are reproduced with the permission of the Mathematical Institute of the University of Oxford.

Hardy's Oxford time-line

1919	Hardy appointed Savilian Professor of Geometry
1920	Comes into office on 19 January
	18 May: Gives inaugural lecture on problems in the theory of numbers
	Srinivasa Ramanujan dies
	Awarded the Royal Medal of the Royal Society
1921	Visits colleagues in Scandinavia and Germany
1922	President of the British Association, Mathematics & Physics section Presidential lecture: *The theory of numbers*
1924	Georg Pólya (first Rockefeller Fellow) visits Oxford
1924	President of the National Union of Scientific Workers
1925–26	Abram Besicovitch visits Oxford
1925–27	President of the Mathematical Association Presidential lectures: *What is geometry?* and *The case against the Mathematical Tripos*
1926	Advocates founding the *Journal of the London Mathematical Society*
1926–28	President of the London Mathematical Society Presidential lecture: *Prolegomena to a chapter on inequalities*
1927	Co-edits Ramanujan's *Collected Papers*
1928	Attends International Congress of Mathematicians in Bologna
1928–29	Spends the academic year in the USA (mainly at Princeton and Caltech)
	Replaced in Oxford by Oswald Veblen
1929	Awarded the London Mathematical Society's De Morgan Medal
1930	Re-launches the *Quarterly Journal of Mathematics*
	Lobbies in the *Oxford Magazine* for a Mathematical Institute
1931	Albert Einstein receives Oxford honorary degree and gives lectures
	Hardy leaves Oxford and returns to Cambridge

Savilian Professor

The Savilian Professorships of Geometry and Astronomy at Oxford University were founded by Sir Henry Savile, Warden of Merton College, in 1619.

From Cambridge to Oxford

From 1897 to 1916 the Geometry chair was occupied by William Esson, following the tenure of J J Sylvester who was appointed to it at the age of 69 after returning from the USA.

William Esson

Esson had been largely responsible for initiating Oxford's intercollegiate lecture system, replacing the arrangement in which the colleges mainly operated independently. He was Sylvester's deputy from 1894 to 1897 when failing eyesight prevented Sylvester from lecturing. Esson died in 1916, during World War I, and the election to the vacant Savilian Chair was suspended until the War was over.

Eventually, in October 1919, an advertisement appeared in the Oxford University Gazette inviting applications. The details of the post included the following: *The Professor will be a Fellow of New College, and will receive a stipend of £900 per year ... It will be the duty of the Professor to lecture and give instruction in Pure and Analytical Geometry ... He is also bound to give not less than forty-two lectures in the course of the academical year.*

Although Hardy was not a geometer, a number of considerations made the Oxford position attractive to him.

By 1919 he had become disillusioned with Cambridge, where he had resided since entering as a scholar in 1896. He was sickened by the War and by the pro-War attitudes of his colleagues at Trinity College, especially in their behaviour towards Bertrand Russell, then a young lecturer.

Bertrand Russell

In 1916 Russell was convicted, and later imprisoned, for his anti-War activities. The College removed Russell's lectureship, reinstating it in 1919, but as Russell's most ardent supporter in the College, Hardy found the atmosphere stifling and needed to get away.

Hardy was also devastated by the illness and departure of his coworker Srinivasa Ramanujan, his researches with J E Littlewood were not going well, and he was suffering from an increased administrative load.

A complete break was called for ...

Hardy is appointed

In December 1919 the following notice appeared in the *Oxford University Gazette:*

> SAVILIAN PROFESSORSHIP OF GEOMETRY
> At a meeting of the Electors held on Friday, December 12, GODFREY HAROLD HARDY, MA, Fellow of Trinity College, Cambridge, was elected Savilian Professor of Geometry, to enter office on January 19, 1920.

The Electors for the Oxford Chair were:
- H Blakiston, *Vice-Chancellor*
- Revd. W Spooner, *Warden of New College* (of 'Spoonerism' fame)
- H H Turner, *Savilian Professor of Astronomy*
- E B Elliott, *Waynflete Professor of Pure Mathematics*
- M J M Hill, *Professor of Mathematics at University College, London*
- Percy MacMahon, *former President of the London Mathematical Society*
- C H Sampson, *Principal of Brasenose College*

Hardy had initially encouraged W H Young, independent inventor of the Lebesgue integral, to apply. Later, when Hardy himself decided to apply, he asked Young to withdraw his name, which he did.

Hardy's inaugural lecture on some famous problems of the theory of numbers was given in the University Observatory on Tuesday 18 May at 5 PM; it was later published.

In his inaugural lecture Hardy discussed Waring's problem: Every positive integer can be *written as the sum of at most 4 perfect squares, 9 cubes or 19 fourth powers; does this continue for nth powers, and how many powers are needed?*

This was Hardy's first involvement with Waring's problem, and during his Oxford years he co-authored six papers with J E Littlewood on it.

SOME FAMOUS PROBLEMS

of the

THEORY OF NUMBERS

and in particular

Waring's Problem

An Inaugural Lecture delivered before the

University of Oxford

BY

G. H. HARDY, M.A., F.R.S.
Fellow of New College
Savilian Professor of Geometry in the University of Oxford
and late Fellow of Trinity College, Cambridge

OXFORD
AT THE CLARENDON PRESS
1920

Hardy in Oxford

Hardy found himself well suited to life at New College, the college attached to the Savilian chairs.

Hardy at New College

At New College the Wardens during Hardy's time were the Revd. William Spooner, who lectured on ancient history, philosophy and divinity, and (from 1925) H A L Fisher, who had served in government and later wrote a celebrated *History of Europe*.

New College Fellows in the 1920s included J S Haldane, H W B Joseph and Hugh Allen. The College then had 300 students, but at most one student took mathematics Finals each year. Among the New College students during Hardy's time were Hugh Gaitskell, Richard Crossman and Lord Longford.

"In the informality and friendliness of New College Hardy always felt completely at home. He was an entertaining talker on a great variety of subjects, and one sometimes noticed everyone in common room waiting to see what he was going to talk about."
E C Titchmarsh
Hardy's successor as Savilian Professor

" … the leader of talk in the senior common room, arranging with mock seriousness complicated games and intelligence tests, his personality gave a quality to the atmosphere quite unlike anything else in my experience…"
Lionel Robbins
Economist

Hardy in his room at New College

Mathematical contemporaries

On the mathematical side, Hardy's Oxford colleagues initially consisted of the Sedleian Professor of Natural Philosophy and the Waynflete Professor of Pure Mathematics, the Savilian Professor of Astronomy, and a number of college tutors. Another mathematics chair was added in 1928.

The Applied mathematics professor was Augustus Love, who occupied the Sedleian Chair for over forty years. He was President of the London Mathematical Society and was awarded its De Morgan Medal. His monumental text on the mathematical theory of elasticity was of great importance, and his lectures to Oxford students were models of clear thinking and style.

Twenty years earlier Love had taught Hardy at Cambridge, as Hardy recalled:

"My eyes were first opened by Professor Love, who taught me and gave me my first serious conception of analysis. But the great debt which I owe to him was his advice to read Jordan's famous *Cours d'Analyse*; and I shall never forget the astonishment with which I read that remarkable work, the first inspiration for so many mathematicians of my generation, and I learnt for the first time what mathematics really meant. From that time onwards I was in my way a real mathematician, with sound mathematical ambitions and a genuine passion for mathematics."

The Waynflete Chair of Pure Mathematics was created in 1892. Its first occupant was E B Elliott. His interests were in algebra and his textbook on invariant theory was highly regarded, but his mathematics was old-fashioned, mainly involving the exploitation of elementary ideas with skill and insight. He had 'no sympathy with foreign modern symbolic methods'.

Elliott retired as Waynflete Professor in 1921 and was replaced by A L Dixon, who studied the applications of algebra to geometry and elliptic functions. Both Elliott and Dixon were Presidents of the London Mathematical Society. In 1928 the Rouse Ball Chair was created in Oxford. Unlike its Cambridge counterpart, where it is reserved for pure mathematics, the Oxford Chair is always held by applied mathematicians.

Its first occupant was E A Milne, a former student of Hardy's in Cambridge. A major figure in astrophysics, Milne revolutionised applied mathematics at Oxford.

Later holders of the Oxford Rouse Ball Chair included C A Coulson and Sir Roger Penrose.

E B Elliott

Teaching and research

While in Oxford Hardy distinguished himself as both a superb lecturer and an inspiring leader of research.

Hardy's teaching

Hardy was a brilliant and widely admired lecturer. One of his Oxford students later wrote:

"Each lecture was carefully prepared, like a work of art, with the intellectual denouement appearing as if spontaneously in the last five minutes of the hour. For me these lectures were an intoxicating joy."

In common with most of his Oxford colleagues Hardy taught in his own college — there was then no Mathematical Institute.

Although not a geometer, Hardy assiduously lectured on geometry each term — *Analytic geometry, Applications of analysis to geometry, Solid geometry and Elements of non-Euclidean geometry* — as well as branching out into nongeometrical topics.

The Hilary Term lecture list for 1920 included a range of other lecture courses, including *Theory of functions* by Elliott, *Mathematical astronomy* by Turner, *Electricity and magnetism* by Love, and *Hydrostatics* by Dixon.

Halfway through the term Hardy commenced a course of lectures introducing the *Analytical geometry of the plane*, given on Tuesdays and Saturdays at 11 AM in New College.

Hardy's lecture courses were always models of good organization and clarity. Fortunately, several of them have been preserved in notes taken by his Oxford research student E H Linfoot. To the left is a page from a number theory course given by Hardy in 1924–25.

Hardy's research

Before Hardy there was no flourishing research tradition in Oxford, although J J Sylvester had tried to initiate one in the 1880s and particular individuals such as Augustus Love were involved in their own researches. In 1925 E B Elliott remarked:

"I still hold soundly that our business as teachers in a University was to educate, to assist young men to make the best use of their powers ... 'But how about research and original work under this famous system of yours? You do not seem to have promoted it much.' Perhaps not! It had not yet occurred to people that systematic training for it was possible."

Hardy determined to change all this. His own research interests were mainly in number theory and mathematical analysis, and his greatest legacy to Oxford mathematics was the internationally renowned research school of analysis that he established there in the 1920s.

Shortly after arriving in Oxford, he introduced Friday evening advanced classes in pure mathematics at New College, at which he, his research students, and any mathematical visitors to Oxford, discussed their current investigations.

J E Littlewood

The fact that Hardy and J E Littlewood were now at different universities made no difference to their productivity; during Hardy's Oxford years they produced about fifty joint papers. Probably the most prolific partnership in mathematical history, they co-authored almost one hundred papers on a wide range of subjects.

A joke of the time held that: *Nowadays, there are only three really great English mathematicians: Hardy, Littlewood and Hardy-Littlewood.*

"The mathematician Hardy-Littlewood was the best in the world, with Littlewood the more original genius and Hardy the better journalist."
                                    Edmund Landau

Hardy's research students included E C Titchmarsh, Mary Cartwright (pictured above), E H Linfoot and his later co-author E M Wright.

Oxford preoccupations

In addition to his many research papers, Hardy was involved with a range of other activities.

Hardy's publishing activities

Hardy and Ramanujan

In the years prior to Hardy's coming to Oxford he produced a number of spectacular joint papers with the brilliant Indian mathematician Srinivasa Ramanujan. But Ramanujan became ill and returned to Madras in March 1919. He died a year later, shortly after Hardy's arrival in Oxford. Hardy was devastated:

"For my part, it is difficult for me to say what I owe to Ramanujan — his originality has been a constant source of suggestion to me ever since I knew him, and his death is one of the worst blows I have ever had."

In 1921 Hardy wrote a beautifully crafted obituary notice, described by the *Manchester Guardian* as 'among the most remarkable in the literature about mathematics.' For the next three years Hardy wrote a number of papers based on, and extending, Ramanujan's work. Meanwhile, he was working through his colleague's extensive notebooks and papers, and in 1927, with two others, he published *Ramanujan's Collected Papers*.

Other Publications

In 1926 Hardy persuaded the London Mathematical Society to found the *Journal of the London Mathematical Society* and in 1930 he was involved in re-launching the *Quarterly Journal of Mathematics*. The first issue included articles by M L Cartwright, P Dienes, W L Ferrar, G H Hardy, E C Titchmarsh, J E Littlewood, E A Milne, L J Mordell, G Pólya, O Veblen, G N Watson, and A Zygmund.

Hardy also found time for the occasional general article. In the *Oxford Magazine* of June 1930, he bemoaned the University's lack of attention to mathematics and encouraged it to build up a fine mathematical school, observing: 'mathematicians are reasonably cheap, but they cannot be had for nothing'. He urged an increase in the number of mathematical lecturers, proposed more research activity as essential for the future, and lobbied for the creation of a Mathematical Institute.

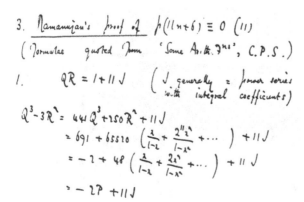

Other Oxford preoccupations

Einstein visits Oxford

In May 1931 a major scientific event took place in Oxford when Albert Einstein received an honorary Doctorate of Science and presented three lectures on the theory of relativity at Rhodes House; a blackboard from one lecture is still preserved in Oxford. Merton College arranged a special dinner for Einstein, which Hardy, Love, Dixon and Milne attended.

Hardy's feud with God

'God-free' Harold Hardy was an ardent atheist who had a permanent feud with God. Above is a bizarre note that he wrote in the form of an imaginary cricket match between himself and God, in which Hardy, of course, won.

Hardy's main relaxation was cricket. Here, during a British Association meeting in Oxford in August 1926, he leads a team of MATHEMATICIANS versus THE REST OF THE WORLD. His team included Titchmarsh, Bosanquet, Linfoot and Ferrar.

From Oxford to Cambridge

After eleven fruitful years in Oxford, Hardy decided to return to Cambridge for the rest of his days.

Hardy leaves Oxford

Although he was very happy at Oxford, Hardy chose to resign his Oxford Chair and return to Cambridge, following the death in 1931 of E W Hobson, Cambridge's Sadleirian Professor of Mathematics.

In spite of all Hardy's efforts to establish an analytical school in Oxford, Cambridge remained the mathematical hub of England and Hardy wished to be considered for its most prestigious mathematical chair. Moreover, he was now aged 54, and if he remained at New College he would have to yield up his rooms upon retirement; if he returned to Trinity College he could live there for life, as indeed did Littlewood.

Hardy was successful in securing the Sadleirian chair, and the following notice duly appeared in the Oxford University Gazette:

> SAVILIAN PROFESSORSHIP OF GEOMETRY
>
> The Electors to this Professorship propose shortly to proceed to an election of a professor in the place of Godfrey Harold Hardy, MA, Fellow of New College, who has resigned as from the beginning of Michaelmas Term, 1931.

By this time the stipend had increased to £1200 per year, and the Statutes now recognised the importance of research: *The duties of every Professor shall include original work by the Professor himself and the general supervision of research and advanced work in his subject and department.*

Hardy's successor as Oxford's Savilian professor was his first research student, E C Titchmarsh, who explicitly asserted that he would not lecture in geometry. Since then, the specification for the Chair has been relaxed and its occupant can now be selected from any area of pure mathematics.

After his departure to Cambridge, Hardy returned to Oxford for several weeks every summer to enjoy the atmosphere, see his college and mathematical colleagues, and captain the New College Senior Common Room cricket team in their annual match against the college servants and the choir school.

He also returned to give the first lecture to the **Invariant Society**, Oxford University's undergraduate mathematics society (still flourishing); this was in Hilary Term 1936 on the subject of Round numbers (those with few different prime factors).

Sayings of G H Hardy

In the 1920s Hardy had many preoccupations besides his teaching and research. He was heavily involved with several national societies, becoming President of the Mathematical Association:

In his first presidential address, *What is geometry?*, Hardy asserted that:

"I do not claim to know any geometry, but I do claim to understand quite clearly what geometry is."

He also joked that the only geometrical result that he had ever proved was:

"If a rectangular hyperbola is a parabola, then it is also an equiangular spiral."

224. [M^1. 8. g.] *A curious imaginary curve.*
    The curve $(x+iy)^2 = \lambda(x-iy)$
is (i) a parabola, (ii) a rectangular hyperbola, and (iii) an equiangular spiral. The first two statements are evidently true. The polar equation is
$$r = \lambda e^{-3i\theta},$$
the equation of an equiangular spiral. The intrinsic equation is easily found to be $\rho = 3is$.
    It is instructive (i) to show that the equation of any curve which is both a parabola and a rectangular hyperbola can be put in the form given above, or in the form
$$(x+iy)^2 = x \text{ (or } y\text{)},$$
and (ii) to determine the intrinsic equation directly from one of the latter forms of the Cartesian equation. G. H. HARDY.

As President of the National Union of Scientific Workers, Hardy's left-wing leanings emerge clearly from the following remarks to an audience of scientists:

"...although our jobs are very different from a coalminer's we are much closer to coalminers than capitalists. At least we and the miners are both skilled workers, not exploiters of other people's work, and if there's going to be a line-up, I'm with the miners."

On visiting America:
During his year in America Hardy wrote to a New College colleague:

"This country is in many ways the only one in the world — it has its deficiencies (tea γ, paper $\beta-$, ... , dinner rather an uninteresting meal, and at 6.30). But American football knocks all other spectacles absolutely flat: the sun shines more or less continuously: and for quietness and the opportunity to be your own master I've never come across anything like it."

7

A Glimpse of J. E. Littlewood

(Excerpts from Robin Wilson's "Hardy and Littlewood" in *Cambridge Scientific Minds*, edited by Peter Harmon and Simon Mitton, Cambridge University Press, 2002, pp. 202–219.)*

From the Editors

In 2002, Robin Wilson contributed an article about Hardy and Littlewood to Cambridge Scientific Minds, *a book celebrating the unmatched scientific legacy of the University of Cambridge. We have here selected portions of Wilson's piece in order to highlight the accomplishments of Hardy's great collaborator, J. E. Littlewood.*

The Hardy-Littlewood Partnership

The mathematical collaboration of Godfrey Harold Hardy and John Edensor Littlewood is the most remarkable and successful partnership in mathematical history. From before the First World War until Hardy's death in 1947 these mathematical giants produced around one hundred joint papers of enormous influence covering a wide range of topics in pure mathematics. Whereas many other mathematicians have collaborated on a short-term basis, there are no other examples of such a long and fruitful partnership.

Hardy and Littlewood dominated the English mathematical scene for the first half of the twentieth century. Throughout the nineteenth century,

---

* Reprinted by kind permission of Robin Wilson and of Cambridge University Press.

mathematical life in England, especially in pure mathematics, had been dwarfed by developments on the Continent and although Cambridge had produced some outstanding applied mathematicians, such as James Clerk Maxwell, George Gabriel Stokes, and William Thomson (Lord Kelvin), there were few pure mathematicians of world class other than Arthur Cayley in Cambridge and James Joseph Sylvester in Oxford. The situation changed with Hardy and Littlewood, who created a school of mathematical analysis unequalled throughout the world. As one contemporary colleague observed: 'Nowadays, there are only three really great English mathematicians, Hardy, Littlewood, and Hardy-Littlewood.'

As frequently happens in collaborative partnerships, the styles and personalities of the two men were very different. Both were mathematical geniuses, completely devoted to their subject, and with many interests in common. But Littlewood was probably the more original of the two, imaginative and immensely powerful and enjoying the challenge of a very difficult problem, while Hardy was the consummate craftsman, a connoisseur of beautiful mathematical patterns and a master of stylish writing. As Littlewood once observed: 'My standard role in a joint paper was to make the logical skeleton, in shorthand—no distinction between r and r^2, 2π and 1, etc., etc. But when I said, "Lemma 17" it stayed Lemma 17.' Hardy, who considered Littlewood as the finest mathematician he had ever known and the more creative partner in the collaboration, always wrote the final draft of their joint papers.

Littlewood's Early Years

J. E. Littlewood was born in Rochester, in Kent, on 9 June 1885, the eldest son of Edward and Sylvia Littlewood. His father, Ninth Wrangler in the 1882 Cambridge Mathematical Tripos, was offered a Fellowship at Magdalene College but turned it down in order to become headmaster of a new school near Cape Town, in South Africa.

Littlewood lived in South Africa from 1892 to 1900, enjoying its beauty but learning little from the school or from tuition at Cape Town University. He returned to England in 1900, and attended St Paul's School in London for three years, where he learned Greek ('mainly conditional sentences about crocodiles'), developed a love for music (particularly Bach, Mozart, and Beethoven), participated actively in a variety of sports, and learned mathematics from the distinguished algebraist F. S. Macauley. Macauley's approach was to encourage his pupils to work independently or with each other, rather than to be spoon-fed with material by the teacher—an approach that Littlewood

7. A Glimpse of J. E. Littlewood

Hardy and Littlewood in New Court, Trinity College, 1920.

was later to adopt with his own research students, urging them to 'Try a hard problem. You may not solve it, but you will prove something else.'

Littlewood obtained a minor Entrance Scholarship to Trinity College, and came into residence in October 1903. He was trained for Part I of the Tripos by the celebrated coach R. A. Herman (once described by Hardy as 'the mildest of the most ferocious of Huns') and became Senior Wrangler, but felt that he had wasted his first two years at Cambridge spending more than half of his time solving exceedingly difficult problems against the clock. In Part II of the Tripos he was placed in Class 1, Division 1.

Littlewood started research during the long vacation of 1906, under the direction of E. W. Barnes, later Bishop of Birmingham. Barnes believed in setting very hard problems in a sink-or-swim manner, and his first problem concerned so-called integral functions of zero order. Littlewood tackled this problem and quickly struck lucky, producing a fifty-page paper in just a few months. Littlewood's next assignment from Barnes was to prove the Riemann hypothesis, a notoriously difficult problem that remains unsettled to this day, relating to the distribution of prime numbers. Littlewood, who relished a challenge, managed to obtain some worthwhile results relating to this problem and wrote them up as a dissertation for a Trinity junior fellowship. The dissertation was well received, but the 1907 fellowship went to a classicist at his final attempt and Littlewood was awarded it the following year. Meanwhile, he won a Smith's prize, and was offered a three-year lectureship at Manchester University which he accepted, feeling that he needed a break from Cambridge. Unfortunately, the workload in Manchester was exceedingly heavy, and Littlewood came to regret his time there.

In 1903 Littlewood returned to Trinity as a college lecturer, succeeding Alfred North Whitehead whose lectures he had enjoyed as an undergraduate. In the following year, he proved a profound converse of a famous theorem on the summation of series by the Norwegian mathematician Niels Henrik Abel. As he later recalled: 'On looking back this time seems to me to mark my arrival at a reasonably assured judgement and taste, the end of my "education". I soon began my 35-year collaboration with Hardy.'

[Hardy moved to Oxford in 1919 and stayed until 1931, while Littlewood remained in Cambridge throughout. In the following selection, Robin Wilson continues with their story during these years apart]

Hardy's Oxford Years

... in Cambridge, Littlewood and Harald Bohr collaborated on a monograph on number theory, but were both so exhausted when the manuscript was complete that they did not have the strength to send it to the printer and it was left for several years. By this time the subject had developed considerably, and they handed it to others to use. In 1926 Littlewood did complete a textbook, *Elements of the Theory of Real Functions*, but he never had Hardy's facility in this direction and preferred to concentrate on research papers.

The 1920s were years of recognition for Hardy and Littlewood. At the Royal Society Hardy was awarded the Royal Medal in 1920, and Littlewood received the same medal nine years later. From 1917 to 1926 Hardy became secretary of the London Mathematical Society, never missing a meeting of the Society or of its Council, and became president from 1926 to 1928; he received the Society's highest honour, the De Morgan Medal, in 1929. Hardy also became president of the Mathematical Association from 1925 to 1927, and the National Union of Scientific Workers in 1926. In 1928 Littlewood was appointed to the newly founded Rouse Ball Chair of Mathematics in Cambridge, a very appropriate appointment, not least because he had been one of Rouse Ball's favourite pupils at Trinity. Littlewood received his first honorary degree in the same year, a Doctorate of Science from Liverpool University.

Both of them found relaxation in sporting activities. Hardy was a keen cricketer and tennis player and after leaving Oxford continued to return every summer to lead the Fellows' cricket team at New College; a well-known picture shows him leading a team of distinguished mathematicians on to the cricket field. Littlewood, an able athlete, enjoyed swimming and took up rock climbing while on holidays in Cornwall and the Isle of Wight. In 1924 he developed a taste for skiing while on vacation in Switzerland.

Reunited Again

In 1931 E. W. Hobson resigned the Sadleirian Chair in Cambridge, a post held for many years by Arthur Cayley. Although Hardy had transformed the Oxford mathematical scene, creating a school of analysis second to none, Cambridge was still the centre for mathematical research in the UK, and Hardy decided to apply. At Trinity he would be allowed to live in college rooms for the rest of his life, whereas no such facility was available in Oxford. Hardy was duly appointed, and spent the rest of his life in Cambridge with Littlewood.

In the late 1920s Hardy and Littlewood had collaborated on two highly original and influential papers, on the rearrangement of series and on a maximal theorem. These papers were to be the foundation of later research by many mathematicians. Their collaboration continued with Hardy's return to Cambridge, and from early 1932 they held weekly mathematical 'conversation classes' in Littlewood's rooms. According to Edward Titchmarsh:

> This was a model of what such a thing should be. Mathematicians of all nationalities and ages were encouraged to hold forth on their own work, and the whole exercise was conducted with a delightful informality that gave ample scope for free discussion after each paper.

During these Cambridge years Hardy and Littlewood continued to collaborate freely with other mathematicians. Hardy wrote his celebrated *An Introduction to the Theory of Numbers* (1938) with his former student

Edward Wright, and his final Cambridge Mathematical Tract, *Fourier Series* (1944), with Werner Rogosinski. Meanwhile Littlewood carried out some far-reaching work on Fourier series in an important series of papers with his brilliant student R. Paley, but this tragically came to an end with the latter's death in a skiing accident at the age of twenty-six. He also collaborated with Cyril Offord on random equations and with Mary Cartwright on some differential equations arising from electrical circuit theory. Later, during the Second World War, Cartwright and Littlewood carried out some important work on the mathematics of radar. He later wrote of this collaboration:

> Two rats fell into a can of milk. After swimming for a time one of them realised his hopeless fate and drowned. The other persisted, and at last the milk was turned to butter and he could get out.

Both Hardy and Littlewood wrote well-known books explaining the nature of mathematics to a general readership. Hardy's *A Mathematician's Apology* (1940), a personal account written at the beginning of the Second World War, is a rather sad book by a mathematician looking back as his powers are waning. In contrast, Littlewood's *A Mathematician's Miscellany* (1953) is a more joyful book, full of mathematical gems and allowing the reader to experience academic life at Trinity through his perceptive eyes.

These years continued to bring external recognition to Hardy and Littlewood. In 1939–41 Hardy again became President of the London Mathematical Society, becoming the only president to serve two terms. Littlewood succeeded him in 1941–43, having received the De Morgan Medal in 1938 and later winning the Senior Berwick Prize (1960). From the Royal Society, Littlewood received the Sylvester Medal in 1943 and the Copley Medal in 1958.

Hardy retired from the Sadleirian Chair in 1942. He continued to write, and his last book, *Divergent Series* (1948) appeared after his death. In 1947 he was elected 'associé étranger' of the Paris Academy of Sciences, one of only ten people from all nations and scientific subjects. He died on 1 December 1947, the very day that he was due to be presented with the Copley Medal by the Royal Society...

Littlewood retired in 1950 at the statutory age of 65. However, he was permitted to continue teaching after the mathematics faculty wrote to the General Board:

> Professor Littlewood is not only exceptionally eminent, but is still at the height of his powers. The loss of his teaching would be irreparable, and it is avoidable. Permission is requested to pay a fee of the order of £100 for each term's course of lectures.

In fact, Littlewood received just £15 for each course.

7. A Glimpse of J. E. Littlewood

Pólya and Littlewood having dinner with a bit of mathematics at Stanford, CA in 1957.

For thirty years Littlewood had suffered from a nervous malady which caused him to function at a fraction of his capacity, and he spent many hours in local cinemas whiling away the tedious hours. But in 1960, a brilliant neurologist experimented with some new drugs and cured the depression. This gave Littlewood a new lease on life and at the age of 75 he started to take up lecturing invitations, particularly in the United States which he visited eight times.

Littlewood continued his mathematical researches until he was well into his eighties. He died on 6 September 1977 at the age of 92.

8

A Letter from Freeman Dyson to C.P. Snow and Letters from Hardy to Dyson

Freeman Dyson, 1948.

From the Editors

Freeman Dyson, a mathematical physicist, was a professor at the Institute for Advanced Study (IAS) in Princeton from 1949 to 1994.

As a child, he was strongly influenced by E. T. Bell's Men of Mathematics. *His training was in mathematics, and he says that he still is a mathematician. He took courses from Hardy in the early 1940s and was a Fellow of Trinity*

College, Cambridge from 1946 to 1949. Dyson has made significant contributions to quantum electrodynamics, solid-state physics, astronomy, and nuclear engineering. He has written several important volumes and has been a frequent contributor to The New York Review of Books. *At age 91 he solved a problem stemming from the "iterated prisoner's dilemma."*

What follows are letters —one from Dyson to C. P. Snow, and two exchanges between Hardy and Dyson. The first letter was written to Snow after Dyson had read Snow's preface to A Mathematician's Apology. *Snow, of course, knew Hardy at Cambridge. He served in several civil service positions and was technical director of the Minister of Labour during World War II. Snow also was a novelist, best known for his* Strangers and Brothers *series. In his nonfiction book* The Two Cultures, *he famously argued that academics were divided into two camps, the humanists and the scientists, where those in one camp could not or would not understand what those in the other were doing. He became Lord Snow upon being knighted in 1947.*

Dyson in his letter to Snow reflected on the great pleasure he had being a student in Hardy's classes, providing insights into his sense of humor and approach to teaching.

May 22, 1967

Lord Snow
199 Cromwell Road
London, SW 5, England

Dear Lord Snow:

I feel impelled to add a postscript to your chapter on Hardy. It answers many questions which I had never expected to have answered, and it brings back memories of events in which you too were peripherally concerned.

I and my friend James Lighthill, whom you probably have met in the course of his distinguished later career, came up to Trinity as undergraduates in 1941. We plunged immediately into the advanced Part III lectures given by Hardy and Littlewood. We knew that our time at Cambridge would be short, and we did not wish to waste any of it. In a way we were lucky, because there were hardly any other advanced students. Hardy and Littlewood lectured alternately in a small room in the Arts School, with the audience sitting around a table. Usually there were four of us at the table, and never more than six. Besides

8. A Letter from Dyson to C.P. Snow and Letters from Hardy to Dyson

Lighthill, there was Aronszajn, then in the uniform of the Pólish Army, and Christine, the future Mrs Bondi.

In that little room we sat within a couple of feet of Hardy, three times a week for two years. He lectured like Wanda Landowska playing Bach, precise and totally lucid, but displaying his passionate pleasure to all who could see beneath the surface. I was always amazed that so great a mathematician (we considered him the best in the world) should spend so much of his time and energy in lecturing to so small a class. And each lecture was carefully prepared, like a work of art, with the intellectual dénouement appearing as if spontaneously in the last five minutes of the hour. For me these lectures were an intoxicating joy, and I used to feel sometimes an impulse to hug that little old man in the white cricket-sweater two feet away from me, to show him somehow how desperately grateful we were for his willingness to go on talking. Only now after reading your chapter, I begin to understand why he put his heart and soul into these lectures in the way that he did.

He very rarely spoke to us as individuals. There was plenty of humour in his lectures, but it tended to be dry and impersonal. Once, he arrived at the lecture-room and began in the most solemn tone, "I have some grave news for those of you who belong to Trinity." He paused while we prepared ourselves to hear of the death of some revered College dignitary, and then continued, "It will be tripe again for lunch." These were days when rationing left us no alternative but to eat whatever the College provided.

I remember only one occasion on which I summoned the courage to visit him in his rooms. I had found a mistake in some of the equations in his book *Ramanujan*. With great pride and trepidation I went in and asked him whether it was possible that he had erred. He looked at it, said in an off-hand way, "Oh yes, you are right, it is a mistake," and the conversation was at an end.

Over us all in those days hung the shadow of the war. My views about the war were much closer to his, as you describe them in your chapter, than to yours. My ethical bible was Aldous Huxley's *Ends and Means*, and I could see clearly enough that even the most just ends did not justify the means that we were employing. But Hardy never spoke to us about these matters, and we would never have dared to speak to him uninvited. I did not imagine that the war was causing him any great concern.

At the end of our two years it was understood that we should go into government service. Somebody (I do not know whether it was Hardy) appealed to the authorities to let Lighthill and me stay another year at Cambridge, to get started in mathematical research. The authorities said they would let one stay, but not both. On these terms neither of us would accept the honour. And so in June 1943 we came before you for placement. You spoke to me in glowing terms about the work I should be doing for the next two years in the Operational Research Section at the headquarters of R.A.F. Bomber Command. I did not speak to you about my moral scruples, and you did not speak to me about your friendship with Hardy.

In September 1946 I came back to Cambridge and started life afresh as a physicist. I felt slightly guilty towards Hardy for having deserted pure mathematics, and I did not seek his company. He was then an invalid, and I was unsettled and depressed, and it did not occur to me that I could have lightened his last days by going to see him occasionally.

Much later, after I had read Hardy's *Apology*, I realized that I probably owe him a debt of gratitude for something more than his two years of superb lectures.

How did it happen that Lighthill and I came to Trinity at the age of 17 knowing enough mathematics to understand him? We both came from Winchester, Hardy's old school. Nothing remotely approaching the Cambridge Part III syllabus was taught there. Hardy's name was never mentioned in the school, and he certainly never came near it. But there was one thing in the school which perhaps was connected with him. There was a copy of Jordan's *Cours D'Analyse* in the library.

When Lighthill and I were fifteen, we discovered the three fat and dusty volumes of Jordan. They had been in the library a long time, and there was no record of how they got there. It is hard to see how anybody, with one exception, would have considered them fit fare for English schoolboys. Looking back, I take great pride in the fact that Lighthill and I understood what a treasure we had found. Slowly and systematically, without help from anybody, we worked through the three volumes, and had our eyes opened to mathematics as Hardy had forty years earlier. I remember most vividly the days spent on the cricket-field, which I hated as much as Hardy loved it, with a volume of Jordan inadequately concealed under my shirt. Occasionally I was lucky enough to play under a captain who would put me to field in the deep, so that I could read openly and undisturbed.

As I read your chapter on Hardy, I felt many regrets. I came just too late into his life to work with him as a collaborator, and I narrowly missed many opportunities of closer contact with him. But perhaps the most poignant of all my regrets is that I never mentioned to him the *Cours D'Analyse* in the library at Winchester. It seems overwhelmingly probable that he must have put it there, with a vague hope that some boys following in his footsteps would some time light upon it. And when after forty years the boys finally found it, were transformed by it magnificently, and came to sit at his feet, they did not know whom to thank for it.

That is the end of my tale. Please excuse my writing at such length. I was just so moved by your chapter that my pen ran away with me.

Yours sincerely,

Freeman Dyson

Hardy to Dyson—1

From the Editors

The first of the Hardy letters to Dyson is a response to a manuscript on Waring's problem that Dyson, then an undergraduate, had sent to Hardy in 1943. Hardy's reply speaks volumes about his concern for students. The impression that Hardy made on Dyson is underscored by the fact that he kept the Hardy letter 72 years after it was written. Dyson observed: "The other thing that is noteworthy about the letter is that it was written in response to a rather puerile attempt by a second-year undergraduate to prove Waring's conjecture for the case of seven cubes. Hardy was remarkably friendly to students at all levels. My attempt failed, and I believe the seven-cube conjecture was afterwards proved by C. A. Rogers."

Trin. Coll. Camb.
June 24 [1943]

Dear Dyson

I am looking through your MS, but it will take some time, and I shall probably have to write to Davenport about it. The truth is that I am completely out of touch with Waring's problem (& actually I never knew this side of it really well, though naturally I have worked through Landau's proof & even done it in lectures). So I don't know what the actual situation is, or whether anybody has succeeded in making any material advance on him. Davenport, I know, got 'almost all' for 4 cubes, quite recently.

Also, I never knew the theory of ternary forms properly, & have very vague ideas about the plausibility of hypotheses like yours. My general reaction is to be rather sceptical about much coming out of it. However, I will do what I can.

Littlewood & I resolved, about 1928 or thereabouts, that we had shot our bolt about W's problem, and would never try to do more with it!

I am inclined to think that you would find Dickson, *Modern elementary theory of numbers*, very useful, more particularly Ch. X. I don't know whether it is difficult to get now.

Yours sincerely
G H Hardy

Hardy to Dyson –2 (1943)

From the Editors

Dyson recalled, "I had enquired about the mathematician [Ernest W. Barnes] who had switched from mathematics to theology and became Bishop of Birmingham." Hardy responded:

Dear Dyson,

Here is the paper. Yes, Barnes is the Bp. [Bishop] all right. He was a really good mathematician (& a colleague of mine for a while here) though incurably unprofessional as judged by "Landau" standards. But he dropped mathematics like a hot brick after getting his FRS [Fellow of the Royal Society].

GHH

Ed. Note —Barnes was Second Wrangler in 1896, won the Smith Prize in 1898, and was elected a Fellow of Trinity College in the same year.

9

Miss Gertrude Hardy

From the Editors

G. H. Hardy famously collaborated with mathematicians Ramanujan and Littlewood, but his foremost personal collaboration was with his sister, Gertrude. They shared the joys—and one terrible accident—of childhood, and she was with him at the end in 1947.

Here we sketch the life of "Miss Gertrude Hardy," as she was known in her professional life. We are most indebted to St. Catherine's School in Surrey for supplying us with this information and doing so with dispatch and good cheer.

Gertrude Edith Hardy was born on April 17, 1878, the year after her brother Godfrey Harold. They were raised in Cranleigh, a Surrey town about 40 miles southwest of London. One imagines a childhood of indoor activities and outdoor play, a pleasant, English upbringing in the latter years of the 19th century.

As we noted in our biographical section (I.1), the darkest event of these youthful days was the moment when Godfrey hit Gertrude's eye with a cricket bat. The eye could not be saved. She would have to deal with the effect of this mishap for the rest of her life. But the trauma did not keep her down, nor, to the credit of all, did it affect her relationship with her brother.

When G. H. Hardy was sent off to Winchester School, Gertrude remained in Surrey. She was a day pupil at St. Catherine's School in Bramley near her home. She graduated with distinction, having shown particular aptitude in French and art, and then went to the University of London where she received her BA degree in 1899. This was far from the common path taken by young ladies at the end of the Victorian period.

Nor did all young women of the day play cricket, but the photo shows young Gertrude with bat in hand, intently awaiting the bowler's offering. Given the terrible price she had paid on the cricket field, this reveals a stoic if not spunky attitude on her part. Or perhaps an obsession with cricket was hard-wired into the Hardy genetic code. In any case, it is an arresting image.

Miss Hardy playing cricket at St. Catherine's School.

After university, Gertrude's *alma mater* called her back, and she began a 42-year career at St. Catherine's. Records indicate that she was the art mistress at school and also taught Latin and coached a bit of mathematics. Her athletic interests led her not only to the cricket field but also to the tennis courts. And, with the help of a small committee, she edited the school magazine.

9. Miss Gertrude Hardy

Students would later characterize Miss Hardy as kind and dedicated. One remembered a woman who was reserved in manner "... but very nice when you got to know her." And, this student continued, she "... had a glass eye which we all found very fascinating." Apparently, faculty were given nicknames by their pupils, and Gertrude's was "Hardicanute." The reference is to a Medieval King of England who shared her surname (sort of). Below is a picture of Miss Hardy/Hardicanute in her heyday at St. Catherine's.

Miss Hardy, Art Mistress.

One has, then, the picture of a dedicated teacher who never married but threw herself into the life of St. Catherine's (a school, by the way, that is still there and still thriving). Her long service carried her through the suffering of both World Wars and might have continued longer, but in 1945 came word of her brother's deteriorating physical condition. Gertrude Hardy knew what she had to do. "It was sad to leave so suddenly," she wrote a few years after the fact, "but there was nothing else to be done. I couldn't possibly leave a semi-invalid brother to fend for himself in college."

And so Gertrude gave up her position and moved to Cambridge to care for G. H. Hardy. She reported that this change of venue had its pros and cons. She found the countryside to be something of a disappointment—she called Cambridgeshire "as flat as a pancake and only properly appreciated when one is on a bicycle." On the other hand, the architecture was spectacular and the flowers in the famous Backs along the Cam were beautiful beyond measure.

In her new life, there were all sorts of domestic matters to be undertaken. "At present," she confided in a letter, "I am occupied in dusting my brother's books. I do about seven shelves a day and hope to be through in a month or so." But as Hardy's condition worsened, her responsibilities assumed a different, more serious character. He died on December 1, 1947, and his last months were surely made easier by the presence of Gertrude at his side.

At that point, she still had a long life ahead. This she spent in Cambridge, enjoying the university town, listening to her wireless (a retirement gift), and attending services in King's College Chapel. She invited her former students to visit her with the hope that "... perhaps some day I shall meet some of you in Trinity Great Court or on King's Parade."

Gertrude Hardy died, in Cambridge, on January 20, 1963. Back in the Chapel at St. Catherine's, a plaque was put up in her honor. And the school magazine published an obituary of this remarkable woman. We end this tribute to Gertrude Hardy by including it in full:

MISS G. E. HARDY (1891–1945)
(Pupil, Assistant Mistress, Housemistress, Senior Mistress)

A host of old St. Catherine's girls will be sad to hear of the death this winter of Miss Hardy.

She came to Bramley as a non-resident teacher in 1903, after attending the school as a pupil, and only left in 1945 to look after her brother who was a Cambridge don. There can be few people who had had such a long perspective of changing fashions in school girls, staff and, indeed, headmistresses.

9. Miss Gertrude Hardy

When Miss Commins left in 1926, Miss Hardy edited the school magazine with considerable success, and when Miss Ireland retired in 1935, she became second mistress.

Even after 40 years my memories of Miss Hardy are very vivid. In those distant days it seemed to us that the staff lived a life almost as circumscribed as our own, but we felt, dimly, that she achieved an affectionate toleration of the régime and its pupils. A teacher of outstanding ability, she combined keen interest in her bright pupils and, as I have good reason to remember, boundless patience with the slow. Her reproofs were astringent, but never undeserved. Her scrupulous fairness gave that feeling of security which is so valuable to the young.

Looking back to my days at St. Catherine's, a jumble of early impressions of Miss Hardy crowd into my mind—her integrity, independence of mind and moral courage; her rather shy and diffident, but most understanding, kindness; her love of walking and the country; her knowledge of wild flowers (most helpful to me, a London chemist, teaching botany to girls, many of whom had grown up in the country!); her scholarship; her power of inspiring the most unlikely people, including myself, with an interest in pictures and architecture; and, not least, her fund of common sense in the practical business of living, going with an almost complete lack of interest in things domestic, especially gadgets.

These wide interests and sympathies remained with Miss Hardy to the end of her life.

After her retirement in 1945, Miss Hardy gave herself up with absolute single-mindedness to her brother, whose many months of illness were cheered and alleviated by her devoted care for him. At the same time she gathered round her a wide circle of good friends of all ages, who loved her for herself and for her thoughtfulness. She was, I think, one of the kindest and most generous people I have ever met; but many of her kindnesses were unknown and unsuspected. During those Cambridge years she enjoyed her surroundings to the full, and appreciated her good fortune in living in such a lovely city with its fine architecture and bright gardens, shape and colour being two of her constant delights. She enjoyed, too, her well-ordered and quietly active life, doing her daily "Times" crossword, reading detective stories and Gerald Durrell, shopping for the old ladies in a nearby home, being visited by Old Girls [i.e., St. Catherine *alumnae*] and their husbands and their children, entertaining and talking to her friends, taking a lively interest in world affairs, choosing clothes, trying to understand modern art, and always ready to help anyone in any way she could. And then, as her sight began to fail, it was her courage that we all admired, and during those last weeks of darkness, with very little hope of sight returning, it was an inspiration to hear her comments on the news of the day and her perpetual interest in her friends and their families. She was always a good companion, and her sense of humour and her unselfishness were with her to the end.

II

Writings by and about G. H. Hardy

This part contains two main chapters.

The first is called "Selections from Hardy's Writings." Here we have combed through his books and letters for passages that touch upon topics ranging from Ramanujan's talent, to the vagaries of the French language, to (believe it or not) football at Princeton. This is meant to be a feast of Hardy's writing.

Then we have a somewhat longer chapter called "Selections from What Others Have Said about Hardy." Here, a variety of authors—Littlewood, Wiener, Halmos, and many others—describe Hardy's approach to mathematics, his boundless love of cricket, his views on the relative merits of Cambridge and Oxford, and a host of other matters.

These excerpts are meant to give a fuller, more compelling picture of G. H. Hardy as seen through his eyes and the eyes of people who knew him.

A COURSE OF PURE MATHEMATICS

Centenary Edition

G. H. Hardy

CAMBRIDGE

10
Hardy on Writing Books

"Young men should prove theorems; old men should write books." was the advice that Hardy gave Freeman Dyson when he was a student at Cambridge in the early 1940s. It appears that he did not follow his own advice. His first book, *Integration of Functions of a Single Variable*, was published in 1905 when he was 28; his second *A Course of Pure Mathematics* was published in 1908 when he was 31. More than one hundred years later both books are still in print! *A Course of Pure Mathematics* is now in its tenth edition, and continues to sell. T. W. Körner of Cambridge notes in his preface to the Centenary Edition that "For the next 70 years [after publication], Hardy's book defined the first analysis course in Britain."

The first edition of *A Mathematician's Apology* was published in 1940 when he was 63. It has been reprinted 19 times and has been widely praised by mathematicians and non-mathematicians as being a beautifully crafted account of the mathematical experience.

In addition to his more than 340 research papers, we also are indebted to him for writing books.

123

11
Selections from Hardy's Writings

We begin with Hardy's characterization—outdated, very British, but nonetheless memorable—of the subject to which he devoted his life: "Mathematics is an ancient, and one would have supposed quite a gentlemanly, study."
—"Mathematics," *The Oxford Magazine*, 48 (1922), p. 819.

Hardy would occasionally reflect on this gentlemanly study, as in the following passage about proof in mathematics:

"At this point I should like to . . . make a few general remarks about mathematical proof as we working mathematicians are familiar with it. It is generally held that mathematicians differ from other people in *proving* things, and that their proofs are in some sense *grounds* for their beliefs. Dedekind said that 'what is provable, ought not to be believed without proof'; and it is undeniable that a decent touch of scepticism has generally (and no doubt rightly) been regarded as some indication of a superior mind. But if we ask ourselves why we believe particular mathematical theorems, it becomes obvious at once that there are very great differences. I believe the Prime Number Theorem because of de la Vallée-Poussin's proof of it, but I do not believe that $2 + 2 = 4$ because of the proof in *Principia Mathematica*. It is a truism to any mathematician that the 'obviousness' of a conclusion need not necessarily affect the interest of a proof."
—"Mathematical Proof," *Mind* (1929), p. 17. Reprinted with permission of the Oxford University Press.

Hardy, who enjoyed a metaphor as much as anyone, likened a mathematician to the first explorer of a mountainous landscape, whose goal ". . . is simply to distinguish clearly and notify to others as many different peaks as he can. There are some peaks which he can distinguish easily, while others are less clear. . . . But when he sees a peak he believes that it is there simply because he sees it. If he wishes someone else to see it, he *points to it*, either directly or

through the chain of summits which led him to recognize it himself. When his pupil also sees it, the research, the argument, the *proof* is finished."
   —"Mathematical Proof," *Mind* (1929), p. 18. Reprinted with permission of the Oxford University Press.

One of Hardy's unquestioned classics was his 1908 text, *A Course of Pure Mathematics*. This was his attempt to bring rigorous, serious analysis to Britain, a nation that had heretofore been satisfied with a more informal treatment of the subject. His mission succeeded. In fact, Hardy could look back almost three decades later and note that *A Course of Pure Mathematics* "... was written when analysis was neglected in Cambridge, and with an emphasis and enthusiasm which seem rather ridiculous... If I were to rewrite it now I should not write (to use Prof. Littlewood's simile) like 'a missionary talking to cannibals'... "
   —Preface to the Seventh Edition, *A Course of Pure Mathematics*, Cambridge University Press, 1937.

"What is essential in mathematics is that its symbols should be capable of some interpretation; generally they are capable of many, and then, so far as mathematics is concerned, it does not matter which we adopt. Mr Bertrand Russell has said that 'mathematics is the science in which we do not know what we are talking about, and do not care whether what we say about it is true,' a remark which is expressed in the form of a paradox but which in reality embodies a number of important truths."
   —*A Course of Pure Mathematics*, Cambridge University Press, 1937.

Of all the branches of mathematics, none captured Hardy's imagination more than number theory. For him, the whole numbers were "... a continual and inevitable challenge to the curiosity of every healthy mind."
   —"The Theory of Numbers," *Nature*, 110 (1922), p. 385.

This same attitude was evident in his Preface to the famous text on number theory that Hardy wrote with E. M. Wright:

"Our first aim has been to write an interesting book, and one unlike other books. We may have succeeded at the price of too much eccentricity, or we may have failed; but we can hardly have failed completely, the subject-matter being so attractive that only extravagant incompetence could make it dull."
   —(with E. M. Wright) Preface to *An Introduction to the Theory of Numbers*, Oxford University Press, 1938, p. v.

In Hardy's inaugural lecture at Oxford on the occasion of his assuming the Savilian Professorship in Geometry, he had to explain why he chose to talk not in philosophical terms about geometry but in mathematical terms about number theory. "I think that a professor should choose, for his inaugural lecture, a subject, if such a subject exists, to which he has made himself some contribution of substance and about which he has something new to say. And about mathematical philosophy I have nothing new to say.... I have therefore finally decided, after much hesitation, to take a subject which is quite frankly mathematical, and to give a summary account of the results of some research which, whether or no they contain anything of any interest or importance, have at any rate the merit that they represent the best that I can do.

"My own favourite subject has certain redeeming advantages. It is a subject, in the first place, in which a large proportion of the most remarkable results are by no means beyond popular comprehension. There is nothing in the least popular about its *methods*; as to its votaries it is the most beautiful, so by common consent it is the most difficult of all branches of a difficult science; but many of the actual results are such as can be stated in a simple and striking form. The subject has also a considerable historical connexion with this particular chair. I do not wish to exaggerate this connexion. It must be admitted that the contributions of English mathematicians to the Theory of Numbers have been, in the aggregate, comparatively slight. Fermat was not an Englishman, nor Euler, nor Gauss, nor Dirichlet, nor Riemann; and it is not Oxford or Cambridge, but Göttingen, that is the centre of arithmetical research to-day. Still, there has been an English connexion, and it has been for the most part a connexion with Oxford and with the Savilian chair." He then went on to cite the contributions by Sylvester and Henry Smith, his predecessors as Savilian Professor.

—"Some Famous Problems of the Theory of Numbers," *Inaugural Lecture*, Oxford University Press, 1920, pp. 6–7.

We leave this topic with a final testimonial to his beloved theory of numbers: "Are there infinitely many primes of the form $2^n - 1$? I find it difficult to imagine a problem more fascinating or more intricate than that."

—"The Theory of Numbers," *Nature*, 110 (1922), p. 383.

G. H. Hardy knew so many of his era's great mathematicians, and he occasionally remembered them in writing. In our next few selections, we sample his reminiscences.

Of course, even among a Who's Who of distinguished colleagues, Srinivasa Ramanujan stood out. The initial letter from Ramanujan was, according to Hardy, "... certainly the most remarkable that I have ever received." After a brief introduction, the letter contained a host of mathematical statements, all without proof. "Some of the formulae were familiar," Hardy noted, "and others seemed scarcely possible to believe."

—"S. Ramanujan, F.R.S.," *Nature*, 105 (June 17, 1920), p. 494–495.

Of Ramanujan, Hardy wrote that [he] "used to entertain his friends with his theorems and formulæ, even in those early days ... ; he could give the values of

$\sqrt{2}, \pi, e, \ldots$ to any number of decimal places ... In manners, he was simplicity itself...."
—*Collected Papers of Srinivasa Ramanujan*, edited by G. H. Hardy, et al., Cambridge University Press, 1927, p. xxi. Reprinted with permission of the Cambridge University Press.

Hardy summed up Ramanujan's contributions as a mathematician: "It is often said that it is much more difficult now for a mathematician to be original than it was in the great days when the foundations of modern analysis were laid; and no doubt in a measure it is true. Opinions may differ as to the importance of Ramanujan's work, the kind of standard by which it should be judged, and the influence which it is likely to have on the mathematics of the future. It has not the simplicity and the inevitableness of the very greatest work; it would be greater if it were less strange. One gift it has which no one can deny, profound and invincible originality. He would probably have been a greater mathematician if he had been caught and tamed a little in his youth: he would have discovered more that was new, and that, no doubt, of greater importance. On the other hand he would have been less of a Ramanujan, and more of a European professor, and the loss might have been greater than the gain."
—*Collected Papers of Srinivasa Ramanujan*, edited by G. H. Hardy, et al., Cambridge University Press, 1927, p. xxxvi. Reprinted with permission of the Cambridge University Press.

Hardy long claimed that he was drawn to mathematics by reading Camille Jordan's *Cours d'Analyse*. In a tribute to Jordan he wrote: "The result was a new book and the rise of a new school; for it is fair to attribute to the inspiration of Jordan the beginnings of the movement which, carried on by Hadamard, Borel and Lebesgue, has revolutionized the foundations of modern analysis." Freeman Dyson—who, like Hardy, was a schoolboy at Winchester College—was also much influenced by Jordan's *Cours* and liked to think that it was Hardy's insistence that Jordan's work was there on the library shelves that made it possible for him to discover these volumes decades later.
—"Camille Jordan, 1838–1922," *Proc. Royal Soc. (A)* 104 (1922), p. xxiii.

In that same tribute, Hardy described [Camille] Jordan as the first mathematician to give a comprehensive account of the theory of Galois and its applications to the study of algebraic equations. Hardy claimed little expertise in this area but was convinced that Jordan showed himself to be "*un grand algébriste*" and added that this is "... at any rate a verdict which no one who knows Jordan as an analyst will be inclined to dispute."
—"Camille Jordan, 1838–1922," *Proc. Royal Soc. (A)* 104 (1922), pp. xxv–xxvi.

The mathematician J. W. L. Glaisher was Hardy's colleague at Cambridge. Glaisher was a colorful character who had been a Second Wrangler and then made his career at his *alma mater*. At one point, he tutored the brilliant Ludwig Wittgenstein. In addition, Glaisher was a natural leader, serving as president of both the Royal Astronomical Society and of the Cambridge Bicycle Club. And he was an especially ardent collector of fine ceramics. In fact, a large room at the University of Cambridge's Fitzwilliam Museum now holds the Glaisher Collection of pottery and porcelain.

But Glaisher was perhaps best known as the sole editor of *Messenger of Mathematics*. This was a wonderful little journal which served not as a forum for earth-shaking research articles but as the perfect place for young scholars to cut their mathematical teeth. "It occupied a comparatively humble position in the mathematical world," wrote Hardy, "but a useful, individual, and honourable position." The journal was very much the enterprise of this one man, a responsibility that Hardy called "thankless." Thus, in Glaisher's obituary, Hardy noted that "Few mathematicians can have expected that the *Messenger* would survive Glaisher's death, and in fact it dies with him." Hardy was, alas, correct.

—"Dr. Glaisher and the '*Messenger of Mathematics*'," *Messenger of Mathematics*, 58 (1929), p. 159.

Another British mathematician of Hardy's time was E. W. Hobson, whom Hardy remembered in a 1934 tribute: "The modern theory of functions of a real variable," he wrote, "was in its infancy when Hobson began his work. In England it was practically unknown, and rather derided. There may, perhaps, have been a little excuse for the people who... regarded it as a monstrosity; for there was still a faint air of mystery hanging about the elements, and much of the superstructure was inelegant and more than a little tiresome. Hobson and Young were the first English mathematicians to see the significance of the new ideas, and fought what must often have been a rather disheartening fight for their recognition. Hobson lived to see real function theory the most highly developed mathematical discipline in Cambridge, a subject recognized even as 'a good Tripos subject'."

—"E. W. Hobson," *J. London Math. Soc.* 9 (1934), p. 236.

As a final testimonial, we include Hardy's assessment of Henri Lebesgue, whom he called "one of the greatest mathematicians of recent times." Acknowledging that Lebesgue did not exhibit the mathematical breadth of a Poincaré or a Hilbert, Hardy nonetheless gave him his due: "He was rather a man with one outstanding claim to fame... all his secondary work, of which there is not much, is overshadowed by his work on integration. There, he was first: the 'Lebesgue integral' is one of the supreme achievements of modern analysis."

—"Prof. H. L. Lebesgue," *Nature*, 153 (1943), p. 685.

In our Introduction, we mentioned that Hardy was a prolific reviewer. Here we present selections from some reviews, beginning with the pithy observation that "... a book on mathematics without difficulties would be worthless."
—"Review of Courant and Robbins' *What Is Mathematics?*" *Nature*, 150 (1942), p. 674.

After a series of not very positive remarks about the precursor of one of the all-time successes in the history of calculus texts, Hardy finished his review of W. A. Granville's *Elements of the Differential and Integral Calculus* with the following weak endorsement: "I could make many other criticisms of detail; but I do not wish to appear ungenerous to a book which, while hardly likely to excite enthusiasm, has solid merits, and is on the whole clear, readable, and reasonably accurate. To profess that I regard its methods as the 'latest and best' [as the preface promised], would, however, be an exaggeration."
—"Review of Granville's *Calculus*," *Math. Gazette* 7 (1913), p. 24.

In considering a new text in function theory he struck a nice balance: "It is not easy to find anything to criticize adversely. The authors cannot lay claim to the conciseness of Jordan or the sprightliness of Picard: but in lucidity, thoroughness, and consistency of purpose they yield to no one, and although the book is not exactly easy reading it is never unreasonably difficult or heavy. It is enlivened with a large number of excellently chosen examples, many of them interesting and important theorems in themselves."
—"Review of O. Stolz and J. A. Gmeiner's *Einleitung in die Funktionentheorie*," *Math. Gazette* 3 (1905), p. 304.

Hardy could be succinct in his reviews. For one book on function theory the whole review was: "The author gives a careful account of certain well known regions of function theory. He makes no pretense to originality; but this thesis is interesting as showing an interest in modern mathematics in Spain."
—"Review of Patricio Reñalver y Bachiller's *Estudio elemental de la prolongación analitica*," *Math. Gazette* 6 (1913), p. 346.

Here Hardy considered a more advanced book on divergent series: "It is a little old-fashioned to say that the equation $1 - 1 + 1 - \cdots = \frac{1}{2}$ has 'no meaning,' although, as the authors mean it, the remark is of course, quite true.... And I may mention here that the examples (it is refreshing to see examples in a German book) are extremely well selected; they contain a large number of important theorems for which there is no space in the text, and nothing which is not of some interest, all 'tricks' being rigidly excluded."
—"Review of O. Stolz and J. A. Gmeiner's *Theoretische Arithmetik*," *Math. Gazette* 2 (1903), pp. 312–313.

Hardy took an especially dim view of *Le Spectre Numérique*, by M. Petrovitch when he observed, "The defect of this book is that there is nothing in it."

Hardy allowed that the book might be redeemed if even one "application of interest" could be found amid all of its mathematical verbiage. But, he continued, "All that appears from the book is that M. Petrovitch has found none."
—"Review of Petrovitch's *Les spectres Numériques*," *Math. Gazette* 10 (1920), p. 77.

On other occasions, however, he was in a remarkably better humor: "This is a charming book, written not only with extreme clearness and precision, but also with a freshness and originality seldom to be found in books which purport to be elementary text-books. When I began reading it I was not acquainted with any of Prof. Kowalewski's writings, but I had not spent an hour over it before I went out and ordered his *Grundzüge der Differential- und Integralrechnung*, of which it is a continuation. I recommend it with confidence to all who are interested in the foundations of the theory of functions of a complex variable."
—"Review of G. Kowalewski's *Die complexen Veränderlichen und ihre Funktionen*," *Math. Gazette* 6 (1912), p. 345.

And one review featured a little French lesson: "This volume seems to me hardly worthy of the firm by whom it is published [Gauthier-Villars]. It is slapdash and inaccurate, and although it has some novel features, the originality is not of so striking an order as to justify an addition to the long list of *Cours* and *Traités* already on the market. Moreover, it is written in what is to me at any rate a very irritating style. French is admittedly the language best adapted for scientific purposes, and the French of the best French mathematicians is unrivalled for lucidity and charm. But it is a language which has the defects of its qualities; its lucidity can wear thin, and its nervous terseness become jerky."
—"Review of R. d'Adhemar's *Leçons sur les principes de l'Analyse*," *Math. Gazette* 8 (1912), pp. 346–347.

Hardy had high praise for a 1905 volume by the great, troubled mathematician René Baire. The book adapted Baire's "... remarkable memoir, *Sur les fonctions de variables réelles*, published as his thesis in 1899." As Hardy noted, "The problem of finding the *necessary and sufficient* conditions that a function whose points of continuity are given should be capable of representation as the sum of a series of continuous functions is one which most mathematicians would have regarded as hopeless if M. Baire had not completely solved it."
—"Review of R. Baire's *Leçons sur les fonctions discontinues*," *Math. Gazette* 3 (1905), p. 232.

We end our survey of Hardy's reviews with his words about a text by J. Edwards. Hardy was not pleased. He observed, "There is always the danger... that a student who reads a textbook may suppose that the statements which it contains are true." Such was not the case with the volume under review. The best Hardy could say was that the book might prove to be of some value to a professor who was "... careful not to allow it to pass into his pupil's hands."

—"Mathematical Analysis," *Nature* 109 (1922), p. 437.

Our next few passages show Hardy in a more reflective, philosophical mood.

On E. W. Hobson's giving the Gifford lectures at Cambridge: "But in Gifford lectures, one looks for something more; one wants, in short, either a theistic or an anti-theistic moral, and Hobson's moral, that religion has nothing to fear from science if only it will leave science alone, may be a very reasonable one but is not at all exciting. The reader itches to know what Hobson himself really thinks about it, and that is just what he will not say."

—"E. W. Hobson," *J. London Math. Soc.* 9 (1934), p. 229.

In critiquing Bertrand Russell's concept of "worship of the non-existent," Hardy offered his own, more down-to-earth view: "Existence means to me much more than it does to him;... I should consider, for example, that a very halting affection for a very imperfect friend, who has at any rate the saving merit of existence, may be far more valuable than the most perfect love of a purely imaginary God."

—"Mr. Russell as a religious teacher," *Cambridge Magazine* (1917), p. 132.

In writing of his own response to external phenomena, Hardy observed, "I know that there are emotions, many, for example, of the emotions inspired by music, which I am almost, if not quite, incapable of experiencing."

—"Mr. Russell as a religious teacher," *Cambridge Magazine* (1917), p. 133.

While on the subject of emotional responses, Hardy described "... a feeling of union, in some sense or other, with the actual world." This, he surmised, is what a poet must experience. For him, such a reaction "... is most commonly excited by the sense of sight, though sometimes by other senses; but always by something which seems mysterious and big, the Indus cutting its way through the desert, the rumble of the Scotch express, the lights of St. Pancras station."

—"Mr. Russell as a religious teacher," *Cambridge Magazine* (1917), p. 133.

Hardy spent a sabbatical at Princeton, and this gave him a chance to comment upon the United States and its peculiar customs.

"I saw my first football match yesterday: much more impressive than I expected, though it was only a runaway match against an obscure university (score 50–0). The stadium is truly magnificent, and the young man who waves his arms and uses the megaphone a marvelous performer—I shall go hysterical when it comes to the big matches. I didn't find it at all hard to follow. I had hoped to get a ticket for the big base-ball matches in NY from a rich merchant I played bridge with on the ship, but unfortunately he seems to have forgotten

it—I read the reports in the papers by the hour and worship Babe Ruth and Lou Gehrig. Batting averages ranging from .387 to .231 are very entertaining to a cricketer."
—Hardy in a letter to Oswald Veblen, 1928, from Princeton.

"This country [the U.S.] is in many ways the only one in the world—it has its deficiencies (... dinner rather an uninteresting meal, and at 6:30). But American football knocks other spectacles absolutely flat; the sun shines more or less continuously; and for quietness and the opportunity to be your own master I've never come across anything like it. Prohibition is a great blessing; the tobacco is good; and at cakes and ice cream it can give any other place a good 30. Also it is a great advantage to be able to have your shoes cleaned only once in 3 weeks. I had mine done today for the first time since I got here. Everybody seems just at first to be merely unsophisticated, but that is a delusion..."
—Hardy in a letter to C. W. M. Cox, 26 October 1928, from Princeton.

Lest one conclude that Hardy's favorable impressions of America were restricted to football, Prohibition, and Babe Ruth, we note that when Hardy visited Harvard in 1936, Garrett Birkhoff reported that he said the United States had become number one in mathematics, "... ahead of Germany, France, or England." Strong words.
—*Men and Institutions in American Mathematics* (Dalton Tarwater, et al., editors), "The Rise of Modern Algebra," by Garrett Birkhoff, p. 62.

We end with some miscellaneous observations from the pen of Professor Hardy.

"It is easy to think, but hard to joke, in symbols."
—"Review of Whitehead and Russell's *Principia Mathematica, Vol. 1*," *Times Literary Supplement* 504 (1911), p. 322.

"I believe in formal lectures, and I believe also in formal examinations."
—"The Case against the Mathematical Tripos," *Math. Gazette* 13 (1926), p. 62.

"There is no word which excites such bitter feeling in a 'teaching university' as the word 'research'."
—"Mathematics," *The Oxford Magazine*, 48 (1922), p. 821.

Hardy was passionate about cricket and also fond of "real" tennis. And he had opinions about chess: "A chess problem is genuine mathematics, but it is in some way 'trivial' mathematics. However ingenious and intricate, however original and surprising the moves, there is something essential lacking. Chess problems are *unimportant*. The best mathematics is *serious* as well as beautiful— 'important' if you like, but the word is very ambiguous, and 'serious' expresses what I mean much better... between 'real mathematics' and chess... we may take it for granted now that in substance, seriousness, significance, the advantage of the real mathematical theorem is overwhelming. It is almost equally obvious, to a trained intelligence, that it has a great advantage in beauty also; but this advantage is much harder to define or locate, since the *main* defect of the chess problem is plainly its 'triviality', and the contrast in this respect mingles with and disturbs any more purely aesthetic judgement. What 'purely aesthetic' qualities can we distinguish in such theorems as Euclid's and Pythagoras's? . . . In both theorems (and in the theorems, of course, I include the proofs) there is a very high degree of *unexpectedness*, combined with 'inevitability' and 'economy.'"

—*A Mathematician's Apology*, Cambridge University Press, 1940, pp. 88–89, 112–113.

"Oxford has produced a good many more distinguished mathematicians than is generally realized, but the majority of English mathematicians are Cambridge men."

—"Mathematics," *The Oxford Magazine*, 48 (1922), p. 819.

In Hardy's day, debate raged about the paradoxes that had appeared in Georg Cantor's set theory. This theory had been touted as a logical foundation for all of mathematics, so something like Russell's Paradox was a troubling discovery that, some thought, undermined this foundation irrevocably. David Hilbert, however, was not about to abandon the ideas of set theory and famously vowed that no one would expel mathematicians from the Paradise that Cantor had created. Hardy was in agreement. "The history of mathematics," he wrote, "shows conclusively that mathematicians do not evacuate permanently ground which they have conquered once." But, picking up on Hilbert's expulsion-from-Paradise imagery, Hardy could not resist the quip, ". . . the worst that can happen to us is that we shall have to be a little more particular about our clothes."

—"Mathematical Proof," *Mind* (1929), p. 5. Reprinted with permission of Oxford University Press.

As mentioned in the Introduction, Hardy wrote a small book, *Bertrand Russell & Trinity/A college controversy of the last war*, which he published privately through the University Press, Cambridge. Russell had been dismissed from Trinity for what were no doubt complicated reasons but mainly as a result of his outspoken stand against World War I and conscription. In the book, Hardy was critical of the dismissal, and his comments to this end demonstrated that it would be unwise to have G. H. Hardy as an adversary. Along with his persuasive arguments were some interesting observations, particularly in light of Hardy's convictions: "It must be remembered that, even if there had been a 'clerical party anxious to frustrate Russell, it would certainly not have been in a position to carry the College with it. There had been one 'religious' controversy in the College a little before, when the Council, in 1912, had attempted to restore compulsory chapel. In this the 'clerical party' had been decisively defeated, and the opposition had been led by the very men (Jackson and McTaggart) who were Russell's most bitter opponents later." When the vote was taken to reinstate Russell, both Hardy and Littlewood had been in favor, though Ramanujan did not join them (probably for reasons beyond College politics). The opposition included the poet A. E. Housman who, explaining in a letter to H. A. Holland that he could not join those voting to invite Russell to return, said: "I am writing this, not to argufy, but only in acknowledgement of your civility in writing to me. I hope I shall not be able to discover 'conscious effort' in the amiability of yourself or Hardy when I happen to sit next to you in the future. I am afraid, however, that if Russell did return he would meet with rudeness from some Fellows of the College, as I know he did before he left. This ought not to be, but the world is as God made it."

As it turned out, the group that argued to reverse the earlier decision won and Russell was invited to return to Trinity. For various reasons, however, he moved on.
—*Bertrand Russell and Trinity*, Cambridge University Press, 1942, pp. 1–3, 31, 54, 60. Reprinted with permission.

In formal committee meetings, when a topic is before the body, a member can move to shelve discussion, essentially killing the motion. In Britain, the terminology for this is that the matter in question "be not put." Hardy weighed in on this parliamentary tactic: "I do not like motions to shelve discussion. I have often longed to move 'that the question "that the question be not put" be not put,' and wondered whether any chairman would accept such a motion."
—*Bertrand Russell and Trinity*, Cambridge University Press, 1942, p. 24. Reprinted with permission.

12
Selections from What Others Have Said about Hardy

G. H. Hardy and E. T. Bell—mathematicians who shared not only a love for number theory but a preference for their first two initials—apparently carried on a long correspondence. According to Bell's biographer Constance Reid, their letters were probably destroyed. But she stated that "of all the important men he [Bell] met, only two in his opinion—Einstein and the English number theorist G. H. Hardy, who had also visited Caltech—could be called geniuses. Outwardly they seemed completely different, and yet 'in complete mastery of their stuff' he found them very much the same—'and there is no thrill quite so great to an American as that of seeing in action a man who is absolute master of his trade.'"
   —Constance Reid, *The Search for E. T. Bell*, Mathematical Association of America (MAA), Washington, DC, 1993, pp. 255, 328.

When Hardy was eight years old, he produced a newspaper containing, among other things, "a speech by Mr. Gladstone, various tradesmen's advertisements, and a full report of a cricket match with complete scores and bowling analysis." Around this time, Hardy also "embarked on writing a history of England for himself, but with so much detail that he never got beyond the Anglo-Saxons."
   —E. C. Titchmarsh "Godfrey Harold Hardy," *J. London Math. Soc.* 25 (April, 1950), p. 81.

While on the subject of his childhood, we note that Hardy's skepticism of all things supernatural had early beginnings. E. C. Titchmarsh observed that Hardy "agonized his nurse with long arguments about the efficacy of prayer and the existence of Santa Claus: 'Why, if he gives me things, does he put the price on?'"
   —E. C. Titchmarsh "Godfrey Harold Hardy," *J. London Math. Soc.* 25 (April, 1950), p. 81.

E. C. Titchmarsh at the Edinburgh Mathematical Society Colloquium in St. Andrews.

H. Burkill in his obituary of Hardy wrote: "Since Hardy published so much, it might be thought that he devoted most of his waking hours to mathematics, but such a conclusion would be quite wrong, for he held that four hours' research was the daily limit for a mathematician... Over breakfast he read *The Times*, starting with the cricket scores—English ones in the summer, the Australian ones in the winter. (J. M. Keynes, the economist, once told [Hardy] that, if he paid the same attention to the stock exchange reports, he would be a rich man.) Then, until lunch, he did mathematics. The afternoon was spent on his favourite games. He might play some real tennis (a complicated indoor game, the ancestor of lawn tennis) or practice at the nets, but above all he liked watching cricket, if possible in the company of a friend with whom to discuss the game or anything else under the sun. Though he enjoyed conversation over dinner, the port and walnuts in the Combination Room, the prospect of which had brought him to Trinity, soon palled. Some evenings may well

have been devoted to literature; Hardy was certainly an exceptionally well-read man."
>—H. Burkill, "G. H. Hardy (1877–1947), *Math. Spectrum* 28 (1995–1996), pp. 25–31. Reprinted by permission of the Applied Probability Trust. First published in *Math. Spectrum*, Copyright © Applied Probability Trust 1996.

"Everyone thought [Hardy] looked appropriately striking. Bertrand Russell said to me (c. 1911) that he had the bright eyes that only very clever men had ('clever' then meant more than now)."
>—*Littlewood's Miscellany*, (ed. Béla Bollobás), Cambridge University Press, 1986, p. 120.

"On the social side, the most distinctive aspect of my contact with Bertrand Russell lay in his Thursday evening parties, or as they were called in view of the number of guests, his 'squashes.' A very distinguished group of men foregathered there. There was Hardy, the mathematician. There was Lowes Dickinson, the author of *Letters from John Chinaman* and *A Modern Symposium*, and the bulwark of the liberal political opinion of the time. There was Santayana, who had left Harvard for good to take up his residence in Europe. Besides these, Russell himself was always an interesting talker. We heard much of his friends, Joseph Conrad and John Galsworthy."
>—Norbert Wiener, *Ex Prodigy*, Simon & Schuster, New York, 1953, p. 194. Reprinted courtesy of the MIT Press.

Hardy "had violent prejudices against food; his vocal obsession with the horror of hot roast mutton was something of a joke."
>—*Littlewood's Miscellany*, (ed. Béla Bollobás), Cambridge University Press, 1986, p. 120.

Titchmarsh recalled that Hardy "has been described as absent-minded, but I never saw any sign of this." However, he did have "a way of passing in the street people whom he knew well without any sign of recognition, but this was due to a sort of shyness, or a feeling of the slight absurdity of a repeated conventional greeting."
>—E. C. Titchmarsh "Godfrey Harold Hardy," *J. London Math. Soc.* 25 (April, 1950), p. 86.

Hardy's concerns were wide-ranging as evidenced by six New Year's resolutions he sent in a postcard to a friend: "(1) prove the Riemann Hypothesis; (2) make 211 not out in the fourth innings of the last Test Match at the Oval; (3) find an argument for the nonexistence of God which shall convince the general public; (4) be the first man at the top of Mount Everest; (5) be proclaimed the first president of the U.S.S.R., of Great Britain and Germany; and (6) murder Mussolini."
>—Paul Hoffman, *The Man Who Loved Only Numbers*, Hyperion, New York, 1998, p. 81 Reprinted by permission of Hachette Books.

The physicist Richard Dawkins, in the first volume of his autobiography, recalled being a new faculty member at Oxford when older colleagues brought out the Senior Common Room Betting Book. In this document, former Oxford dons had recorded wagers made among themselves over port and conversation at high-table. Dawkins learned that "... the most assiduous betting man was the eccentrically brilliant G. H. Hardy." For instance, one colleague bet his entire fortune versus Hardy's bet of a ha'penny that the sun would rise the next morning. Hardy bet another that he would *not* become the President of Magdalen College (he didn't). And Dawkins read this peculiar entry: "Professor Hardy bets Mr Creed 2/6 to 1/6 that the New Prayer Book would go phut."
—Richard Dawkins, *An Appetite for Wonder*, Harper Collins, New York, 2013, p. 217.

Hardy made up other favorite cricket teams, beyond those in our Introduction. He wrote them down on slips of paper. Here are a few others. Challenge to the reader: Can you identify the players? You might note that the first column is made up of a team for Hardy, the second a team for Pólya. The third might be an all-Jewish team.

Hayward (T)	Poincaré (H)	D. [sic] Spinoza
Hannibal	Porsena	A. Einstein
Haydn	Pontius Pilate	David
Hakon	Poe	B. Disraeli (Capt.)
Hamilton (Sir WR)	Poisson	God (F)
Hardy (T)	Potiphar (Mrs.)	H. Stinnes
Hafiz	Poincaré (R)	P. G. H. Fender
Hapsburg (R von)	Poushkin	E. Lasker
Harmodius	Pond	Paul
Hamlet	Poinsot	God (S)
Hadamard	Polycrates	God (HG)

Hint: P. G. H. Fender was a cricketer. And a reminder: players here include God the Father, God the Son and God the Holy Ghost.
—G. L. Alexanderson, *Random Walks of George Pólya*, MAA, 2000, pp. 65–67.

Littlewood reported that "I stood by, while Hardy wrote a telegram to New College in the little Post Office in Rose Crescent in Cambridge. 'Inexpressibly shocked pusillanimous surrender reactionary proposal Council stop wanton abandonment sacrosanct principle liberty, etc' The postmistress fulfilled her duty under the regulations, and read it back with a perfectly straight face."
—*Littlewood's Miscellany*, (ed. Béla Bollobás), Cambridge University Press, 1986, p. 165.

12. Selections from What Others Have Said about Hardy

Ralph Boas as a Harvard student in 1933. He spent 1938 at Cambridge taking courses from Hardy, Littlewood, and Besicovich. He also attended the Conversation Class which was given by Hardy in Littlewood's rooms, but without Littlewood.

"G. H. Hardy did not particularly esteem the Ph.D. degree, or take it very seriously (he didn't have a Ph.D. himself), and was not above writing a thesis for a candidate. I was told that he once did this for a foreign student, who then asked Hardy to write a letter saying that he had written a good thesis, so that he would get a better job in his home country. Hardy balked at doing this, but he said, 'Show your thesis to Littlewood, and *he* will write you a letter.' Littlewood wrote the letter and the student got the job. What makes me confident about the validity of this tale is that I repeated it to a compatriot of the student, and he said 'Oh, I know that man,' but he wouldn't tell me who it was. 'Never criticize the sonatas of archdukes—you never know who wrote them.'"
—Ralph P. Boas, Jr., in *Lion Hunting and Other Mathematical Pursuits*, MAA, Washington, DC, 1995, p. 128.

Hardy "did not believe in wasting time reviewing his earlier lectures—the students were expected to retain in their minds the material already covered. It

chanced that a series of mathematical lectures given by . . . Hardy at Cambridge University was interrupted by the long vacation. On the first class meeting after the vacation, Professor Hardy advanced to the blackboard with chalk in hand and said, 'It thus follows that . . .'"
> —Stanley B. Jackson, in Howard Eves, *Mathematical Circles Adieu*, Prindle, Weber and Schmidt, Boston, 1977, p. 55. (Reissued by the Mathematical Association of America, 2003.)

A very common story about mathematicians, told, no doubt, by many about their not so favorite teachers in graduate school, or about their dissertation advisors, might actually be true in the case of Hardy. Ralph P. Boas, Jr., who spent time in England, before becoming the leader of the department at Northwestern University, tells us that the "story is told of G. H. Hardy (and of other people) that during a lecture he said 'It is obvious . . . *Is* it obvious?' left the room, and returned fifteen minutes later, saying 'Yes, it's obvious.' I was present once when Rogosinski asked Hardy whether the story were true. Hardy would admit only that he might have said 'It's obvious . . . *Is* it obvious?' (brief pause 'Yes it's obvious).'"
> —G. L. Alexanderson and D. H. Mugler, *Lion Hunting and Other Mathematical Pursuits*, MAA, Washington, DC, 1995, p. 128.

At Cambridge, mathematics students were able to attend the "conversation class" run by Hardy and Littlewood. Titchmarsh called this seminar "memorable" and regarded it as a "model of what such a thing should be." In particular, "the audience was always amazed by the sure instinct with which Hardy put his finger on the central point and started the discussion with some illuminating comment, even when the subject seemed remote from his own interests."
> —E. C. Titchmarsh "Godfrey Harold Hardy," *J. London Math. Soc.* 25 (April, 1950), p. 87.

Hardy was widely admired among a generation of American mathematicians for his brilliant lectures, his mentoring (though he probably would not have used the word), and for the quality of his mathematics. He was often cited for his influence. Garrett Birkhoff said, "At Cambridge University in 1932–33 Hardy was by far my most stimulating teacher." David Blackwell said that he fell in love with mathematics when he used Hardy's *Pure Mathematics* as a text. "That's the first time I knew that serious mathematics was for me. It became clear that it was not simply a few things that I liked. The whole subject was just beautiful." Peter Hilton said, "I owe a tremendous debt to Hardy, who exercised such a strong influence on British mathematics." And George Pólya, who often wrote about Hardy, summed it up with "Hardy had a very great personal

12. Selections from What Others Have Said about Hardy

David Blackwell, distinguished statistician at the University of California, Berkeley. He was a member of the National Academy of Sciences and received the National Medal of Science.

influence on me." And those are but a few such expressions of the debt owed to Hardy."
—D. J. Albers and G. L. Alexanderson (eds.), *Mathematical People*, Birkäuser Boston, 1985, pp. 7, 21, 143–144, 250.

Titchmarsh addressed this same matter when he remembered that "[Hardy] was an extremely kind-hearted man, who could not bear any of his pupils to fail in their researches." And, drawing from personal experience, Titchmarsh contributed perhaps the ultimate accolade: "There could have been no more inspiring director of the work of others."
—E. C. Titchmarsh "Godfrey Harold Hardy," *J. London Math. Soc.* 25 (April, 1950), p. 86.

"Ralph P. Boas, Jr. noted that: 'Even research in mathematics is, to a considerable extent, a teachable skill. A student of G. H. Hardy's once described to me how it was done. If you were a student of Hardy's, he gave you a problem that he was sure you could solve. You solved it. Then he asked you to generalize it in a specific way. You did that. Then he suggested another generalization, and

so on. After a certain number of iterations, you were finding (and solving) your own problems. You didn't necessarily learn to be a second Gauss that way, but you could learn to do useful work.'"
—G. L. Alexanderson and D. H. Mugler, *Lion Hunting and Other Mathematical Pursuits*, MAA, Washingon, DC, 1995, p. 235.

When Norbert Wiener as a young man moved from Harvard to Cambridge he "loved the new freedom of mind and personality he encountered at Cambridge, where for once he did not feel like a freak set apart from the crowd. He contrasted the local scene with that of Harvard, which, as he asserted, 'has always hated the eccentric and the individual, while... in Cambridge eccentricity is so highly valued that those who do not really possess it are forced to assume it for the sake of appearances.' ... One Trinity man Wiener held in the highest regard was Godfrey Harold Hardy, England's foremost pure mathematician of that period. With his fair features and sporting passion, Hardy had the air of a sophomore, but he was in his mid-thirties and a fragile, shy man. Wiener thanked Hardy, not Russell, for grounding him in the new tools of modern mathematics."
—Flo Conway and Jim Siegelman, *Dark Hero of the Information Age/In Search of Norbert Wiener/The Father of Cybernetics*, Basic Books, New York, 2005, pp. 31–32.

"Hardy in his thirties held the view that the late years of a mathematician's life were spent most profitably in writing books; I remember a particular conversation about this, and though we never spoke of the matter again, it remained an understanding. The level below his best at which a man is prepared to go on working at full stretch is a matter of temperament; Hardy made his decision, and while of course he continued to publish papers, his last years were mostly devoted to books; whatever has been lost, mathematical literature has greatly gained. All his books gave him some degree of pleasure, but this one, his last, was his favourite. When embarking on it he told me that he believed in its value (as he well might), and also that he looked forward to the task with enthusiasm.... The title holds curious echoes of the past, and of Hardy's past. Abel wrote in 1828: 'Divergent series are the invention of the devil, and it is shameful to base on them any demonstration whatsoever.' In the ensuing period of critical revision they were simply rejected. Then came a time when it was found that something after all could be done about them. This is now a matter of course, but in the early years of the century the subject, while in no way mystical or unrigorous, *was* regarded as sensational, and about the present title, now colourless, there hung an aroma of paradox and audacity."
—John E. Littlewood, Preface to *Divergent Series*, Oxford University Press, 1949, p. [vii].

The classic book, *Inequalities*, by Hardy, Littlewood and Pólya, remains popular after almost 80 years. Ralph P. Boas, Jr., recalled, however, that "Hardy was once heard to complain that whenever he needed an inequality the precise one that he wanted was not there."
—G. L. Alexanderson, *Random Walks of George Pólya*, MAA, Washington, DC, 2000, pp. 209–210.

In an interview with Freeman Dyson, Donald J. Albers asked Dyson why, when he was offered his choice of a book for a prize, he chose Hardy and Wright's *Introduction to the Theory of Numbers*. Dyson responded, "It wasn't my first prize book. I got it in my second year in high school. A few years later I got to know Hardy in Cambridge and asked him why he spent so much time and effort writing that marvelous book when he might have been doing serious mathematics. He answered, 'Young men should prove theorems, old men should write books.'"
—Donald J. Albers, "Freeman Dyson: Mathematician, Physicist and Writer," *College Math. J.*, 25:1 (1994), p. 3.

"It was Hardy's *Course in* [sic] *Pure Mathematics* (1908), together with the reform of the Cambridge Tripos system the following year, which really marked the turning point in British university-level education. From then on, analysis would be a fundamental component... Another reason for the prominence of Hardy's name in any account of British mathematics in the early 20th century is his sheer productivity during this period, especially in comparison with his contemporaries. Turning again to publication in the London Mathematical Society's two journals, we find that among the most frequent contributors between 1900 and 1940 were Burnside (31 papers), Hobson (32 papers), and Young (69 papers). But it was Hardy who was by far the most prolific, with a yield of 106 papers, a figure that amounts to nearly 10% of the Society's entire output over this whole period. When the sheer quantity of this work is considered together with its high quality, it is hardly surprising that Hardy demands so much attention."
—Adrian C. Rice and Robin J. Wilson, "The Rise of British Analysis in the Early 20th Century: the Role of G. H. Hardy and the London Mathematical Society," *Hist. Math.* 30 (2003), pp. 178, 190. Reprinted with permission of the publisher.

"I know of no writing—except perhaps Henry James's introductory essays—which convey so clearly and with such an absence of fuss the excitement of the creative artist."
—(Henry) Graham Greene, "An Austere Art" (a review of *A Mathematician's Apology), The Spectator* 165 (December 19, 1940), p. 682.

Ralph P. Boas, Jr., wrote that "at that time Hardy was an editor of the *Journal of the London Mathematical Society*. He used to tell referees to ask three questions: Is it new? Is it true? Is it interesting? The third was the most

important. I would now add: Is it decently written? I think Hardy took that as a given. If he got a paper that was interesting but badly written, he would ask a graduate student to rewrite it. I know, because I did a couple of such rewrites for Hardy, for authors who subsequently became very well known."
  —G. L. Alexanderson and D. H. Mugler, *Lion Hunting and Other Mathematical Pursuits*, MAA, Washington, DC, 1995, p. 10.

"Littlewood read the first proof-sheets of an article by Hardy on the remarkable Hindu mathematician Ramanujan. In the article Hardy had written, 'as someone said, each of the positive integers was one of his personal friends.' On reading this, Littlewood commented, 'I wonder who said that; I wish I had.' In the final proof-sheets Littlewood read (and this is how it came out in the printed article), 'It was Littlewood who said, each of the positive integers was one of his personal friends.'"
  —Howard Eves, *Mathematical Circles Squared*, Prindle, Weber and Schmidt, Boston, 1972, p. 83. (Reissued by the Mathematical Association of America, 2003.)

"In his lifetime Hardy had assigned the royalties on his books to the society [London Mathematical Society] and when he died it was found he had left it the remainder of his estate after reserving a life interest for his sister Gertrude. After her death the bequest was partly used to fund a Hardy Visiting Lectureship, which involves a lecture tour of a number of British universities, and to endow a Hardy Junior Research Fellowship at New College, which supports a young researcher. Although he was not a great traveller Hardy made several visits to America [mainly to Princeton University and the California Institute of Technology], and had many friends there. On one of these visits he had tried with some success, to interest the officers of the Rockefeller Foundation in the idea of helping to establish a Mathematical Institute at Oxford, as had been done in Göttingen and Paris."
  —Ioan James, *Remarkable Mathematicians from Euler to von Neumann*, Cambridge-MAA, 2002. p. 304. Reprinted with the kind permission of the Cambridge University Press.

Hardy's assertion that as mathematicians get to a certain age, they can no longer do really creative work but must settle for something less, was challenged by L. J. Mordell, a most capable number theorist who was a student of Titchmarsh's. He quoted from *A Mathematician's Apology* where Hardy claimed "if then I find myself writing not mathematics but 'about' mathematics, it is a confession of weakness, for which I may rightly be scorned or pitied by younger and more vigorous mathematicians. I write about mathematics because, like any other mathematician who has passed sixty, I have no longer the freshness of mind, the energy, or the patience to carry on effectively with my proper job." Mordell responded: "It seems almost nonsense to say that anyone would scorn or pity [Hardy], and the use of the term 'rightly' is even more nonsensical." He particularly objected to Hardy's citing "sixty" as the age beyond which a mathematician cannot be expected to be creative and cites counterexamples: "Great activity among octogenarians is shown by Littlewood, his lifelong collaborator, Sydney Chapman, his former pupil and collaborator, and myself. There is also Besicovitch in the seventies. Davenport, who had passed sixty, was as active and creative as ever, and his recent death is a very great loss to mathematics since he could have been expected to continue to produce beautiful and important work."

Mordell added that "no mathematician can always be producing new results. There must inevitably be fallow periods during which he may study and perhaps gather ideas and energy for new work. In the interval, there is no reason why he should not occupy himself with various aspects of mathematical activity, and every reason why he should. The real function of a mathematician is the advancement of mathematics." One such activity of Hardy's cited by Mordell was his hard work on the reform of the Tripos. Mordell claims that before Hardy, the Tripos was "looked upon as a sporting event, reminding one of the Derby."

—L. J. Mordell, "Hardy's 'A Mathematician's Apology,'" *Amer. Math. Monthly* 77:8 (1970), pp. 832–833.

L. J. Mordell pointed out that while Hardy was concerned about "ugly mathematics" he didn't define it. Mordell gave it a try: "Among such, I would mention those involving considerable calculations to produce results of no particular interest or importance; those involving such a multiplicity of variables, constants, and indices, upper, lower, right, and left, making it very difficult to gather the import of the result; and undue generalization apparently for its own sake and producing results with little novelty. I might also mention work which places a heavy burden on the reader in the way of comprehension and verification unless the results are of great importance."

—L. J. Mordell, "Hardy's 'A Mathematician's Apology,'" *Amer. Math. Monthly*, 77:8 (1970), pp. 834–835.

Hardy in his *A Mathematician's Apology* seemed, according to L. J. Mordell, to "denigrate the usefulness of 'real' mathematics." Quoting Hardy, he said: "The 'real' mathematics of the 'real' mathematicians, the mathematics of Fermat and Euler and Gauss and Abel and Riemann is almost wholly 'useless'" Mordell went on: "This statement is easily refuted. A ton of ore contains an almost infinitesimal amount of gold, yet its extraction proves worthwhile. So if only a microscopic part of pure mathematics proves useful, its production would be justified. Any number of instances of this come to mind... the

investigation of the properties of the conic sections by the Greeks and their application many years later to the orbits of the planets. Gauss' investigations in number theory led him to the study of complex numbers. This is the beginning of abstract algebra, which has proved so useful for theoretical physics and applied mathematics. Riemann's work on differential geometry proved of invaluable service to Einstein for his relativity theory. Fourier's work on Fourier series has been most useful in physical investigations."
>—L. J. Mordell, "Hardy's 'A Mathematician's Apology,'" *Amer. Math. Monthly*, 77:8 (1970), p. 835.

The Hardy-Weinberg law was, perhaps unfortunately for Hardy, not his only contribution to applicable mathematics. "In 1973, the British mathematician Clifford Cocks used the theory of numbers to create a breakthrough in cryptography—development of codes. Cocks's discovery made [a] statement of Hardy obsolete. In his famous book *A Mathematician's Apology* ... he pronounced: 'No one has yet discovered any war-like purpose to be served by the theory of numbers.' ... Codes have been absolutely essential for military communications."
>—Mario Livio, *Is God a Mathematician?*, Simon and Schuster, New York, 2009, p. 5. Reprinted with the permission of Simon & Schuster, Inc., from *Is God a Mathematician?*, by Mario Livio, Copyright © 2009, by Mario Livio.

Laurence Young, son of the distinguished mathematicians William Henry and Grace Chisholm Young wrote: "In mathematics, in science generally, Cambridge is considered more important than Oxford. This is not entirely fair, and as a Cambridge man I should be the first to admit it: I do not believe in ranking universities any more than in ranking individuals. Oxford, on the whole, I believe to be ahead of Cambridge in ideas; the hard work that new ideas demand is then mainly carried out in Cambridge. The relationship is not unsimilar to that of Hardy and Littlewood: in Cambridge, Hardy was, in Littlewood's words, a rebel—he was happier in the more intellectual atmosphere of Oxford. Littlewood was more typically a Cambridge man, but they collaborated very well: they had on the whole the same way of looking at things. Oxford and Cambridge have likewise a same underlying structure, unique in today's world: they are *museums of learning.*"
>—Laurence Chisholm Young, *Mathematicians and Their Times/History of Mathematics and Mathematics of History*, North-Holland, Amsterdam, 1981, p. 266.

Because Hardy was perhaps among a few first-class scholars to hold chairs at both Oxford and Cambridge, "this moving back and forth between Cambridge and Oxford led to a question about his loyalty to the two universities. First one must know that Hardy loved cricket but despised rowing. Pólya remarked that when someone asked Hardy, 'For which university are you in sports?',

Hardy replied, 'It depends. In cricket I am for Cambridge, in rowing I am for Oxford.'"
  —G. L. Alexanderson, *The Random Walks of George Pólya*, MAA, Washington, DC, 2000, p. 64.

Littlewood weighed in on Hardy's divided loyalties between Oxford and Cambridge, noting that Hardy "liked to pose as Oxford at Cambridge and vice versa." Regarding cricket, Hardy's preference for Cambridge mellowed a bit over the years. Littlewood remembered that Hardy "still wanted a Cambridge victory by innings and 200, but no longer wanted the Oxford Captain to be hit in the stomach by the fast bowler."
  —*Littlewood's Miscellany*, (ed. Béla Bollobás), Cambridge University Press, 1986, pp. 119–120.

"It was with the arrival of Esson's successor G. H. Hardy as Savilian professor in 1920 that the modern era of mathematics in Oxford really began. In many ways Hardy was a natural heir of Henry Smith in number theory, and he arrived in Oxford at the height of his mathematical powers, picking up essentially where Sylvester had left off by creating a fully fledged research school in mathematics for the first time in Oxford. One needs only to look at the list of subsequent professors, including E. C. Titchmarsh, Sydney Chapman, E. A. Milne, Henry Whitehead, C. A. Coulson, Graham Higman, Sir Michael Atiyah, Sir Roger Penrose, Daniel Quillen, and Simon Donaldson, to see that Oxford has maintained its research momentum in both pure and applied mathematics ever since."
  —Raymond Flood, Adrian Rice, and Robin Wilson, *Mathematics in Victorian Britain*, Oxford, 2011, p. 50.

"From 1920 to 1931, Hardy was the outstanding influence in pure mathematics at Oxford. I felt that I joined a rather very special group when I began to attend Hardy's class at New College in January 1928. He had just ceased to be Secretary and was beginning his term as President of the London Mathematics [sic] Society. He was obviously strained and working very hard. Looking back, I think that we knew that we were studying under a very great man who had not yet been recognized fully."
  —J. Tattersall and S. McMurran, "An Interview with Dame Mary L. Cartwright, D. B. E., F. R. S.," *College. Math. J.* 32:4 (2001), p. 248.

When at Oxford, Hardy "also lectured occasionally on mathematics for philosophers, and drew large audiences of Oxford philosophers to whom ordinary mathematics made no appeal."
  —E. C. Titchmarsh "Godfrey Harold Hardy," *J. London Math. Soc.* 25 (April, 1950), p. 85.

"In Hardy's presidential address to the London Mathematical Society in 1928 he was able to boast that he had sat through every word of every lecture of every meeting of every paper since he became secretary in 1917."
—T. W. Körner, Foreword to Hardy's *A Course of Pure Mathematics, Cambridge*, 1908, p. x.

Though Hardy seemed to enjoy his 11 years at Oxford, between 1920 and 1931, he did return to his Fellowship at Trinity. Ioan James (one of Hardy's successors in the Savilian Chair at Oxford) claimed that "one of the reasons he gave for the move was that eventually, when he reached the mandatory retirement age, he would be able to continue to live in college, something not normally permissible at Oxford. However, as an Honorary Fellow of New College, he was able to retain his connection with Oxford and frequently spent his weekends, and vacations there."
—Ioan James, *Remarkable Mathematicians from Euler to von Neumann*, Cambridge-MAA, 2002, pp. 303–304. Reprinted by kind permission of the Cambridge University Press.

"The diffident cricket-loving G. H. Hardy would plead not to be seated next to Mary Cartwright at the dinners of the Cambridge Mathematical Society, because 'her fast ball', he complained, 'is so very devastating."
—Walter Gratzer, *Eurekas and Euphorias/The Oxford Book of Scientific Anecdotes*, Oxford, 2002, p. 273.

In his memoirs Norbert Wiener wrote: "Father went up with me to Cambridge. We looked up Bertrand Russell in his rooms at Trinity, and he helped us to orient ourselves. While we were in Russell's rooms a young man came in whom my father took to be an undergraduate and who excited no particular attention in us. It was G. H. Hardy, the mathematician who was to have the greatest influence on me in later years."
—Norbert Wiener, *Ex Prodigy*, Simon & Schuster, New York, 1953, p. 183. Reprinted by courtesy of the MIT Press.

Hardy's youthful appearance was also noted by Littlewood, who recalled that "Until he was about 30, he looked incredibly young; there was a legend that when he was a Prize Fellow the Kitchen Office wouldn't let him order beer."
—*Littlewood's Miscellany*, (ed. Béla Bollobás), Cambridge University Press, 1986, p. 120.

Norbert Wiener, describing his stay in Cambridge, said: "In laying out my course, Russell had suggested to me the quite reasonable idea that a man who was going to specialize in mathematical logic and in the philosophy of mathematics might just as well know something of mathematics. Accordingly, I took at various times a number of mathematical courses, including one by Baker, one by Hardy, one by Littlewood, and one by Mercer. I did not continue

12. Selections from What Others Have Said about Hardy

Baker's course long, as I was ill prepared for it. Hardy's course, however, was a revelation to me. He proceeded from the first principles of mathematical logic, by way of the theory of assemblages, the theory of the Lebesgue integral, and the general theory of functions of a real variable, to the theorem of Cauchy and to an acceptable logical basis for the theory of functions of a complex variable. In content it covered much the same ground that I had already covered with Hutchinson of Cornell, but with an attention to rigor which left me none of the doubts that had hindered my understanding of the earlier courses. In all my years of listening to lectures in mathematics, I have never heard the equal of Hardy for clarity, for interest, or for intellectual power. If I am to claim any man as my master in my mathematical training, it must be G. H. Hardy."

—Norbert Wiener, *Ex Prodigy*, Simon & Schuster, New York, 1953, p. 190. Reprinted courtesy of the MIT Press.

On the question of the Cambridge Tripos: "In every respect but one, in fact, the old Mathematical Tripos seemed perfect. The one exception, however, appeared to some to be rather important. It was simply—so the young creative mathematicians such as Hardy and Littlewood kept saying—that the training had no intellectual merit at all. They went a little further and said that the Tripos had killed serious mathematics in England stone dead for a hundred years."

—C. P. Snow, *The Two Cultures*, Cambridge, 1993, pp. 20–21.

"At the suggestion of G. H. Hardy, Pólya was awarded the first international Rockefeller Fellowship in 1924 . . . Hardy was in the midst of his campaign to reform the mathematics Tripos and asked Pólya to take the exam unofficially. Hardy expected Pólya's poor showing would demonstrate that most of the questions on the Tripos were irrelevant to 'modern continental mathematics.' Unfortunately for Hardy's plan, Pólya's performance was the best on the examination, and he would have been named Senior Wrangler if he had been a student."

—Halsey Royden, "A History of Mathematics at Stanford," in *A Century of Mathematics in America, Part II*, American Mathematical Society, Providence, RI, 1989, pp. 250–251. Reprinted with the kind permission of the American Mathematical Society.

"When I visited Hardy in his rooms in Trinity College [Cambridge] in 1933, I saw three pictures on his mantelpiece, namely, one of Lenin, one of Jesus Christ, and one of Hobbs (the then star cricketer in England). Asked why there were exactly those three, Hardy said they were, in his opinion, the only important personalities who had achieved a hundred percent of what they wanted to achieve."

—Helmut Hasse, in Howard Eves, *Mathematical Circles Adieu*, Prindle, Weber and Schmidt, Boston, 1977, p. 55. (Reissued by the Mathematical Association of America, 2003.)

George D. Birkhoff discovered the Ergodic Theorem and was the winner of numerous awards, including the Bôcher Prize of the American Mathematical Society. He also has a crater on the Moon named in his honor.

"G. H. Hardy, a later successor to Cayley's Cambridge chair, had decided views, as Lawrence Young recorded: 'I once asked Hardy whether he thought Cayley was a great mathematician: Hardy simply glared at me' Hardy was perhaps influenced by the mood for debunking the Victorians, which existed after the World War; he was after all, a friend and contemporary of Lytton Strachey, author of *Eminent Victorians*."
—Tony Crilly, *Arthur Cayley, Mathematician Laureate of the Victorian Age*, Johns Hopkins University Press, Baltimore, 2006, p. 439.

The Harvard mathematician George David Birkhoff had made a name for himself with his work on ergodic theory, dynamical systems, and combinatorics, but in the 1930s his focus unexpectedly broadened to include a mathematical assessment of aesthetics. This was not met with universal delight. When Birkhoff's son was in Cambridge, Hardy asked him how his father's

aesthetic research was coming along. When he heard that Birkhoff had completed a book on the subject, Hardy replied, "Good. Now your father can get back to real mathematics."
—Steve Nadis and S. T. Yau, *A History in Sum*, Harvard University Press, 2013, p. 78.

H. S. M. Coxeter, the eminent geometer, later at the University of Toronto, was a fellow at Trinity College, Cambridge in the 1920s and at one point sent a query on evaluating the volume of a spherical tetrahedron to the *Mathematical Gazette*. He received back "a registered letter from G. H. Hardy, then a professor of geometry at Oxford and recognized as the greatest mathematician in England. 'I tried very hard *not* to spend time on your integrals,' Hardy scribbled around the perimeter on one of several pages of calculations, 'but to me the challenge of a definite integral is irresistible.' With that Donald Coxeter performed a rite of passage."
—Siobhan Roberts, *King of Infinite Space*, Walker, Toronto, 2006, pp. 57–58. © Siobhan Roberts, 2006, *King of Infinite Space*, by permission of Walker Books, an imprint of Bloomsbury Publishing Inc.

Howard Eves recalled that during his "year of graduate study at Princeton University, Professor Hardy came over from England to spend a semester lecturing on some Tauberian theorems. He soon had us all charmed. Hardy wore heavy tweed suits, and even in the coldest weather augmented his dress with only a very long dirty white scarf that he wrapped round and round his neck. It wasn't long before I coveted that scarf for my mathematical museum. So, one day, after Hardy had gone into the great Gothic dining hall where we ate suppers, I took his long scarf off the hook where he had hung it and replaced it with the nicest and most expensive scarf I had been able to purchase in the village. After supper, when Hardy came out to reclaim his scarf, he found the beautiful one in its place. He began grumbling, but I soothed him by pointing out that whoever had taken his scarf had certainly left a considerably nicer one in its place. He conceded this, and finally took the new scarf in place of the old one. I waited several days to be sure he was satisfied with the surreptitious exchange. Finding him quite content, I entered his scarf as a new acquisition into my mathematical museum—in its original dirty and unkempt condition, of course."
—Howard Eves, *Mathematical Circles Revisited*, Prindle, Weber and Schmidt, Boston, 1971, p. 132. (Reissued by the Mathematical Association of America, 2003.)

"Hardy is sold [on] American football. He even came over to my house to listen in on the Princeton-Ohio State game broadcast from Columbus by WJZ and the blue network. Hasn't missed a game in Princeton and knows more

Harold Scott MacDonald Coxeter was a geometer who made important contributions to polytopes, and non-euclidean geometry. He was a Fellow of the Royal Society.

about the plays, rules, etc. than I did. Very much worried for fear the Australian cricket matches will not be fully reported in the N.Y. papers."
—J. W. Alexander, in a letter to Oswald Veblen, 17 November 1928, from Princeton.

When Hardy visited Princeton in the 1930s, his place at Cambridge was taken by Oswald Veblen. Hardy described him as having "all the American, all the British, all the Scandinavian virtues."
—G. L. Alexanderson, *The Random Walles of George Pólya*, MAA, Washington, DC, 2000, p. 83.

In a letter to his mother dated 6 October, 1936, [Alan Turing] wrote of meeting [G. H.] Hardy at Princeton: "At first he was very standoffish or possibly shy. I met him in Maurice Pryce's rooms the day I arrived, and he didn't say a word to me. But he is getting much more friendly now."
—Andrew Hodges, *Alan Turing: The Enigma* (Centenary Edition), Princeton University Press, 2012, p. 117.

12. Selections from What Others Have Said about Hardy

Paul Halmos of *Finite-Dimensional Vector Spaces* fame.

The collaboration between Hardy and Littlewood was not without complications. At one point Hardy wrote in a letter to Russell "I wish you could find some tactful way of stirring up Littlewood to do a little writing. Heaven knows I am conscious of my huge debt to him. But the situation which is gradually stereotyping itself is very trying for me. It is that, in our collaboration, he will contribute ideas and ideas only: and that *all* the tedious part of the work has to be done by me. If I don't, it simply isn't done, and nothing would ever get published... At the moment I am committed to write out two joint papers for publication in Germany, inside about two months. And I can get absolutely no help from him at all: not even an enquiry as to how I am getting on! The effect on morale is most disheartening."
  —Robin Wilson in Peter Harman and Simon Mitton (eds.), *Cambridge Scientific Minds*, Cambridge, 2002, pp. 204–205. Reprinted with permission of the Cambridge University Press.

For the outside observer, though, the collaboration between Hardy and Littlewood seemed quite without friction or unpleasantness. Paul Halmos commented on the "marriage principle of collaboration, a principle strongly supported by Hardy and Littlewood, the famous pair of great collaborators. The principle is that a collaboration once joined together shall not be put asunder. Once a conversation that might lead to a collaborative result is begun, then anything it leads to, for better, for worse, is a collaborative result. The contributions of the partners to the final result might be numerically equal—the

same number of definitions, the same number of theorems, the same number of proofs—or they might not; it must not matter, the count must not be made. Perhaps one partner contributes the insight and the other the technique; perhaps one partner asks the questions and the other knows the literature well enough to avoid the waste of time that comes from trying to re-discover already known answers; or, possibly, one is active and the other is the foil needed to keep up his morale and inspiration. No matter—once a collaboration is begun (which is not necessarily the same thing as formally declared), anything that comes out of it is, must be, called a collaborative product."
—Paul R. Halmos, *I Want To Be a Mathematician/An Automathography*, Springer, New York, 1985, p. 98, reissued by the Mathematical Association of America in 2005.

"When G. H. Hardy wrote after the First World War to the effect that he had not been a fanatical anti-German, and felt confident that Landau would want to resume further relations, Landau replied: 'As a matter of fact my opinions were much the same as yours, with trivial changes of sign.'"
—*Littlewood's Miscellany*, (ed. Béla Bollobás), Cambridge, 1986, p. 125.

Hardy had always felt close to mathematicians in the German school—Landau and others—so he also felt close to Mittag-Leffler [from Sweden]. After World War I the French led a move to keep German mathematicians from attending the first mathematical congress following the war, the International Congress of Mathematicians that had been scheduled for Stockholm in 1920. It turned out it was rescheduled for Strasbourg and had to be renamed because if Germans were not allowed to attend, it could no longer be labeled a congress of mathematicians. So it was called the International Mathematical Congress. J. W. Dauben described the situation thus: "Despite the anti-German vehemence expressed by mathematicians like Emile Picard and embodied in the exclusionary statutes of the new Academic Unions of the allied powers after the war, there were nevertheless those like G. H. Hardy and Gösta Mittag-Leffler who were anxious to put politics aside, to normalize relations, and to reestablish the free exchange of research that had, at least under Mittag-Leffler's careful editorship, preserved his *Acta Mathematica* as an open forum for mathematicians throughout those dark and difficult years of World War I."
—Joseph W. Dauben, "Mathematicians and World War I: The International Diplomacy of G. H. Hardy and Gösta Mittag-Leffler as Reflected in their Personal Correspondence," *Hist. Math.* 7 (1980), p. 277. Reprinted with the permission of the publisher.

"Hardy, along with Mittag-Leffler, fought against the exclusionary rules at the Strasbourg International Congress in 1920. Their argument was that the

12. Selections from What Others Have Said about Hardy

congress had to change its name to an International Congress of Mathematics, not of Mathematicians, to reflect the fact that German mathematicians were not being invited to the congress. They did get the name changed by the organizing committee but by the time the proceedings came out they were the *Comptes Rendus du Congrès International des Mathématiciens*. Hardy did not appear in Toronto in 1924, but was back on the rosters of delegates and members at Bologna in 1928 and, in Zurich in 1932. He does not appear on the list of members at that small congress in Oslo in 1936 where the first Fields Medals were awarded, but there he was appointed to head up the committee to choose the Fields Medalists by the next Congress, but a month later Hardy resigned from the committee, to be replaced by Lefschetz. It did not matter very much. Because of the outbreak of World War II, that congress was never held."
—Guillermo P. Curbera, *Mathematicians of the World, Unite!*, AK Peters, Wellesley, MA, 2009, pp. 72–74, 106.

"The late 1930s were a difficult time for European and American mathematicians who had close friends and professional colleagues in Germany. In 1938, after the Editor of Springer's *Zentralblatt für Mathematik*, Otto Neugebauer, left Germany for Copenhagen, Hardy joined Harald Bohr, Courant, Tamarkin, and Veblen in resigning from its Editorial Board. The move was not all in one direction, however. Helmut Hasse, a long-time collaborator of Harold Davenport's at Cambridge, took that opportunity to join the Board. That ended the collaboration."
—Sanford L. Segal, *Mathematicians under the Nazis*, Princeton University Press, 2003, p. 160.

"The more serious side of [Hardy's] nature showed in his efforts to promote reconciliation with the German mathematicians after the First World War, and later, to help the academic refugees from Nazi-occupied Europe. In Britain, although much was also done by individuals, including Hardy himself, the relevant organization for this purpose was the Academic Assistance Council, formed in 1933. This body raised a fund to provide maintenance for displaced university scholars, from whatever country, and acted as a centre of information, to put them in touch with institutions that could best help. The government was sympathetic, regarding it as in the public interest to 'try and secure for [the United Kingdom] prominent Jews who were being expelled from Germany and who had achieved distinction whether in pure science, applied science . . . music or art.' Hardy, who was usually consulted where mathematicians were concerned, saw this as an opportunity to strengthen mathematics in the United Kingdom, especially in Oxford and Cambridge. At the end of

the first year, four displaced mathematicians had settled in Britain; seven more came later. Hardy spent much time and effort on this exacting work."
—Ioan James, *Remarkable Mathematicians from Euler to von Neumann*, Cambridge-MAA, 2002, p. 303. Reprinted by permission of the Cambridge University Press.

"Hardy had opposed the ill treatment of conscientious objectors, and denounced the dismissal and imprisonment of Russell. Much later, in a privately circulated little book . . . he described the Russell case and the storms that raged over it. Russell was reinstated in Trinity in 1944, and made to feel at home. On one occasion he kept the Fellows spellbound with his anecdotes in the Combination Room, so late that next morning the manciple came to the Master with the resignations of the whole Kitchen Staff. Trevelyan, the Master and himself a Labour man, was not above appealing to the innate conservatism of College Servants: 'But you wouldn't want me to interrupt an Earl?' The resignations were withdrawn."
—Laurence Chisholm Young, *Mathematicians and Their Times/History of Mathematics and Mathematics of History*, North-Holland, Amsterdam, 1981, p. 280.

For Hardy, mathematics was a harmless enterprise. Indeed, it is difficult to imagine nations going to war over the Riemann hypothesis or the Goldbach conjecture. David Leavitt put it particularly well when he wrote, "Hardy saw mathematics as fundamentally innocent, even neutral—as a sort of Switzerland of the Sciences . . . "
—David Leavitt, *The Man Who Knew Too Much*, W. W. Norton, 2006, p. 132.

"When G. H. Hardy—as the reader may easily find in his book, *A Mathematician's Apology*—values number theory precisely for its lack of practical application, he is not fully facing the moral problem of the mathematician. It takes courage indeed to defy the demands of the world and to give up the fleshpots of Egypt for the intellectual asceticism of the pure mathematician, who will have no truck with the military and commercial assessment of mathematics by the world at large. Nevertheless, this is pure escapism in a generation in which mathematics has become a strong drug for the changing of science and the world we live in rather than a mild narcotic to be indulged by lotus eaters. When I returned to Cambridge as a mature mathematician after working with engineers for many years, Hardy used to claim that the engineering phraseology of much of my mathematical work was a humbug, and that I had employed it to curry favor with my engineering friends at the Massachusetts Institute of Technology. He thought that I was really a pure mathematician in disguise, and that these other aspects of my work were superficial. This, in fact, has not been the case."
—Norbert Wiener, *Ex Prodigy*, Simon & Schuster, New York, 1953, p. 189. Reprinted courtesy of the MIT Press.

12. Selections from What Others Have Said about Hardy

Norbert Wiener, a child prodigy, who began graduate school at Harvard at age fourteen. He took courses from Hardy in 1914 in Cambridge, and was strongly influenced by him. He worked in communication theory and invented the field of cybernetics. Crater Wiener on the Moon is named in his honor.

Constance Reid in her definitive book on David Hilbert noted that in 1909 he published a paper "settling Waring's century-old conjecture [that for a positive integer n there exists a function $g(n)$ so that n can always be written as a sum of at most $g(n)$ kth powers of integers]. I should classify the latter paper among his most original ones, but we can forego considering it more closely because a decade later Hardy and Littlewood found a different approach which yields asymptotic formulas for the number of representations, and it was the Hardy-Littlewood 'circle method' which has given rise in recent times to a considerable literature on this and related subjects."
—Constance Reid, *Hilbert*, Springer, New York, 1970, pp. 263–264.

In 1949, Norbert Wiener wrote an obituary of Hardy for the *Bulletin of the American Mathematical Society* in which he described Hardy's mathematics. This was followed up quickly by a short article by J. E. Littlewood, G. Pólya, L. J. Mordell, E. C. Titchmarsh, and H. Davenport, followed by a response to them from Wiener, "Two Statements Concerning the Article on G. H. Hardy," pointing out disagreements with Wiener's assessment, that (1) Hardy was not a conscientious objector (Wiener responds with apologies that he had been

misinformed); (2) Neither Bohr nor Landau could be described as students of Hardy's (Wiener claimed that they had "received a maximum influence from his work so that I consider that my statement although false in detail is correct in implications"), and (3) "To say that 'Hardy chose as his field the analytic theory of numbers' is to leave out of account his important work on Tauberian Theorems, on Diophantine approximation, and on Fourier series" (Wiener pointing out that "I may be quite wrong, but I feel that Hardy's very great work on Tauberian theorems, Diophantine approximation, and Fourier series is very largely an outgrowth of his interest in analytical number theory and even tools for its development".)
    —Norbert Wiener, "Godfrey Harold Hardy—1877–1947," *Bull. Amer. Math. Soc.* 55 (1949), pp. 72–77.

Hardy's work in the foundations of mathematics, especially his long paper "Mathematical Proof" of 1929, was in some ways the last in a series because the world of foundations changed in 1931 with the works of Kurt Gödel. Grattan-Guinness pointed out that "like other mathematicians of that time interested in foundational issues, especially those for whom proof was an important aspect of their practice, Hardy appears to have become most attracted to metamathematics among the three positions; early in 1929, during a sabbatical year in the USA, he lectured twice on 'Hilbert's logic'. It would have been fascinating to see him expand his lectures and the recent article... especially after Kurt Gödel... was to show in 1931 that neither logicism nor metamathematics could be executed in the manner in which their proponents envisioned it." This point was also noted by the editors of Hardy's *Collected Papers*, when they appended to his "Mathematical Proof" the comment: "Some of the later sections would probably have been modified if the paper had been written after Gödel's incompleteness results became known."
    —Ivor Grattan-Guinness, "The Interest of G. H. Hardy, F.R.S., in the philosophy and the history of mathematics," *Notes Rec. R. Soc. Lond.* 55:3 (2001), pp. 411–424. Reprinted with permission of the publisher.

"Bertrand Russell remarked to his colleague, the mathematician G. H. Hardy, that, if he had calculated that Russell would die in five minutes, his regret at losing a friend should his prediction be borne out, would be outweighed by his satisfaction at being proved right."
    —Walter Gratzer, *Eurekas and Euphorias/The Oxford Book of Scientific Anecdotes*, Oxford, 2002, p. 4.

M. H. A. Newman, writing an obituary about Hardy in *The Mathematical Gazette* in 1948, wrote that "In the *Gazette* it is fitting that we should record with special emphasis the debt which teachers of mathematics in this country owe to Hardy for the vast improvements in the teaching of analysis during the

12. Selections from What Others Have Said about Hardy   161

past forty years, from vagueness to precision, from obscurity to clarity, along lines mapped out and laid down for us by him."

Newman then ends his obituary with

> MATHEMATICIS QUOTQUOT UBIQUE SUNT
> OPERUM SOCIETATEM NUNC DIREMPTAM
> MOX UT OPTARE LICET REDINTEGRATURIS
> D. D. D. AUCTORES
> HOSTES IDEMQUE AMICI,

the dedication that Hardy and Marcel Riesz used for their 1915 book, *The General Theory of Dirichlet's Series*. (Translation: To the mathematicians (how many and wherever they may be): that they may soon again take up, as is to be hoped, the confraternity of their works which is currently disrupted, we, the authors, friends and foes at the same time, present and dedicate this book.) It was a suitable dedication for a joint effort by Hardy and Riesz in the early years of the First World War, Riesz at that time being in Stockholm but having come from Hungary, part of the "German" side in the war. Landau used this in his address on the occasion of the opening of the Hebrew University in Jerusalem in 1925. [11]

—M. H. A. Newman, *et al.*, "Godfrey Harold Hardy, 1877–1947," *Math. Gazette* 32 (1948), p. 50.

III
Mathematics

13
An Introduction to the Theory of Numbers

(*Bulletin of the American Mathematical Society* 35:6 (1929), pp. 773–818)*

From the Editors

This essay on number theory was originally a Josiah Willard Gibbs Lecture sponsored by the American Mathematical Society and the American Association for the Advancement of Science, in New York City, December 28, 1928. In 1932 this paper was awarded the Chauvenet Prize for expository writing by the Mathematical Association of America.

Hardy on an excursion at the 1932 International Congress of Mathematicians in Zurich.

---

* Reprinted with the kind permission of the publisher, The American Mathematical Society.

Part I

1. *Farey Series*. The theory of numbers has always occupied a peculiar position among the purely mathematical sciences. It has the reputation of great difficulty and mystery among many who should be competent to judge; I suppose that there is no mathematical theory of which so many well-qualified mathematicians are so much afraid. At the same time it is unique among mathematical theories in its appeal to the uninstructed imagination and in its fascination for the amateur. It would hardly be possible in any other subject to write books like Landau's *Vorlesungen* or Dickson's *History*, six great volumes of overwhelming erudition, better than the football reports for light breakfast table reading.

The excursions of amateur mathematicians into mathematics do not usually produce interesting results. I wish to draw your attention for a moment to one very singular exception. Mr. John Farey, Sen., who lived in the Napoleonic era, has a notice of twenty lines in the *Dictionary of National Biography*, where he is described as a geologist. He received as a boy "a good mathematical training". He was at one time agent to the Duke of Bedford, but afterwards came to London, where he acquired an extensive practice as a consulting surveyor, which led him to travel much about the country and "collect minerals and rocks". His principal work was a geological survey of Derbyshire, undertaken for the Board of Agriculture, but he also wrote papers in the *Philosophical Magazine*, on geology and on many other subjects, such as music, sound, comets, carriage wheels and decimal coinage. As a geologist, Farey is apparently forgotten, and, if that were all there were to say about him, I doubt that he would find his way into the *Dictionary of National Biography* today.

It is really very astonishing that Farey's official biographer should be so completely unaware of his subject's one real title to fame. For, in spite of the *Dictionary of National Biography*, Farey is immortal; his name stands prominently in Dickson's *History* and in the German encyclopaedia of mathematics, and there is no number-theorist who has not heard of "Farey's series". Just once in his life Mr. Farey rose above mediocrity and made an original observation. He did not understand very well what he was doing, and he was too weak a mathematician to prove the quite simple theorem he had discovered. It is evident also that he did not consider his discovery, which is stated in a letter of about half a page, at all important; the editor of the *Philosophical Magazine* printed a very stupid criticism in the next volume, and Farey, usually a rather acrid controversialist, ignored it completely. He had obviously no idea that this casual letter was the one event of real importance in his life. We may be tempted to think that Farey was very lucky; but a man who has made an observation that has escaped Fermat and Euler deserves any luck that comes his way.*

Farey's observation was this. The *Farey series of order n* is the series, in order of magnitude, of the irreducible rational fractions between 0 and 1 whose denominators do not exceed n. Thus

$$\frac{0}{1}, \frac{1}{7}, \frac{1}{6}, \frac{1}{5}, \frac{1}{4}, \frac{2}{7}, \frac{1}{3}, \frac{2}{5}, \frac{3}{7}, \frac{1}{2},$$
$$\frac{4}{7}, \frac{3}{5}, \frac{2}{3}, \frac{5}{7}, \frac{3}{4}, \frac{4}{5}, \frac{5}{6}, \frac{6}{7}, \frac{1}{1}$$

is the Farey series of order 7. There are two simple theorems about Farey series; (1) if p/q and p'/q' are two consecutive terms, then

$$p'q - pq' = 1,$$

and (ii) if p/q, p'/q', p''/q'' are three consecutive terms, then

$$\frac{p'}{q'} = \frac{p+p''}{q+q''}.$$

The second theorem (which is that actually stated by Farey) is an immediate consequence of the first, as we see by solving the equations

$$p'q - pq' = 1, \quad p''q' - p'q'' = 1,$$

for p' and q'.

The theorems are not of absolutely first class, importance, but they are not trivial, and all of the many proofs have some feature of real interest. One of the simplest uses the language of elementary geometry. We consider the *lattice* or *Gitter L* in a plane formed by drawing parallels to the axes at unit distance from each other; the intersections, the points (x, y) with integral coordinates, are called the *points of the lattice:* It is obvious that the properties of the lattice are independent of the particular lattice point O selected as origin and symmetrical about any origin. The lattice is transformed into itself by the linear substitution

$$x' = \alpha x + \beta y, \quad y' = \gamma x + \delta y,$$

where $\alpha, \beta, \gamma, \delta$ are integers and $\Delta = \alpha\delta - \beta\gamma = 1$, since then there is a pair x, y which give any assigned integral values for x', y'.

---

* It should be added that Farey's discovery had been anticipated 14 years before by C. Haros: see Dickson's *History*, vol. 1, p. 156. Cauchy happened to see Farey's note and attributed the theorem to him, and everyone else has followed Cauchy's example.

The area of the parallelogram P based on the origin and two lattice points (x_1, y_1), and (x_2, y_2), not collinear with O, is

$$\delta = \pm(x_1 y_2 - x_2 y_1).$$

We can construct a lattice L' (an oblique lattice) by producing and drawing parallels to the sides of P. A necessary and sufficient condition that L' should be equivalent to L, that is, that they should contain the same lattice points, is that $\delta = 1$, that is, that δ should have its smallest possible value. It is clear that this is also a necessary and sufficient condition that *there should be no lattice point inside P*, and it is easy to see that if there is such a point inside P, there is one inside, or on the boundary of, the triangular half of P nearer to O.

We may call the lattice point (q, p) which corresponds to a fraction p/q in its lowest terms a *visible* lattice point; there is no other lattice point which obscures the view of it from O. Let us consider all the visible lattice points which lie inside, or on the boundary of, the triangle bounded by the lines $y = 0$, $x = n, y = x$. It is plain, that these points correspond one by one to the fractions of the Farey series of order n. When the ray R from O to (q, p) rotates from the x-axis to the line $y = x$, it passes through each of these points in turn. If we take two consecutive positions of R, corresponding to the points $(q, p), (q', p')$, the parallelogram based on these two points contains no lattice point inside it, since otherwise there would be a lattice point inside its nearer triangle, and therefore a Farey fraction between p/q and p'/q'. It follows that

$$\delta = p'q - pq' = 1,$$

which proves Farey's theorem.

2. *Purpose of this Lecture.* So much then for Farey's discovery; it is a curious theorem, and its history is still more curious; but I have no doubt allowed myself to dwell upon it a little longer than its intrinsic importance deserves. My discussion of it will, however, help me to explain what I am trying to do in this lecture.

I shall imagine my audience to be made up entirely of men like Farey. I know that most of them are very much better mathematicians, but I shall not assume so; I shall assume only that they possess the common school knowledge of arithmetic and algebra. But I shall also assume that, like Farey, they are curious about the properties of integral numbers; one need after all be no Ramanujan for that.

Let us then imagine such a man playing about with numbers (as so many retired officers in England do) and puzzling himself about the curious properties which they seem to possess. What odd properties would strike him? What are the first questions he would ask? We must not try to be very systematic; if

13. An Introduction to the Theory of Numbers

we do, we shall make no progress in an hour. We must aim merely at a rough preliminary survey of the ground. If in the course of our survey, we find the opportunity for any illuminating remark, we may delay to make it, as I have already delayed over Mr. Farey, even if it does not seem to fall in quite its proper logical place. Then, if time permits, we may return to examine a little more closely any important difficulties which our preliminary survey has revealed.

3. *Congruences to a Modulus*. There is no doubt that the first general idea which we should have to explain is that of a *congruence*. Two numbers a and b are *congruent to modulus m* if they leave the same remainder when divided by m, that is, if m is a divisor of $a - b$. We write

$$a \equiv b \pmod{m}, \quad m | a - b.$$

It is obvious that congruences are of immense practical importance. Ordinary life is governed by them; railway time tables and lists of lectures are tables of congruences. The absolute values of numbers are comparatively unimportant; we want to know what time it is, not how many minutes have passed since the creation.

A great many problems both of arithmetic and of common life depend upon the solution of congruences involving an unknown x, such as

$$a_0 x^n + a_1 x^{n-1} + \cdots + a_n \equiv 0 \pmod{m}.$$

Such congruences may be classified like algebraical equations, as linear, quadratic,..., according to the value of n. Our first instinct in dealing with congruences is to follow up the analogy with algebra. In algebra a linear equation has one root, a quadratic two, and so on. We find at once that there are obvious and striking contrasts; even the linear congruence suggests a whole series of problems, and a full discussion of quadratic congruences involves quite an imposing body of general ideas.

Let us take the simplest case, the linear congruence, and suppose first that we are concerned only with one particular modulus, such as 7 or 24. We have then an example of a genuinely finite mathematics. Congruent numbers have exactly the same properties and cannot be distinguished, and our mathematics contains only *a finite number of things*. In such a mathematics any problem can be solved by enumeration; we can solve $2x \equiv 5 \pmod{7}$ by trying all possible values of x, and we find there is a unique solution, $x \equiv 6$. If we try to solve $2x \equiv 5 \pmod{24}$, we find that there is no solution; if I lecture every other day, I shall sooner or later lecture on Thursday, but if I lecture every other hour, I may never lecture at 5 P.M.

The difference is of course accounted for by the fact that 7 is *prime* and 24 is not. Here we encounter the notion of a prime, a number without factors, and all kinds of speculations suggest themselves. Can we tell, by any method short of trial of all possible divisors, whether any given number is prime or not? Are there formulas for primes? Are the primes infinite in number, and if so, what is the law of their distribution?

Again, it appears that all numbers are composed of primes, that primes are the ultimate material out of which the world of numbers is built up. We are bound to ask *how;* and here we meet our first big theorem, the "fundamental theorem of arithmetic," the theorem that factorization is unique. But we shall probably be wise to allow our enquirer to take this theorem for granted until he has acquired a little of the sophistication which comes with wider knowledge.

We may observe, however, before passing on, that the contrast between arithmetic and algebra becomes much more marked as soon as we consider congruences of higher degree. An equation of the fourth degree has, with appropriate conventions, just four roots. But

$$x^4 \equiv 1 \qquad (\mathrm{mod}\ 13)$$

has 4 roots, 1, 5, 8, and 12;

$$x^4 \equiv 1 \qquad (\mathrm{mod}\ 16)$$

has 8 roots, 1, 3, 5, 7, 9, 11, 13, and 15; and

$$x^4 \equiv 2 \qquad (\mathrm{mod}\ 16)$$

has none.

4. *Regarding Decimals.* I pass to another subject that has an irresistible fascination for amateurs, the subject of *decimals*. Some decimals are finite and some recurring, but it is easy to write down decimals, such as

(a) 0.10100100010...   (b) 0.11010001000...

which are neither. Here (a) the number of 0's increases by one at each stage, (b) the ranks of the 1's are 1, 2, 4, 8, ... More amusing examples are

(c) 0.01101010001010...

(in which the 1's have prime rank) and

(d) 0.23571113171923...

(formed by writing down the prime numbers in order). The proof for (c) demands the knowledge that there is an infinity of primes, and that for (d) rather more.*

---

* See Pólya and Szegö's *Aufgaben und Lehrsätze aus der Analysis*, vol. 2, pp. 160, 383.

13. An Introduction to the Theory of Numbers

The answer to some of the obvious questions is immediate. A finite decimal represents a rational fraction $p/(2^\alpha 5^\beta)$, a pure recurring decimal a fraction p/q, where q is not divisible by 2 or 5, and a mixed recurring decimal a fraction in which q is divisible by 2 or 5 and also by some other number. The converses of these theorems are also true, but the proof demands a little genuinely arithmetical reasoning. I shall state the proof in the simplest case, since it depends upon the logical principle which is perhaps our most effective weapon in the elementary parts of the theory, where we are dealing with so simple a subject matter that our choice of arguments is naturally very restricted.

Suppose $p < q$ and q prime to 10. If we divide all powers 10^ν by q, there are only q possible remainders, and one at least must be repeated. It follows that there are a ν_1 and a $\nu_2 > \nu_1$ such that

$$10^{\nu_2} \equiv 10^{\nu_1}, \quad 10^{\nu_1}(10^{\nu_2-\nu_1} - 1) \equiv 0$$

to modulus q. It follows that, if we write $\nu_2 - \nu_1 = N$, we have $10^N \equiv 1$, so that $q | 10^N - 1$ and

$$\frac{p}{q} = \frac{P}{10^N - 1} = P \cdot 10^{-N} + P \cdot 10^{-2N} + \cdots.$$

Since $P < 10^N$, this is a pure recurring decimal with a period of at most N. The principles which we have used are (a) that *if there are more than q things of at most q kinds, there must be two of them of the same kind*; (b) that if $10^\nu Q$ is divisible by q, and q is prime to 10, then Q is divisible by q. In the second we are of course appealing to the "fundamental theorem". The first is the general logical principle to which I referred just now.

Let us take a slightly more complicated variant of this principle. *If there are two sets of objects*

$$a_1, a_2, \ldots, a_m, \quad b_1, b_2, \ldots, b_m,$$

no two of either set being the same; and if every b is equal to an a; then the b's are the a's arranged in a different order. We may apply this principle to obtain further information about the period of our recurring decimal. I suppose now that q is prime. If q and a are given, and a is not a multiple of q, it is impossible that

$$ra \equiv sa \qquad (\mod q)$$

unless $r \equiv s$. If (ra) is the remainder when ra is divided by q, the two sets

$$r, \quad (ra) \qquad\qquad (r = 1, 2, \ldots, q - 1)$$

satisfy the conditions of our principle and are therefore the same except in order. It follows that

$$(q-1)!a^{q-1} \equiv \Pi(ra) \equiv \Pi r = (q-1)! \quad (\bmod q),$$

and therefore that

$$a^{q-1} \equiv 1 \quad (\bmod q);$$

Fermat's Theorem. In the particular case in which we are interested, a is 10, and Fermat's Theorem shows that we may take $N = q - 1$, so that the period of p/q cannot exceed $q - 1$ figures. Observe that we have appealed to the fundamental theorem twice in the proof.

It is familiar to everyone that $\frac{1}{7}$ has 6 figures, the maximum, number. We are bound to ask what other primes q possess this, property; the values of q less than 50 are in fact 7, 17, 19, 23, 29, and 47, but here we begin to get into deeper water. I cannot stop to discuss this question now, but before passing on I must mention another familiar text-book theorem which I shall have to quote later. This is Wilson's Theorem, that

$$(q-1)! + 1 \equiv 0 \quad (\bmod q)$$

if and only if q is prime. Of the mass of proofs catalogued by Dickson, that of Dirichlet depends most directly on principles which we have used already. It is an immediate consequence of these principles that, if x is any one of the set $1, 2, \ldots, q-1$, there is just one other, y, such that $xy \equiv 1 \pmod{q}$; we call y the *associate* of x. It is plain that 1 and $q - 1$ are associated with themselves; and no other number can be, since $x_1^2 \equiv x_2^2$ implies $x_1 \equiv x_2$ or $x_1 \equiv q - x_2$. It follows that the numbers $2, 3, \ldots, q-2$ are composed of $\frac{1}{2}(q-3)$ distinct pairs the product of each of which is congruent to 1. Hence

$$2 \cdot 3 \cdots (q-2) \equiv 1^{(q-3)/2} = 1,$$
$$(q-1)! \equiv q - 1 \equiv -1,$$

which is one half of Wilson's Theorem. The converse half is practically obvious, since $(q - 1)!$ would be divisible by any factor of q.

5. *Algebraic and Transcendental Numbers*. The study of decimals leads directly to problems concerning *rationality and irrationality*. Our decimals such as 0.1010010001 ... must represent irrational numbers. What criteria are there for deciding whether a given number is rational or irrational? To ask this question is to go a little outside the theory of numbers proper, which is concerned first with integers, and then with rationals or irrationals of special forms, such as the form $a + b\sqrt{2}$, and not with irrationals as a whole or general

13. An Introduction to the Theory of Numbers

criteria for irrationality. The problem is, however, one about which an amateur will certainly demand information.

The famous' argument of Pythagoras shows that $\sqrt{2}$ is irrational; if a/b is in its lowest terms and $a^2 = 2b^2$, then a and b must both be even, a contradiction. It is obvious to us now that the Pythagorean argument extends at once to $\sqrt{3}, \sqrt{5}, \ldots, 2^{1/3}, \ldots,$ and generally to $N^{1/m}$, where N is any number which is not a perfect mth power. There is a curious and very instructive historical puzzle connected with this argument. There is a passage in Plato's *Theaetetus*, discussed at length by Heath in his *History of Greek Mathematics*, about the attempt of Theodorus to generalize Pythagoras's proof. Theodorus, working some 50 years after Pythagoras, proved the irrationality of $\sqrt{N}$ for all values of N (except square values) up to 17 inclusive. Why, ask the historians did he stop? Why in any case should it have taken mathematicians like the Greeks 50 years to make so obvious an extension? Zeuthen in particular expended a great deal of ingenuity upon this question, but I think that the ingenuity was misplaced, and that the answer is obvious.

Theodorus *did not know the fundamental theorem of arithmetic;* there is something of a puzzle about the history of that theorem, but it cannot have been known to the Greeks before Euclid's time. The triviality of the generalization to us is due entirely to our knowledge of this theorem. Suppose, for example, we wish to prove that

$$a^2 = 60b^2,$$

where a and b are integers without common factor, is impossible. We argue that a^2 cannot be divisible by 3 unless a is divisible by 3; hence $a = 3c$, $a^2 = 9c^2$, $3c^2 = 20b^2$, and a repetition of the argument shows that b also is divisible by 3. We can prove that $3|a^2$ implies $3|a$ *without the fundamental theorem,* by enumeration of possible cases, considering separately the cases in which $a \equiv 0, 1, 2 \pmod{3}$. If it were 17 instead of 3, the process would be a little tedious; and in any case such a classification of numbers would have been very novel in Theodorus's time. I am, so far from being puzzled by the limitations of his work that I regard what he did as a very remarkable achievement.

There are very few types of numbers which present themselves at all naturally in analysis and which can be proved to be irrational. It is obvious that a number like $\log_{10} 2$ is irrational, for a power of 2 cannot be a power of 10. The proof for e, from the exponential series, is quite easy, and that for e^2 not very much more difficult. That for π is decidedly more so, and when we come to numbers like e^3 and π^2, it ceases to be worth while to worry about elementary proofs; we may as well go the whole way and prove e and π are transcendental. The most famous constant in analysis, after e and π, is Euler's constant γ; and the proof of the irrationality of γ is one of the classical unsolved problems of mathematics. It has never been proved that $2^{\sqrt{2}}$, $3^{\sqrt{2}}$, and similar numbers are irrational; no plausible method for attacking such problems has even been suggested. I am inclined to think that the number which holds out the best hopes for new discovery is the number e^{π}, which presents itself so naturally in the formulas of elliptic functions.

I said just now that e and π were "transcendental". I must not stop to talk at length about this famous theorem of Lindemann,* which contains the final proof that the quadrature of the circle, in the classical sense, is impossible; but the *statement* of the theorem introduces a notion that we shall require, that of an *algebraic number*. An algebraic number is the root of an equation

$$a_0 x^n + a_1 x^{n-1} + \cdots + a_n = 0,$$

where the a's are integers. An *algebraic integer* is an algebraic number whose characteristic equation has unity for its leading coefficient. Thus $\sqrt{2}$ and $1 + \sqrt{(-5)}$ are algebraic integers. A *transcendental* number is a number which is not algebraic; and Lindemann's Theorem is that π *is transcendental*. It is easy to show that all lengths which can be constructed by euclidean methods are algebraic, and indeed algebraic numbers of a quite special kind. If follows that the quadrature of the circle by any euclidean construction is impossible.

There is another direction in which we may be tempted to digress at this point, the theory of the approximation of irrationals by rationals, what is now called "diophantine approximation". There is just one theorem in this field that I shall mention, because it is connected so directly with what I have just been saying, and because it depends upon another of the stock arguments of number theory, the principle that *an integer numerically less than* 1 *is* 0. This is Liouville's theorem, that *there are transcendental numbers*. It is naturally much easier to prove this than to prove that a given number such as π is transcendental.

Liouville proves first that *it is impossible to approximate rationally to an algebraic number with more than a certain accuracy*. It is quite easy to see why. Suppose that ξ is an algebraic number defined by

$$f(\xi) = a_0 \xi^n + a_1 \xi^{n-1} + \cdots + a_n = 0.$$

We may suppose that the equation is irreducible, that is to say that $f(\xi)$ cannot be resolved into simpler algebraic factors of similar form; in this case we say

---

* See for example Hobson's *Trigonometry*, third edition, p. 305, or the same author's *Squaring the Circle*.

that ξ is *of degree n*. We can obviously find a number M, depending only on ξ, such that

$$|f'(x)| < M$$

for x near ξ. Suppose now that p/q is a rational, near ξ. Then

$$f\left(\frac{p}{q}\right) = \frac{N}{q^n},$$

where N is an integer not zero. It follows from our general principle that $|N| \geq 1$ and

$$\left|f\left(\frac{p}{q}\right)\right| \geq \frac{1}{q^n}.$$

But

$$f\left(\frac{p}{q}\right) = f\left(\frac{p}{q}\right) - f(\xi) = \left(\frac{p}{q} - \xi\right) f'(\eta),$$

where η lies between p/q and ξ. Hence, for all q,

$$\left|\frac{p}{q} - \xi\right| = \left|\frac{f(p/q)}{f'(\eta)}\right| > \frac{1}{Mq^n}.$$

It is impossible to approximate rationally to an algebraic number of degree n with an order of accuracy higher than q^{-n}.

On the other hand it is easy to write down numbers which have rational approximations of much higher accuracy than this; we have only to take a decimal of 0's and 1's in which the 1's are spaced out sufficiently widely. Thus

$$\xi = \frac{1}{10^{1!}} + \frac{1}{10^{2!}} + \frac{1}{10^{3!}} + \cdots = .11000100000\cdots$$

is approximated by its first k terms, that is, by a fraction

$$\frac{p}{q} = \frac{p}{10^{k!}}$$

with an error of order $10^{-(k+1)!} = q^{-k-1}$. Hence it is not an algebraic number of degree k and, since k is arbitrary, it must be transcendental. Obviously Liouville's argument enables us to *construct* transcendental numbers as freely as we please.

6. *Arithmetic. Forms.* The theory of irrationals starts from Pythagoras, and there is another great branch of the theory of numbers which also starts from him and about which I must now say something. This is the theory of *forms*.

Our interest in the theory of forms begins when we observe that there are Pythagorean triangles with integral sides; thus $3^2 + 4^2 = 5^2$. The first problem which suggests itself is that of determining all such triangles, and the solution given in substance by Diophantus, is easy. All the integral solutions of

$$x^2 + y^2 = z^2$$

are given by

$$x = \lambda(\xi^2 - \eta^2), \quad y = 2\lambda\xi\eta, \quad z = \lambda(\xi^2 + \eta^2),$$

where the letters are integers and ξ and η are coprime and of opposite parity. This problem is trivial, but it suggests an infinity of others.

It is natural to begin by a generalization of the problem. Let us discard the hypothesis that the hypotenuse z is integral; then

$$n = z^2 = x^2 + y^2$$

is the sum of two squares, and we are led to ask what numbers n possess this property. This is the first and simplest problem in the theory of *quadratic forms*, and the answer to it shows that no such problem can be quite easy. Even linear forms are not quite trivial; the solution of $ax + by = n$ in integers is a quite interesting elementary problem. When we consider quadratic forms, we come up against difficulties of a different order.

The first theorem in the subject is another theorem of Fermat, that $x^2 + y^2 = n$ is soluble when n is a prime $p = 4m + 1$ and, apart from trivial variations of the sign and order of x and y, uniquely. It is to be observed that the equation is plainly insoluble when n is $4m + 3$, since any square is congruent to 0 or 1 to modulus 4. This theorem is one of the most famous in the theory of numbers, and very rightly so, since it was the first really difficult theorem in the subject proved by any mathematician. There is no really simple proof, and the most natural, that which depends on the Gaussian numbers $a + bi$, introduces a whole series of ideas of revolutionary importance.

The first stage of the proof consists in proving that *there is a number x such that*

$$x^2 \equiv -1 \qquad (\mathrm{mod}\ p),$$

or $p | 1 + x^2$. Let us go back for a moment to the proof I sketched of Wilson's Theorem. Let us associate the numbers $x = 1, 2, \ldots, p - 1$ in pairs x, y not, as then, so that $xy \equiv 1$, but so that

$$xy \equiv -1 \qquad (\mathrm{mod}\ p).$$

If any x is associated with itself, our proposition is established. If not, we have arranged the numbers from 1 to $p - 1$ in $\frac{1}{2}(p - 1)$ pairs of different numbers

13. An Introduction to the Theory of Numbers

each satisfying the condition. Hence

$$(p-1)! \equiv \Pi xy \equiv (-1)^{(p-1)/2} = 1;$$

which is false, since, by Wilson's Theorem,

$$(p-1)! \equiv -1.$$

We thus obtain our proposition by *reductio ad absurdum*.

The second stage of the proof depends on much more novel ideas. We are concerned with the simplest case of an *algebraic field*. The field $K(i)$ is the aggregate of numbers

$$\xi = r + si = r + s\sqrt{(-1)},$$

where r and s are rational. This number satisfies the equation

$$\xi^2 - 2r\xi + r^2 + s^2 = 0,$$

and is an algebraic integer, in the sense I defined, before, when $2r$ and $r^2 + s^2$ are integers, that is, when r and s are integers. We may denote by $K^*(i)$ the aggregate of all the integers

$$\alpha = a + bi$$

of $K(i)$; a and b are ordinary integers. The numbers of $K^*(i)$ reproduce themselves by addition and multiplication, and we can define division in this field just as we define it in ordinary arithmetic. We can also define a *prime* of $K^*(i)$, and factorization of numbers into primes. There are four numbers, ± 1 and $\pm i$, which play a part in the new arithmetic similar to that of 1 and -1 in ordinary arithmetic. These are the "unities" or divisors of 1. If we define the *norm* of $\alpha = a + bi$ as

$$N(\alpha) = a^2 + b^2,$$

then the unities are characterized by the fact that their norm is 1. We do not count them as primes, just as, in the ordinary theory, we do not count 1 as a prime.

We now make an assumption, namely that *the analog of the fundamental theorem holds in the field $K^*(i)$*, that is to say that, apart from any trivial complications which may be introduced by the unities, *the factorization of a number of $K^*(i)$ into primes is unique*. This assumption is in fact correct. Returning now to the first stage of our proof, there is an x such that

$$p | 1 + x^2 = (1 + ix)(1 - ix).$$

It is obvious that p does not divide $1 + ix$ or $1 - ix$, so that p *divides the product of two numbers without dividing either of them.* Hence p cannot be a prime in $K^*(i)$. We may therefore write

$$p = \pi\lambda,$$

where $N(\pi) > 1$ and $N(\lambda) > 1$. But

$$N(\pi)N(\lambda) = N(p) = p^2,$$

so that $N(\pi)$ and $N(\lambda)$ must each be p. If we write

$$\pi = a + ib,$$

it follows that

$$p = N(\pi) = a^2 + b^2,$$

which is Fermat's theorem.

We may be tempted by our success to further efforts in the same direction. It is easy to satisfy ourselves, by considering particular cases, that *any prime $p = 20m + 1$ is of the form $a^2 + 5b^2$*: thus $61 = 4^2 + 5 \cdot 3^2$. Let us try to prove this theorem by a similar method. We must evidently consider now the field $K^*[\sqrt{(-5)}]$ formed of the algebraic integers of the form

$$\alpha = a + b\sqrt{(-5)};$$

it is easy to show that such a number is an algebraic integer if and only if a and b are ordinary integers. There is no difficulty in defining divisibility and primality in this field also.

The first step, in our proof must plainly be to prove the existence of an x for which $p|1 + 5x^2$. This is not difficult, but it demands a little more knowledge of quadratic congruences than I can assume, and I must take it for granted.

We define the norm $N(\alpha)$ of a number of this field as $a^2 + 5b^2$. We then argue as before; we have

$$p|1 + 5x^2 = (1 + x\sqrt{(-5)})(1 - x\sqrt{(-5)}),$$

so that p divides a product without dividing either factor and is therefore not a prime. Hence, as before $p = \pi\lambda$, where $N(\pi) > 1$ and $N(\lambda) > 1$, and $N(\pi)$ and $N(\lambda)$ must each be p. It follows that

$$p = N(\pi) = a^2 + 5b^2,$$

the theorem we set out to prove.

At this point, however, there is a shock in store for us; we find that we can prove *too much*. The number

$$q = (2 + \sqrt{(-5)})(2 - \sqrt{(-5)})$$

is divisible by 3, while neither factor is so. Hence 3 is not a prime. Hence

$$3 = \pi\lambda, \quad 9 = N(\pi)N(\lambda),$$

and $N(\pi)$ and $N(\lambda)$ are each 3. It follows that

$$3 = N(\pi) = a^2 + 5b^2.$$

Similarly we can prove that

$$7 = a^2 + 5b^2;$$

and both of these theorems are obviously false.

There must therefore be a mistake somewhere in our argument, and if you examine it, and are prepared to believe that I have not been misleading you wilfully, you will see that there is only one step which can be questioned. In all three cases I concluded the argument by an appeal to the same theorem; *a number which divides the product of two numbers without dividing either of them cannot be prime*. This is true in ordinary arithmetic, because of the fundamental theorem; if 7 were a divisor of $15 = 3 \cdot 5$, 15 would be factorable into primes in two distinct manners. It follows that *the analog of the fundamental theorem in the field $K^*[\sqrt{(-5)}]$ must be false*; and this is easily verified when once our suspicions have been excited; thus

$$2 \cdot 3 = (1 + \sqrt{(-5)})(1 - \sqrt{(-5)}),$$
$$3 \cdot 7 = (1 + 2\sqrt{(-5)})(1 - 2\sqrt{(-5)}),$$

and all of these numbers are prime in $K^*[\sqrt{(-5)}]$. The proof which I gave of the theorem concerning primes $20m + 1$ was therefore fallacious, although the theorem is true. The proof of Fermat's theorem, on the other hand, was correct, since factorization *is* unique in $K^*(i)$.

7. *Further Problems*. It is clear that we must go back to the beginning and study the theory of primes a little more closely; but before I do this I should like to call your attention to a series of further problems suggested by Fermat's theorem. We know now when a *prime* is the sum of two squares, and we have to consider the same problem for general n. Here in fact there are three different problems.

The first and most obvious problem is that of determining the necessary and sufficient conditions that n should be representable. This problem may be solved quite easily with the aid of the Gaussian numbers; n must be $2^\alpha M^2 N$, where α is 0 or 1 and N contains prime factors of the form $4m + 1$ only. We are then led naturally to the corresponding problem for other forms, first for the general binary quadratic form

$$ax^2 + bxy + cy^2,$$

then for quadratic forms in a larger number of variables, such as

$$x^2 + y^2 + z^2, \quad x^2 + y^2 + z^2 + t^2,$$

and then for forms of higher degree, such as $x^3 + y^3$ and $x^4 + y^4$. There is a highly developed theory of the general quadratic form; the most famous theorem is perhaps Lagrange's theorem, that *every number is the sum of four squares*. But as soon as we begin to consider cubic or higher forms we find ourselves on the boundary of knowedge. There is for example no criterion analogous to Fermat's by which we can decide whether a given number is the sum of two cubes.

The second problem about the form $x^2 + y^2$ suggested by Fermat's theorem is that of determining the *number of representations*. This problem may be interpreted in two different ways. We may want an exact formula, in terms of the factors of n, and in this case the Gaussian theory again gives what we want; $r(n)$, the number of representations, is given by the formula

$$r(n) = 4\{d_1(n) - d_3(n)\},$$

where $d_1(n)$ and $d_3(n)$ are the numbers of divisors of n of the forms $4m + 1$ and $4m + 3$ respectively. This is, however, not the most interesting interpretation of the problem. We may want, not a formula like this, but information concerning the *order of magnitude* of $r(n)$, whether $r(n)$ is generally large when n is large, whether numbers are usually representable freely or with difficulty. In this case our formula gives us very little help, and the solution of the problem requires quite different methods.

It is here that we come into contact for the first time with a new branch of the theory, the modern "analytic" theory. This theory has two special characteristics. The first is one of method; it uses, besides the methods of the classical theory, the methods of the modern theory of functions of a complex variable. The second is that it is concerned primarily with problems of order of magnitude and asymptotic distribution. The distinction is not a perfectly sharp one; there are "exact", "finite" theorems which have only been solved by "analytic" methods. For example, *every number greater than $10^{10^{10}}$ is expressible as the sum of 8 cubes*; this theorem includes no reference to "order of magnitude", and is a

13. An Introduction to the Theory of Numbers

"finite" theorem in just the same sense as Fermat's theorem about the squares, but the only known proof is analytic. On the whole, however, it is the problems of asymptotic distribution which dominate the theory.

The answer given by the analytic theory to the special question which I raised is roughly as follows. The average value of $r(n)$ is π. It must be observed that representations which differ only trivially, that is, in the sign or order of x and y, are reckoned as distinct. If we allow for this, the average number of representations is rather less than a half; this is explained by the fact that, as we shall see, *most* numbers are not representable. On the other hand $r(n)$ tends to infinity with n with tolerable rapidity for numbers of appropriate forms, more rapidly for example than any power of log n. The corresponding problems for cubes or higher powers present difficulties which are at present quite insuperable, and all that I can do is to mention a few curiosities. The smallest number representable by two cubes in two really distinct ways is

$$1729 = 1^3 + 12^3 = 9^3 + 10^3,$$

and the smallest representable in three ways is probably

$$175959000 = 70^3 + 560^3 = 198^3 + 552^3 = 315^3 + 525^3.$$

It can be proved that there exist numbers with as many different representations as we please. A. E. Western has carried out very heavy computations concerning representations by cubes; he has for example found 6 numbers, of which the smallest is 1,259,712, representable as the sum of *three* cubes in *six* different ways. The smallest number doubly representable by two fourth powers is probably

$$635318657 = 59^4 + 158^4 = 133^4 + 134^4;$$

there is, so far as I know, no known example of a number with three such representations, nor any proof that such a number exists.

The nature of the problems of the analytic theory becomes clearer when we consider the third problem suggested by Fermat's theorem. This is the problem of determining the *distribution* of the representable numbers. We want to know *how many numbers are representable*, or, to put it more precisely, how many numbers less than a large assigned number x are representable. If $Q(x)$ is the number of such numbers, what is the order of magnitude of $Q(x)$? Are nearly all numbers representable, or just a majority, or only a few? The answer is in fact that $Q(x)$ is approximately

$$\frac{Ax}{(\log x)^{1/2}},$$

where A is a constant; to put it roughly, quite a lot of numbers are representable, but strictly an infinitesimal proportion of the whole. This explains why the average number of representations turned out to be less than one.

This problem about $Q(x)$ is a very interesting one, but there is another of the same kind which is obviously still more interesting and much more fundamental. This is the problem of the distribution of the primes themselves; *how many primes are there less than x?* I shall say something about this problem in a moment; it is in any case time for us to return to the theory of primes, since all our enquiries have ended in questions about them, and it is obviously impossible to make serious progress until we know more both of their elementary properties and of the laws which govern their distribution.

Part II

8. *The Fundamental Theorem.* The *fundamental theorem* of arithmetic is the beginning of the theory of numbers, and it is plain that our first task must be to make this theorem secure.

There is another historical puzzle about the fundamental theorem. Who first stated the theorem, explicitly and generally? The natural answer is *Euclid*, since the *Elements* contain all the materials for the proof. Everything rests on Euclid's famous algorithm for the greatest common divisor. Given two numbers a, b, of which a is the greater, we form the table

$$a = bc + b_1, \quad b = b_1c_1 + b_2, \quad b_1 = b_2c_2 + b_3, \ldots$$

where $b_1, b_2, \ldots$, are the remainders in the ordinary sense of elementary arithmetic. Since

$$b > b_1 > b_2 > \cdots,$$

b_n must sooner or later be zero. The last positive remainder δ has the properties implied by the words *greatest common divisor*, and it follows from the process by which δ is formed that any number which divides both a and b divides δ.

Let us note in passing that there is an analogous process in $K^*(i)$, but that the analogy fails in $K^*[\sqrt{(-5)}]$. In ordinary arithmetic, given a and b, we can find a number congruent to a mod b and less than b. There is a similar theorem for the Gaussian numbers. Here there is no strict order of magnitude between different numbers, and we have to use the order of magnitude of their norms. Given α and β, there is a number, congruent to α mod β, whose norm is less than that of β. There is no such theorem in $K^*[\sqrt{(-5)}]$, and the process analogous to Euclid's fails.

When the existence of δ is once established, the proof of the fundamental theorem is easy. We write
$$\delta = (a, b)$$
and we say that a is prime to b when $(a, b) = 1$. The crucial lemma is that *if $(a, b) = \delta$ and $b|ac$, then $b|c$*; in particular, *a prime cannot divide a product without dividing one or other of the factors*. This once granted, anybody can construct the proof of the fundamental theorem for himself; and you will remember that it was just this proposition which led to our troubles in $K^*[\sqrt{(-5)}]$.

The lemma itself may be proved as follows. We construct the euclidean algorithm for a and b, with the final remainder 1. If we multiply it throughout by c, we have the algorithm for ac and bc, and the final remainder is c. It follows that
$$(ac, bc) = c.$$
Since b divides ac, by hypothesis, and also bc, it divides c.

This is Euclid's own argument, and with it he had proved what is essential in the fundamental theorem. It is a very singular thing that he should then omit to state the magnificent theorem that he has proved. He is over the line and free, but apparently disdains the formality of touching down. I do not know of any formal statement of the theorem earlier than Gauss. The substance of the theorem, however, is in the *Elements;* it was plainly unknown, as I explained before, to the Greeks from 50 to 100 years before Euclid's time; and I see no particular reason for questioning the obvious view that it is Euclid's own.

As soon as we have proved the fundamental theorem our elementary knowledge falls into line. The theory of linear congruences, the theorems of Fermat and Wilson and all their consequences, the elementary theory of decimals and of the divisors of numbers, may be developed straight-forwardly and without the introduction of essentially new ideas. I can now say something about the more modern side of the theory of primes.

9. *Problems Concerning Primes.* What are the most natural questions to ask about primes? I say deliberately the most *natural;* we must remember that a natural question does not always seem, on fuller reflection, to have been a *reasonable* one. It is natural to an engineer to ask us for a finite formula for
$$\int e^{-x^2} dx,$$
or for a solution of some simple looking differential equation in finite terms. If we fail to satisfy him, it is not because of our stupidity, but because the world does not happen to have been made that way.

So, if any one asks us (1) *to give a general formula for the nth prime p_n*, a formula in the sense in which

$$p_n = n^2, \quad p_n = n^2 + 1, \quad p_n = [e^n],$$

where $[x]$ denotes the integral part of x, would be a formula, I can only reply that it is not a reasonable question. It is, I will not say demonstrably impossible, but wildly improbable, that any such formula exists. The distribution of the primes, is not like what it would have to be on any such hypothesis. I should make the same reply to a good many other questions which an amateur might be likely to ask, for example if he asked me (2) *to give a rule for finding the prime which immediately follows a given prime*. It would of course be perfectly reasonable that he should press me for the reasons why I gave so purely a negative a reply. On the other hand the problem (3) *to find the number of primes below a given limit* is, if interpreted properly, an entirely reasonable and a soluble problem. The problems (4) *to prove that there are infinitely many pairs of primes differing by 2*, and (5) *to prove that, there are infinitely many primes of the form $n^2 + 1$*, are also entirely reasonable, and if (as is the case) we cannot solve them, it is quite reasonable to condemn our lack of ingenuity.

10. *The Distribution of Primes.* If we wish to classify these problems and to decide which of them are reasonable and which are not, the first essential is to understand broadly the present state of knowledge about the distribution, the distribution *in the large* or *asymptotic* distribution, of the primes. It is this theory which gives the solution of problem (3).

We denote by $\pi(x)$ the number of primes not exceeding x. The first step is to prove that (a) *the number of primes is infinite*; $\pi(x)$ tends to infinity with x. This is another of Euclid's great contributions to knowledge, and, Euclid's proof is perhaps the classical example of proof by *reductio ad absurdum*. If the theorem is false, we may denote the primes by $2, 3, 5, \ldots, P$, and all numbers are divisible by one of these. On the other hand the number

$$(2 \cdot 3 \cdot 5 \ldots P) + 1$$

is obviously not divisible by any of $2, 3, \ldots, P$, and this is a contradiction.

Another very interesting proof is due to Pólya.* It is easy to see that any two of the numbers

$$2 + 1, 2^2 + 1, 2^4 + 1, \ldots, u_n = 2^{2^n} + 1$$

---

* See Pólya and Szegö, loc. cit., pp. 133, 342.

13. An Introduction to the Theory of Numbers

are prime to each other. For suppose that p is an odd prime and that $p|u_n$, $p|u_{n+k}$. Then also

$$p|2^{2^{n+k}} - 1 = u_{n+k} - 2,$$

since

$$x^{2^{km}} - 1$$

is algebraically divisible by $x^m + 1$, and therefore

$$p|u_{n+k} - (u_{n+k} - 2) = 2,$$

which is absurd. It follows that the number of primes less than u_n is at least n, and therefore that the number of primes is infinite. In fact the argument shows not merely that $\pi(x) \to \infty$ but that

$$\pi(x) > A \log \log x,$$

where A is a constant. Something in this direction, though a little less, can be proved by a refinement of Euclid's argument.

There is a third line of argument which is a little less elementary but may be made to prove a good deal more.*

If $2, 3, 5, \ldots, P$ were the only primes, then every number would be of the form

$$2^a 3^b 5^c \cdots P^k.$$

If this number is less than, x, then *a fortiori* 2^a is less than x, so that a is less than a constant multiple of $\log x$, and the same argument applies to $b, c, \ldots, k$. The number of possible choices of $a, b, \ldots, k$ is therefore less than a multiple of $(\log x)^\pi$, where π is the total number of primes. In other words the number of numbers less than x is less than

$$A(\log x)^\pi,$$

where A is a constant, and this is impossible, since x tends *to* infinity more rapidly than any power of $\log x$. A refinement of the argument leads to the inequality

$$\pi(x) > A \frac{\log x}{\log \log x};$$

and the underlying principle may be stated roughly thus, that *if the number of primes were finite, there would not be enough numbers to go round.*

---

* See Dickson's *History*, vol. 1, p. 414, where the proof is attributed to Auric.

We are still a very long way from the ultimate truth. It is in fact possible to prove, and by comparatively elementary methods, that *the order of magnitude of $\pi(x)$ is $x(\log x)^{-1}$*. This theorem, conjectured by Legendre and Gauss, was first proved by Tchebycheff in 1848.

There are two much earlier theorems of Euler which point in this direction. The first is the theorem that (b) *the series*

$$\sum \frac{1}{p}$$

extended over all prime numbers p, is divergent. The proof of this theorem depends upon an identity, also due to Euler, upon which the whole of the modern theory of primes is founded. The identity is

$$1^{-s} + 2^{-s} + 3^{-s} + \cdots = \sum n^{-s}$$
$$= \frac{1}{(1-2^{-s})(1-3^{-s})(1-5^{-s})\cdots}$$
$$= \prod \left(\frac{1}{1-p^{-s}}\right)$$

and is valid for $s > 1$; it is at bottom merely the analytical expression of the fundamental theorem, and its importance arises from the fact that it asserts the equivalence of two expressions of which one contains the primes explicitly while the other does not. From Euler's identity we deduce (b) roughly as follows: if $\sum p^{-1}$ were convergent, then

$$\prod \left(\frac{1}{1-p^{-1}}\right)$$

would be convergent, and therfore $\sum n^{-1}$ would be convergent, which is false. Of course the proof really needs a rather more careful statement.

Euler's second theorem is (c) *the quotient of $\pi(x)$ by x tends to zero*; or in symbols

$$\frac{\pi(x)}{x} \to 0,$$

or, as we write it now

$$\pi(x) = o(x).$$

The proportion of primes is ultimately infinitesimal, *"almost all"* numbers are composite. The theorem is a quite simple corollary of (b); roughly, if we remove from the numbers less than x all multiples of the primes $2, 3, \ldots, p$, other than these primes themselves, we are left something like

$$x\left(1-\frac{1}{2}\right)\left(1-\frac{1}{3}\right)\left(1-\frac{1}{5}\right)\cdots\left(1-\frac{1}{p}\right)$$

numbers. The product multiplying x tends to zero when $p \to \infty$, because of (b), and from this we can deduce Euler's second theorem.

It is rather curious that, although Euler's second theorem is a corollary of the first, the lessons which we learn from the two theorems concerning the distribution of the primes have exactly opposite tendencies. The second theorem tells us that the number of primes below a given limit is *not too great*, that the primes are in the end rather liberally spaced out; it is in fact exactly equivalent to the theorem that (d) *the nth prime p_n has an order of magnitude greater than n*, or

$$\frac{p_n}{n} \to \infty.$$

If on the other hand the order of magnitude of p_n were *much* greater than n, if it were for example n^2 or $n^{10/9}$ or $n(\log n)^2$, then the series $\sum p_n^{-1}$ would be *convergent*, which is just what Euler's first theorem denies. What we learn from the two theorems together is something like this. If, as we hope, the true order of magnitude of p_n can be measured by some simple function $\phi(n)$, then that function must be of order higher than n, but somewhere near the boundary of convergence of the series

$$\sum \frac{1}{\phi(n)}.$$

The most obvious function which satisfies these requirements is $n \log n$, and to say that p_n is of order $n \log n$ is the same thing as to say that $\pi(x)$ is of order $x(\log x)^{-1}$. This is just what is asserted by Tchebycheff's theorem.

11. *Tchebycheff's Theorem.* The formal statement of Tchebycheff's theorem is (e) *the order of magnitude of $\pi(x)$ is $x(\log x)^{-1}$*; there are constants A and B such that

$$\frac{Ax}{\log x} < \pi(x) < \frac{Bx}{\log x}.$$

This theorem is precisely equivalent to (f) *the order of magnitude of p_n is $n \log n$*; there are constants A and B such that

$$An \log n < p_n < Bn \log n.$$

The proofs of these theorems given by Tchebycheff have been simplified a good deal by Landau, and I can give you a sketch of one half of the proof which should enable you to understand without much difficulty the general character of the whole.

We begin by replacing $\pi(x)$ by another function. We can write $\pi(x)$ in the form

$$\pi(x) = \sum_{p \leq x} 1;$$

count one for every prime up to x. A more convenient and really a more natural function is

$$\theta(x) = \sum_{p \leq x} \log p,$$

the logarithm of the product of all primes up to x. This function seems at first sight a more complicated function, but it is easy enough to see why it is more convenient to work with. The most natural operation to perform on primes is *multiplication*, and this is the operation which we employ in forming $\theta(x)$. It is because it is natural to multiply primes and not to add or subtract them that problems like the problem of the prime pairs $(p, p+2)$, or Goldbach's problem of expressing numbers as sums of primes, turn out to be so terribly difficult.

Since $x/x^{1-\delta}$ tends to infinity, for any positive value of δ, we may expect that nearly all the primes which contribute to $\theta(x)$ will lie in the interval $(x^{1-\delta}, x)$, so that their logarithms lie between $(1-\delta) \log x$ and $\log x$. Hence we may expect $\theta(x)$ to be very much the same function as $\pi(x) \log x$, and in fact there is no difficulty in proving that

$$\theta(x) \sim \pi(x) \log x,$$

that is, that the ratio of the two functions tends to 1. It follows that the inequalities in (e) are equivalent to

$$Ax < \theta(x) < Bx.$$

I shall sketch the proof of the second inequality, which is rather the simpler.

Suppose that x is a power of 2, say 2^m. The primes between $x/2$ and x divide $x!$ but not $(x/2)!$, so that

$$\prod_{x/2 < p \leq x} p \;\Big|\; \frac{x!}{(x/2)!(x/2)!}.$$

The expression on the right is *one term* in the binomial expansion of $(1+1)^x = 2^x$, and therefore

$$\prod_{x/2 < p \leq x} p \leq 2^x.$$

Replacing x by

$$x/2, x/4, x/8, \ldots$$

and multiplying the results, we find that

$$\prod_{p \leq x} p \leq 2^{x+x/2+x/4+\cdots} \leq 2^{2x},$$

and

$$\theta(x) \leq 2 \log 2 \cdot x.$$

This proves the theorem when $x = 2^m$. If

$$2^m < x < 2^{m+1}$$

we have

$$\theta(x) \leq \theta(2^{m+1}) \leq 4 \log 2 \cdot 2^m < 4 \log 2 \cdot x.$$

Hence we may take $B = 4 \log 2$. The proof of the second inequality is, as I said, not quite so simple, but does not involve essentially more difficult ideas. We have thus determined the order of magnitude of $\pi(x)$ and of p_n, and it is perhaps a little astonishing that a problem which sounds so abstruse should have so comparatively simple a solution.

12. *The Prime Number Theorem.* Tchebycheff's solution of the problem is, however, one with which it is impossible to remain content for long, since the whole trend of our discussion has been to suggest that much more is true than we have proved. In fact Tchebycheff's work, fine as it is, is the record of a failure; it is what survives of an unsuccessful attempt to prove what is now called the *Prime Number Theorem*.

This is the theorem that $\pi(x)$ *and* $x(\log x)^{-1}$ *are asymptotically equivalent; the ratio of the two functions tends to unity.* We express this by writing

$$\pi(x) \sim \frac{x}{\log x}.$$

The Prime Number Theorem is equivalent to

$$p_n \sim n \log n,$$

and we may express it very roughly by saying that *the odds are* $\log x$ *to* 1 *that a large number x is not prime.*

The Prime Number Theorem, the central theorem of the analytic theory of numbers, was proved independently by Hadamard and by de la Vallée-Poussin in 1896. The empirical evidence for its truth had for long been overwhelming, and I suppose that every number-theorist since Legendre had tried to prove it. The theorem differs from all those which I have discussed so far in that it is apparently impossible to prove it by properly elementary methods; there is no proof known which does not depend essentially on complex function theory. I do not mean to imply that there is any terrible difficulty in the proof; there are considerable difficulties of detail, but the fundamental ideas on which it depends are tolerably straightforward. They are, however, quite unlike any of those of which I have spoken, and I should require a whole lecture to explain them even to a strictly mathematical audience. Actually, a good deal more is known; it can be proved that $\pi(x)$ is approximated still more closely by the "logarithm-integral" of x,

$$\text{Li}\, x = \int_2^x \frac{dt}{\log t},$$

that in fact

$$\pi(x) = \text{Li}\, x + O\left\{\frac{x}{(\log x)^k}\right\}$$

for every k, the error being of lower order than the quotient of x by *any* power of $\log x$; and it is probable, though not yet proved, that the order of the error does not very materially exceed that of $\sqrt{x}$.

13. *Formulas for Primes.* I return now for a moment to a question which I discussed shortly before, the question whether it was reasonable to expect an "elementary formula" for the nth prime p_n. Let us imagine that my questioner was obstinate in his desire for such a formula; how could I refute his successive suggestions? If he suggested

$$p_n = n \log n,$$

I should have the obvious reply that $n \log n$ is not an integer. Suppose then that he modified his formula to

$$p_n = [n \log n].$$

I should reply that his formula did not agree with the known facts of the asymptotic theory. It agrees with $p_n \sim n \log n$, the first and most obvious deduction from the Prime Number Theorem itself; but the theory carries us much further; it enables us, for example, to show that

$$p_n = n \log n + n \log \log n + O(n),$$

which contradicts the formula. If, becoming more cautious, he asked me what ground I had for denying that p_n might be *some* elementary combination of

$$n, \log n, \log \log n, \ldots,$$

I should naturally find it harder to refute him, but I could advance three arguments which are enough in the aggregate to make up a tolerably convincing case. (i) Since Li x is a very good approximation to $\pi(x)$, the inverse function $\mathrm{Li}^{-1} n$ must be a very good approximation to p_n. Now it is demonstrable that neither the logarithm integral nor its inverse* is an elementary function. It is therefore very unlikely that there should be an elementary formula for p_n. (ii) If the "elementary formula" does not involve the symbol [...] of the "integral part", the function which it defines will generally not be integral for integral n. If it does, it loses all its simplicity and all its plausibility, (iii) An elementary function may be expected to behave with tolerable regularity at infinity, and so may all its *differences*. Now extremely little is known about the difference $p_{n+1} - p_n$ of two successive primes, but everything that is known, or seems probable from the evidence of the tables, suggests *extreme irregularity* in its behavior. The Prime Number Theorem shows that the *average* value of $p_{n+1} - p_n$ must be $\log n$, and tend to infinity with n. On the other hand there is overwhelming evidence that the smallest possible values of $p_{n+1} - p_n$, namely, 2, 4, 6, ..., recur indefinitely. It seems practically certain, not merely that there are infinitely many prime pairs $(p, p + 2)$ but that there are infinitely many triplets $(p, p + 2, p + 6)$, and so with any combination of successive primes that is arithmetically possible; such a combination as $(p, p + 2, p + 4)$ is naturally not possible, since one of these numbers must be divisible by 3. All this seems hopelessly inconsistent with the existence of such a formula as was suggested, and it is clear that speculation in this direction is a waste of time.

There are, however, questions which have a somewhat similar tendency and which cannot be dismissed so summarily. There is one, for example, mentioned in Carmichael's little book. The problem, as he states it, is *"to find a prime greater than a given prime,"* which might be interpreted as meaning either *"to find an elementary function $\phi(n)$ such that $\phi(n) \to \infty$ and $\phi(n)$ is prime for every n, or for all n beyond a certain limit"* or as meaning *"to find an elementary function $\phi(p)$ such that $\phi(p) > p$ and $\phi(p)$ is prime whenever p is prime."* With either interpretation, it is a reasonable challenge, and the problem has not been solved.

Let us take the first form of the problem, which is perhaps the more natural, and let us begin by demanding *less*, namely that $\phi(n)$ shall be prime only for *an infinity of values of n*. In this case the problem becomes trivial, since n is a solution, by Euclid's theorem. It is, however, very interesting to observe that even then n, and certain simple linear functions such as $4n - 1$ and $6n - 1$, are the *only* trivial solutions. Dirichlet proved that *any* linear function $an + b$

---

* This may be deduced from general theorems proved recently by J. F. Ritt.

has the property required, provided only that b is prime to a, or in other words that *every arithmetical progression* (subject to the last reservation) *contains an infinity of primes*. This theorem is quite difficult, except in a few special cases such as those which I mentioned, and it exhausts our knowledge in this particular direction. No one has ever proved that any of the functions

$$n^2 + 1, \quad 2^n - 1, \quad 2^n + 1$$

is prime for an infinity of values of n. With functions of *two* variables we can progress a good deal farther; we know for example that every quadratic form $am^2 + bmn + cn^2$ contains an infinity of primes, provided of course that a, b, c have no common factor and that $b^2 \neq 4ac$, and we can study the law of their distribution.

To find a $\phi(n)$ prime for *every* n is naturally still more difficult. Here linear functions are obviously useless, and no solution of any kind is known. Fermat conjectured that

$$2^{2^n} + 1,$$

is always prime, but Euler proved that this is false, since

$$2^{32} + 1 = 4294967297 = 641 \cdot 6700417.$$

So far as I know, no one else has ever advanced any other suggestion which is even plausible.

In view of the apparently insuperable difficulties of this problem, there is a certain interest in *negative* results. It is plain, first, that $an + b$ cannot be prime for all n, or all large n. More generally, no polynomial

$$f(n) = a_0 n^k + a_1 n^{k-1} + \cdots + a_k$$

can be prime for all or all large n; for if $f(m) = M$ then $f(rn + m)$ is divisible by M for all r. There are entertaining curiosities in this field; thus

$$n^2 - n + 41$$

is prime for the first 41, and

$$n^2 - 79n + 1601$$

for the first 80 values of n. It is obvious that forms like

$$a^n - 1, \quad a^n + 1$$

cannot be prime for all large n, since, for example, $a^{3m} - 1$ is divisible by $a^m - 1$, and it is natural to suppose that the same is true for

$$P(n, 2^n, 3^n, 4^n, \ldots, k^n),$$

13. An Introduction to the Theory of Numbers

where P is any polynomial with integral coefficients.*

14. *The Fundamental Theorem in an Algebraic Field.* I must not allow myself to succumb to the temptation of talking too long about the theory of the distribution of primes, which is after all only one chapter in arithmetic. There are other topics about which our imaginary enquirer will certainly demand more information, and of these I think one stands out; it is certain that he will want fuller explanations about the field $K^*[\sqrt{(-5)}]$ and the other algebraic fields in which the analog of the fundamental theorem fails. All ordinary arithmetic depends, it seems, upon the fundamental theorem; how then can there *be* an arithmetic in a field in which it is false? It would seem that the arithmetic of such a field can bear no real resemblance to ordinary arithmetic. I shall spend the rest of my time in an attempt to explain, in the very broadest outline, how order is restored.

I shall begin by quoting a remark of Hilbert which is trivial in itself but which shows us at once the direction in which we must look for a solution. Consider the numbers

$$1, 5, 9, 13, 17, 21, \ldots$$

of the form $4m + 1$. These numbers form a group for multiplication (though naturally not for addition), and we can define divisibility and primality in the group. The "primes" are the numbers

$$5, 9, 13, 17, 21, 29, 33, 37, 41, 49, \ldots$$

which are greater than 1, of the form $4m + 1$, and not decomposable into factors of this form. Thus 21, 57, 77, and 209 are "primes"; but

$$4389 = 21 \cdot 209 = 57 \cdot 77,$$

so that a number of the group may be resolved into "prime" factors in different ways.

In this case the solution of the mystery is obvious. The "fundamental theorem" fails *because of the absence from the group of the numbers* $4m + 3$ *of ordinary arithmetic*. In fact

$$21 = 3 \cdot 7, \quad 57 = 3 \cdot 19, \quad 77 = 7 \cdot 11, \quad 209 = 11 \cdot 19$$

and

$$21 \cdot 209 = (3 \cdot 7)(11 \cdot 19) = (3 \cdot 19)(7 \cdot 11) = 57 \cdot 77.$$

---

* Morgan Ward of Pasadena has found a very simple proof of this theorem.

We cannot give a proper account of the properties of the numbers $4m+1$ so long as we insist on excluding the numbers $4m+3$; *the numbers $4m+1$ do not form by themselves an adequate basis for arithmetic.* This observation has of course no intrinsic interest, since no reasonable person would expect that they would do so. It is trivial in itself, but it is not at all trivial in its suggestion, since it suggests that the troubles of the field $K^*[\sqrt{(-5)}]$ may be remedied *by considering the field as part of some larger field.*

This is in fact the solution found by Kummer. We consider the field $L[\sqrt{(-5)}]$ of numbers

$$\xi = \sqrt{(a + b\sqrt{(-5)})},$$

where a and b are ordinary integers. This is only an approximate statement; we do not actually consider all such numbers, but only those satisfying certain further conditions; the greatest common divisor of a and b must be a square or five times a square, and $a^2 + 5b^2$ must be a square. The field L includes K^*. The numbers of L form a group for multiplication, and we can define divisibility and primality in the field. Finally, the analog of the fundamental theorem is valid; *factorization is unique in L.* The proof of this is quite simple, but requires a little attention to detail, and I must refer you for the details to Mordell's tract on *Fermat's Last Theorem*.

We can now give a simple account of the equations in $K^*(\sqrt{(-5)})$ which puzzled us before. Consider for example the equation

$$3 \cdot 7 = (1 + 2\sqrt{(-5)})(1 - 2\sqrt{(-5)}).$$

It is easily verified that

$$3^2 = (2 + \sqrt{(-5)})(2 - \sqrt{(-5)}),$$
$$7^2 = (2 + 3\sqrt{(-5)})(2 - 3\sqrt{(-5)}),$$
$$(1 + 2\sqrt{(-5)})^2 = -19 + 4\sqrt{(-5)} = -(2 + \sqrt{(-5)})(2 - 3\sqrt{(-5)}),$$
$$(1 - 2\sqrt{(-5)})^2 = -19 - 4\sqrt{(-5)} = -(2 - \sqrt{(-5)})(2 + 3\sqrt{(-5)}).$$

Hence, if we write

$$\alpha = \sqrt{(2 + \sqrt{(-5)})}, \quad \alpha' = \sqrt{(2 - \sqrt{(-5)})},$$
$$\beta = \sqrt{(2 + 3\sqrt{(-5)})}, \quad \beta' = \sqrt{(2 - 3\sqrt{(-5)})},$$

we have

$$3 = \alpha\alpha', 7 = \beta\beta', 1 + 2\sqrt{(-5)} = -\alpha'\beta, 1 - 2\sqrt{(-5)} = -\alpha\beta',$$
$$3 \cdot 7 = \alpha\alpha' \cdot \beta\beta' = \alpha'\beta \cdot \alpha\beta' = (1 + 2\sqrt{(-5)})(1 - 2\sqrt{(5)});$$

13. An Introduction to the Theory of Numbers

and all of these equations are entirely natural. *In order to obtain a satisfactory theorem of factorization in K^*, we must conceive K^* as immersed in the larger field L.* The logic of the solution is exactly the same as that of the solution of the corresponding, but trivial, problem for the numbers $4m + 1$.

On the other hand there is an obvious contrast between the two solutions. It is *natural* to think of the field "$4m + 1$" as part of the field "m"; "m" is the more obvious and simpler field. It is not natural to think of K^* as part of L; K^* is a much simpler and more natural field than L, and we should like to do without the reference to the latter if we could. It will be very tiresome if, whenever we consider an algebraic field, we are to be compelled to construct some more elaborate field of which it is a part. We should prefer to tidy up the house without going out of doors.

We may look for a hint once more in the numbers $4m + 1$. Some of these numbers are divisible by 7, a number outside the field; and these numbers stand in certain specific relations to one another inside the field. Could we give a rational account of these relations without explicit reference to the number 7? It is a very unnatural thing to try to do, since what is important about the numbers is precisely that they are divisible by 7, but we could do it; we could define the class

$$21, 49, 77, 105, \ldots,$$

of numbers $4m + 1$ divisible by 7 in terms of the field $4m + 1$ itself. For example, we could take the first two numbers 21 and 49, and say "the class in question is the class which begins with these two numbers and whose members recur at regular intervals in the field." It is of course an artificial definition, and it is impossible to conceal from ourselves what we are really doing.

It is often a very profitable exercise for a mathematician to force himself to solve some simple problem without the weapon obviously appropriate to the occasion, to throw away the key of the front door and insist on forcing himself in somehow through the window. The forced and unnatural solution of one problem will often turn out to contain the germ of a quite natural solution of another. So it proves in this case; it is natural to try to define the numbers of K^* divisible by ξ without going outside K^*; it is natural, and possible, and it gives us the key to what is, in the general case, the established method of constructing a satisfactory arithmetic.

It is obvious that, if α and β belong to K^*, and $\xi|\alpha$ and $\xi|\beta$, then $\xi|\lambda\alpha + \mu\beta$, where λ and μ are any numbers of K^*. The converse proposition is not true; it is not true that if I is any set of numbers of the field K^* which has the property "if α and β belong to I, then $\lambda\alpha + \mu\beta$ belongs to I, for every λ and μ, of the field", then there exists a number ξ, belonging to K^* which divides every number of I. What *is* true is that every number of I is divisible by a ξ which belongs to L but not in general to K^*. The set I is identical with the set of numbers of K^* divisible by ξ. Such a set I, or the more general set based on any finite number of numbers $\alpha, \beta, \gamma, \ldots$, of K^*, is called an *ideal*, the numbers ξ, underlying K^* but not belonging to it, having been described by Kummer as "ideal numbers". In ordinary arithmetic ideals are simply the sets of numbers divisible by some special number such as 3, and there is nothing in particular to be gained by their introduction. In an algebraic field they are not, in general, the sets of numbers divisible by a number *of the field*, and their introduction is essential before arithmetic can get properly started. We can define multiplication and division of ideals, prime ideals, and so on, and when we have done this we find that the arithmetic of ideals has all the properties of ordinary multiplicative arithmetic. In particular, *every ideal can be resolved uniquely into prime ideals*; the fundamental theorem is true when stated in terms of ideals.

The proof of the fundamental theorem is not particularly difficult; Landau presents it, with all the preliminary definitions, in about a dozen pages of quite simple reasoning. But I would not commit the impertinence, even if I had the time, of assuming the airs of an expert in the algebraic theory of numbers, a subject which I admire only at a distance and in which I have never worked. It is ordinary rational arithmetic which attracts the ordinary man, and I have digressed outside it only because there is a good deal in it which it is impossible to appreciate properly without a little knowledge of the larger theory. It is impossible, for example, to appreciate Euclid's arithmetical achievements until we realize that there are arithmetics in which the most obvious analogs of his theorems are false.

15. *Conclusion. Pedagogy.* There are few things in the world for which I have less taste than I have for mathematical pedagogics, but I cannot resist the temptation of concluding with one pedagogic lesson. There was, and I fear still is, a popular English text book of algebra which I used at school and which contained a chapter on the theory of numbers. It might be expected that such a chapter would be among the most instructive in the book; we might suppose, for example, that Euclid's algorithm, with its elegance, its simplicity, and its far reaching consequences, would be an ideal text for the instruction of a bright young mathematician. In fact the algorithm was never mentioned; one was to find the highest common factor of 12091 and 14803, I suppose, by "trial"; and all that the authors had to say of the fundamental theorem was that "it is so evident that it may be regarded as a necessary law of thought." It is possible of course that all this may have been expunged from later editions. It is certain, however, that chapters on number theory in textbooks of algebra

are usually quite intolerably bad, and it is conceivable that Oxford University may have been right in erasing the subject altogether from its more elementary examination schedule.

The elementary theory of numbers should be one of the very best subjects for early mathematical instruction. It demands very little previous knowledge; its subject matter is tangible and familiar; the processes of reasoning which it employs, are simple, general and few; and it is unique among the mathematical sciences in its appeal to natural human curiosity. A month's intelligent instruction in the theory of numbers ought to be twice as instructive, twice as useful, and at least ten times as entertaining as the same amount of "calculus for engineers". It is after all only a minority of us who are going to spend our lives in engineering workshops, and there is no particular reason why most of us should feel any overpowering interest in machines; nor is it in the least likely that, on those occasions when machines are of real importance to us, we shall require the power of dealing with them by methods more elaborate than the simplest rule of thumb. It is not engineering mathematics that is wanted for the understanding of modern physics, and still less is it wanted by most of us for the ordinary needs of life; we do not actually drive cars by solving differential equations. There may be a case for subordinating mathematics to the linguistic and literary studies which are so much more obviously useful to ordinary men, but there is none for sacrificing a splendid subject to meet a quite imaginary need.

14

Prime Numbers

(*British Association Report* 10 (1915), pp. 350–354)

From the Editors

The British Association was established in 1831 to encourage an interest in science, somewhat the counterpart of the American Association for the Advancement of Science (AAAS), established 17 years later. The name of this organization was changed to the British Science Association in 2009. Though the Association arranged for reports on mathematics at its meetings—Hardy spoke on various occasions—it is clear that mathematics has not been a central area of concern for members, just as it is for AAAS, whose Section A is for mathematics, but it is not the primary scientific society for mathematicians. Hardy was president of the London Mathematical Society in 1926–28 and 1939–41 (Littlewood in 1941–43), but Hardy never appeared on the roster of presidents of the British Association, whose lists of officers, though they may occasionally include someone with a mathematical position, are more often filled with people more widely recognized as astronomers or physicists, or these days, economists. Hardy did, however serve in 1922 as president of the Association's Section A (Mathematics).

THE Theory of Numbers has always been regarded as one of the most obviously useless branches of Pure Mathematics. The accusation is one against which there is no valid defence; and it is never more just than when directed against the parts of the theory which are more particularly concerned with primes. A science is said to be useful if its development tends to accentuate the existing inequalities in the distribution of wealth, or more directly promotes the destruction of human life. The theory of prime numbers satisfies no such criteria. Those who pursue it will, if they are wise, make no attempt to justify their interest in a subject so trivial and so remote, and will console themselves with the thought that the greatest mathematicians of all ages have found in it a mysterious attraction impossible to resist.

The foundations of the theory were laid by Euclid. Among Euclid's theorems two in particular are of fundamental importance. The first (Euc. vii. 24) is that *if a and b are both prime to c, then ab is also prime to c*. This theorem is the basis of the whole theory of the factorisation of numbers, systematised later by Euler and by Gauss, and in particular of the theorem that *every number can be expressed in one and only one way as a product of primes*. The second theorem (Euc. ix. 20) is that *the number of primes is infinite:* to this theorem I shall return in a moment.

In modern times the theory has developed in two different directions. In the first place there is what may be called roughly the theory of *individual* or *isolated* primes, a theory which it is difficult to define precisely, but of which a general idea may be formed by considering a few of its characteristic problems. How can we determine whether a given number is prime? What conditions are necessary and what sufficient? Can we define forms which represent prime numbers only? Are there infinitely many pairs of primes which differ by 2? Is (as Goldbach asserted) every even number the sum of two primes? This theory I shall dismiss very briefly. We know a number of very beautiful theorems of this character. I need only mention Wilson's theorem, Fermat's theorem, and the extensions of the latter by Lucas. But on the whole the record of research in this direction is a record of failure. The difficulties are too great for the methods of analysis at our command, and the problems remain unsolved.

Very different results are revealed when we turn to the second principal branch of the modern theory, the theory of the *average or asymptotic distribution of primes*. This theory (though one of its most famous problems is still unsolved) is in some ways almost complete, and certainly represents one of the most remarkable triumphs of modern analysis. The theory centres round one theorem, the *Primzahlsatz* or *Prime Number Theorem*; and it is to the history of this theorem, which may almost be said to embody the history of the whole subject, that I shall devote the remainder of this lecture.*

The problem may be stated crudely as follows: *How many primes are there less than a given number x?* More precisely, let $\pi(x)$ denote the number of primes[†] not exceeding x: then *what is the order of magnitude of $\pi(x)$?* The Prime Number Theorem provides a complete answer to this last question. It

---

* A full account of the history of the theorem will be found in Landau's *Handbuch der Lehre, von der Verteilung der Primzahlen* (Teubner, 1909).

† It proves most convenient not to count 1 as a prime.

14. Prime Numbers

asserts that

$$\pi(x) \sim \frac{x}{\log x},$$

that is to say, that $\pi(x)$ and $x/(\log x)$ *are asymptotically equivalent*, or that their ratio tends to 1 when x tends to infinity.

The first step towards the proof of this theorem was made by Euclid, when he proved that the number of primes is infinite, or that

$$\pi(x) \to \infty.$$

Euclid's proof is classical, and can hardly be repeated too often. If the number of primes is finite, let them be $2, 3, 5, \ldots, P$. The number $2.3.5\ldots P + 1$ is not divisible by any of $2, 3, 5, \ldots, P$. It is therefore prime itself, or divisible by some, prime greater than P; and either alternative contradicts the hypothesis that P is the greatest prime. It is worth remarking that Euclid's reasoning may he used to prove rather more, viz. that the order of $\pi(x)$ is at least as great as that of $\log \log x$.*

The next advances were made by Euler, probably about 1740. It was Euler to whom we owe the introduction into analysis of the Zeta-function, the function on whose properties, as later research has shown, the whole theory depends.

Let $s = \sigma + it$. Then the function $\zeta(s)$ is defined, when $\sigma > 1$, by the equations

$$\zeta(s) = \sum n^{-s} = 1^{-s} + 2^{-s} + 3^{-s} + \cdots;$$

and Euler's fundamental contribution to the theory is the formula

$$\zeta(s) = \Pi\left(\frac{1}{1-p^{-s}}\right),$$

where the product extends over all prime values of p. Euler, it is true, considered $\zeta(s)$ as a function of a *real* variable only. But his formula at once indicates the existence of a deep-lying connection between the theory of $\zeta(s)$ and the theory of primes.

Euler deduced from his formula that the series $\sum p^{-s}$, obviously convergent when $s > 1$, is divergent when $s = 1$; and from this it is easy to deduce important consequences as to the order of $\pi(x)$. It is evident that $\pi(x) < x$, so that the order of $\pi(x)$ certainly does not exceed that of x, or, in the notation which is usual now, $\pi(x) = O(x)$.† It is an easy corollary of Euler's result

---

* This was pointed out to me by Prof. H. Bohr of Copenhagen.
† $f = O(\phi)$ means that the absolute value of f is less than a constant multiple of ϕ: thus $\sin x = O(1), 100x = O(x)$.

that the order of $\pi(x)$ is *not very much less than that of x*; that, for example, $\pi(x) \neq O(x^a)$ for any value of *a* less than 1; or again, more precisely, that

$$\pi(x) \neq O\left\{\frac{x}{(\log x)^{1+a}}\right\}$$

for any value of *a* greater than 1.

It is also easy to prove that the order of $\pi(x)$ is *definitely less than that of x*, or that, as we should express it now, $\pi(x) = o(x)$.* This theorem, when read in conjunction with those which precede, is, I think, enough to suggest the Prime Number Theorem as a very plausible conjecture, or at any rate to suggest that the true order is that of $x/(\log x)$. The theorem was in fact conjectured first by Gauss (1793) and by Legendre (1798); and it is in Legendre's *Essai sur la théorie des nombres* that the conjecture first appears in print.

In this state the problem remained for fifty years, until the publication (1849–1852) of the researches of the Russian mathematician Tschebyschef. I have no time to speak of Tschebyschef's work as fully as it deserves, but his chief results, in so far as they bear directly on the problem now before us, were as follows:—

(1) Tschebyschef showed that the problem is simplified if we take as fundamental not the function $\pi(x)$ itself, but the closely related function

$$\theta(x) = \sum_{p \leq x} \log p$$

(the sum of the logarithms of all primes not exceeding *x*). He showed that the order of $\theta(x)$ is the same as that of $\pi(x) \log x$, and that the Prime Number Theorem itself is equivalent to the theorem that

$$\theta(x) \sim x.$$

(2) He showed that $\theta(x)$ is actually of order *x*, and $\pi(x)$ of order $x/(\log x)$, in fact that positive constants A and B exist such that

$$A\frac{x}{\log x} < \pi(x) < B\frac{x}{\log x}.$$

(3) He showed that *if $\theta(x)/x$ tends to a limit, then that limit must be unity*.

What Tschebyschef could not prove is that the limit does in fact exist, and, as he failed to prove this, he failed to prove the Prime Number Theorem. And about Tschebyschef's methods (interesting as they are), I shall say nothing; for

---

* $f = o(\phi)$ means that $f/\phi \to 0$. Thus $\sin x = o(x)$. This theorem also was stated by Euler, hut without satisfactory proof.

14. Prime Numbers

later research has shown that it was the essential inadequacy of his methods which was responsible for his failure, and that the theorem lies deeper in analysis than any of the ideas on which he relied.

The next great step was taken by Riemann in 1859, and it is in Riemann's famous memoir *Ueber die Anzahl der Primzahlen unter einer gegebenen Grösse* that we first find the ideas upon which the theory has now been shown really to rest. Riemann did not prove the Prime Number Theorem: it is remarkable, indeed, that he never mentions it. His object was a different one, that of finding an explicit expression for $\pi(x)$, or rather for another closely associated function, as a sum of an infinite series. But it was Riemann who first recognised that, if we are to solve any of these problems, we must study the Zeta-function as a function of the *complex* variable $s = \sigma + it$, and in particular study the distribution of its zeros.

Riemann proved

(1) that $\zeta(s)$ is an analytic function of s, regular all over the plane except for a simple pole at the point 1;
(2) that $\zeta(s)$ satisfies the functional equation

$$\zeta(1-s) = 2(2\pi)^{-s} \cos \tfrac{1}{2} s\pi \, \Gamma(s) \zeta(s);$$

(3) that $\zeta(s)$ has zeros at the points $-2, -4, -6\ldots$, and no other zeros *except possibly complex zeros whose real parts lie between 0 and 1 inclusive.*

To these propositions he added certain others of which he could produce no satisfactory proof. In particular he asserted that there is in fact an infinity of complex zeros, all naturally situated in the 'critical strip' $0 \le \sigma \le 1$; an assertion now known to be correct. Finally he asserted that it was 'sehr wahrscheinlich' that all these zeros have the real part $\tfrac{1}{2}$: the notorious 'Riemann hypothesis', unsettled to this day.

We come now to the time when, a hundred years after the conjectures of Gauss and Legendre, the theorem was finally proved. The way was opened by the work of Hadamard on integral transcendental functions. In 1893 Hadamard proved that the complex zeros of Riemann actually exist; and in 1896 he and de la Vállée-Poussin proved independently that *none of them have the real part* 1, and deduced a proof of the Prime Number Theorem.

It is not possible for me now to give an adequate account of the intricate and difficult reasoning by which these theorems are established. But the general ideas which underlie the proofs are, I think, such as should be intelligible to any mathematician.

In the first place Euler's formula shows that $\log \zeta(s)$ behaves, throughout the half-plane $\sigma > 1$, much like the series Σp^{-s}. But $\zeta(s)$ has a simple pole for $s = 1$, and so the sum of the series $\Sigma p^{-1-\delta}$ tends logarithmically to $+\infty$ when $\delta \to 0$ through positive values. Suppose now that (if possible) $\zeta(1 + ti) = 0$. Then the real part of $\log \zeta(1 + \delta + ti)$, and therefore the real part of the series $\Sigma p^{-1-\delta-ti}$, tends, also logarithmically, to $-\infty$ when $\delta \to 0$. It follows that the series

$$\sum p^{-1-\delta}, \quad -\sum p^{-1-\delta} \cos(t \log p)$$

tend to $+\infty$ *with equal rapidity* when $\delta \to 0$. As the first series is a series of positive terms, while the signs of the terms in the second series change with a certain regularity, it is natural to suppose that our last conclusion is impossible; and this is in fact not particularly difficult to prove.

I come now to the proof of the Prime Number Theorem itself. If we differentiate Euler's formula logarithmically, we obtain

$$\frac{\zeta'(s)}{\zeta(s)} = \sum \left(\frac{\log p}{p^s} + \frac{\log p}{p^{2s}} + \cdots \right) = \sum_{p,m} \frac{\log p}{p^{ms}};$$

or

(1) $$-\frac{\zeta'(s)}{\zeta(s)} = \sum \frac{\Lambda(n)}{n^s}$$

where p assumes all prime values, m and n all positive integral values, and $\Lambda(n)$ is equal to $\log p$ if n is of the form p^m and to zero otherwise. Let

$$\psi(x) = \sum_{n \leq \infty} \Lambda(n)$$

Then $\psi(x)$ is, for our present purpose, equivalent to $\theta(x)$; it is easy to show that the difference between the two functions is of order $\sqrt{x}$. We have therefore to prove that $\psi(x) \sim x$.

The series on the right-hand side of the equation (1) is what is called a 'Dirichlet's series'; and the theory of such series resembles the more familiar theory of Taylor's series in one very important respect. *We can express the coefficients by contour integrals in which the function represented by the series appears under the sign of integration.* In particular we can show that

(2) $$\psi(x) = -\frac{1}{2\pi i} \int \frac{\zeta'(s)}{\zeta(s)} \frac{x^s}{s} ds,$$

where the path of integration is a line parallel to the imaginary axis and passing to the right of the point $s = 1$.

14. Prime Numbers

The general idea of the proof is now easy enough to grasp. Every element of the-integral (2) is of order x^σ where $\sigma > 1$: we can therefore draw no *direct* conclusion as to the behaviour of $\psi(x)$ when x is large. But it is at once suggested that we should try to make use of Cauchy's theorem. The subject of integration has a simple pole at the point 1, corresponding to the pole of $\zeta(s)$ itself, and the residue at the pole is precisely x; and there are no other singularities on the line $\sigma = 1$, since $\zeta(s)$, as we have seen, has no poles *or zeros* on that line. Suppose then that we can move the path of integration across to the left of the line, introducing the appropriate correction due to the pole. Plainly we shall then have an expression for $\psi(x) - x$ in the form of an integral *in which every element is of order less than that of x*. And if we can show that the same is true of the integral itself, we shall have proved that $\psi(x) \sim x$, that is to say, we shall have proved the Prime Number Theorem. It will be observed that, if $\zeta(s)$ had zeros whose real part is equal to 1, then the result would be definitely false, since there would be additional residues of order x. It thus becomes clear why the older attempts to prove the theorem, without using the theory of functions of a complex variable, were unsuccessful.

The arguments which I have advanced are not exact: I have merely put forward a chain of reasoning which seems likely to lead to the desired result. The achievement of Hadamard and de la Vallée-Poussin was to replace these plausibilities by rigorous proofs. It might be difficult for me to make clear to you how great this achievement was. Some branches of pure mathematics have the pleasant characteristic that what seems plausible at first sight is generally true. In this theory anyone can make plausible conjectures, and they are almost always false. Nothing short of absolute rigour counts; and it is for this reason that the Analytic Theory of Numbers, while hardly a subject for an amateur, provides the finest possible discipline in accurate reasoning for anyone who will make a real effort to understand its results.

15

The Theory of Numbers

(*British Association Report* 90 (1922), pp. 16–24; reprinted in *Nature* 110:2759 (September 18, 1922), pp. 381–385)

From the Editors
This is the transcript of an address given at the beginning of Hardy's term as president of Section A, the mathematical wing of the British Association (later known as the British Science Association). It was delivered in Hull, September 8, 1922.

I find myself to-day in the same embarrassing position in which a predecessor of mine at Oxford found himself at Bradford in 1875 [*sic*], the president of a Section, probably the largest and most heterogeneous in the Association, which is absorbed by a multitude of divergent professional interests, none of which agree with his or mine.

There are two courses possible in such circumstances. One is to take refuge, as Prof. Henry Smith did then, with visible reluctance, in a series of general propositions to which mathematicians, physicists, and astronomers may all be expected to return, a polite assent. The importance of science and scientific method, the need for better organisation of scientific education and research, are all topics on which I could no doubt say something without undue strain either on my own honesty or on your credulity. That there is no finer education and discipline than natural science; that it is, as Dr. Campbell has said, "the noblest of the arts"; that the crowning achievements of science lie in those directions with which this Section is professionally concerned: all this I could say with complete sincerity, and, if I were the head of a deputation approaching a Government Department, I suppose that I would not shirk even so unprofitable a task.

It is unfortunate that these essential and edifying truths, important as it is that they should be repeated as loudly as possible from time to time, are, to the man whose interest in life lies in scientific work and not in propaganda, unexciting, and in fact quite intolerably dull. I could, if I chose, say all these things, but, even if I wanted to, I should scarcely increase your respect for mathematics and mathematicians by repeating to you what you have said yourselves, or read in the newspapers, a hundred times already. I shall say them all some day; the time will come when we shall none of us have anything more interesting to say. We need not anticipate our inevitable end.

I propose therefore to adopt the alternative course suggested by my predecessor, and try to say something to you about the one subject about which I have anything to say. It happens, by a fortunate accident, that, the particular subject which I love the most, and which presents most of the problems which occupy my own researches, is by no means overwhelmingly recondite or obscure, and indeed is sharply distinguished from, almost every other branch of pure mathematics, in that it makes a direct, popular, and almost irresistible appeal to the heart of the ordinary man.

There is, however, one preliminary remark which I cannot resist the temptation of making. The present is a particularly happy moment for a pure mathematician, since it has been marked by one of the greatest recorded triumphs of pure mathematics. This triumph is the work, as it happens, of a man who probably would not describe himself as a mathematician, but who has done more than any mathematician to vindicate the dignity of mathematics, and to put that obscure and perplexing construction commonly described as "physical reality," in its proper place.

There is probably less difference between the methods of a physicist and a mathematician than is generally supposed. The most striking among them seems to me to be this, that the mathematician is in much more direct contact with reality. This may perhaps seem to you a paradox, since it is the physicist who deals with the subject-matter to which the epithet "real" is commonly applied. But a very little reflection will show that the "reality" of the physicist, whatever it may be (and it is extraordinarily difficult to say), has few or none of the attributes which common-sense instinctively marks as real. A chair may be a collection of whirling atoms, or an idea in the mind of God. It is not my business to suggest that one account of it is obviously more plausible, than the other. Whatever the merits of either of them may be, neither draws its inspiration from the suggestions of common-sense.

Neither the philosophers, nor the physicists themselves, have ever put forward any very convincing account of what physical reality is, or of how the physicist passes, from the confused mass of fact or sensation with which he starts, to the construction of the objects which he classifies as real. We cannot be said, therefore, to know what the subject-matter of physics is; but this need not

15. The Theory of Numbers

prevent us from understanding the task which a physicist is trying to perform. That, clearly, is to correlate the incoherent body of facts confronting him with some definite and orderly scheme of abstract relations, the kind of scheme, in short, which he can borrow only from mathematics.

A mathematician, on the other hand, fortunately for him, is not concerned with this physical reality at all. It is impossible to prove, by mathematical reasoning, any proposition whatsoever concerning the physical world, and only a mathematical crank would be likely now to imagine it his function to do so. There is plainly one way only of ascertaining the facts of experience, and that is by observation. It is not the business of a mathematician to suggest one view of the universe or another, but merely to supply the physicists with a collection of abstract schemes, which it is for them to select from, and to adopt or discard at their pleasure.

The most obvious example is to be found in the science of geometry. Mathematicians have constructed a very large number of different systems of geometry, Euclidean or non-Euclidean, of one, two, three, or any number of dimensions. All these systems are of complete and equal validity. They embody the results of mathematicians' observations of *their* reality, a reality far more intense and far more rigid than the dubious and elusive reality of physics. The old-fashioned geometry of Euclid, the entertaining seven-point geometry of Veblen, the space-times of Minkowski and Einstein, are all absolutely and equally real. When a mathematician has constructed, or, to be more accurate, when he has observed them, his professional interest in the matter ends. It may be the seven-point geometry that fits the facts the best, for anything that mathematicians have to say. There may be three dimensions in this room and five next door. As a professional mathematician, I have no idea; I can only ask some competent physicist to instruct me in the facts.

The function of a mathematician, then, is simply to observe the facts about his own intricate system of reality, that astonishingly beautiful complex of logical relations which forms the subject-matter of his science, as if he were an explorer looking at a distant range of mountains, and to record the results of his observations in a series of maps, each of which is a branch of pure mathematics. Many of these maps have been completed, while in others, and these, naturally, are the most interesting, there are vast uncharted regions. Some, it seems, have some relevance to the structure of the physical world, while others have no such tangible application. Among them there is perhaps none quite so fascinating, with quite the same astonishing contrasts of sharp outline and mysterious shade, as that which constitutes the theory of numbers.

The number system of arithmetic is, as we know too well, not without its applications to the sensible world. The currency systems of Europe, for example, conform to it approximately; west of the Vistula, two and two make something approaching four. The practical applications of arithmetic, however, are tedious beyond words. One must probe a little deeper into the subject if one wishes to interest the ordinary man, whose taste in such matters is astonishingly correct, and who turns with joy from the routine of common life to anything strange and odd, like the fourth dimension, or imaginary, time, or the theory of the representation of integers by sums of squares or cubes.

It is impossible for me to give you, in the time at my command, any general account of the problems of the theory of numbers, or of the progress that has been made towards their solution even during the last twenty years. I must adopt a much simpler method. I will merely state to you, with a few words of comment, three or four isolated questions, selected in a haphazard way. They are seemingly simple questions, and it is not necessary to be anything of a mathematician to understand them; and I have chosen them for no better reason than that I happen to be interested in them myself. There is no one of them to which I know the answer, nor, so far as I know, does any mathematician in the world; and there is no one of them, with one exception which I have included deliberately, the answer to which any one of us would not make almost any sacrifice to know.

1. *When is a number the sum of two cubes, and what is the number of its representations?* This is my first question, and first of all I will elucidate it by some examples. The numbers $2 = 1^3 + 1^3$ and $9 = 2^3 + 1^3$ are sums of two cubes, while 3 and 4 are not: it is exceptional for a number to be of this particular form. The number of cubes up to 1,000,000 is 100, and the number of numbers, up to this limit and of the form required, cannot exceed 10,000, one-hundredth of the whole. The density of the distribution of such numbers tends to zero as the numbers tend to infinity. Is there, I am asking, any simple criterion by which such numbers can be distinguished?

Again, 2 and 9 are sums of two cubes, and can be expressed in this form in one way only. There are numbers so expressible in a variety of different ways. The least such number is 1729, which is $12^3 + 1^3$ and also $10^3 + 9^3$. It is more difficult to find a number with *three* representations; the least such number is

$$175{,}959{,}000 = 560^3 + 70^3 = 552^3 + 198^3 = 525^3 + 315^3.$$

One number at any rate is known with *four* representations, namely,

$$19 \times 363510^3$$

(a number of 18 digits), but I am not prepared to assert that it is the least. No number has been calculated, so far as I know, with more than four, but

15. The Theory of Numbers

theory, running ahead of computation, shows that numbers exist with five representations, or six, or any number.

A distinguished physicist has argued that the possible number of isotopes of an element is probably limited because, among the ninety or so elements at present under observation, there is none which has more isotopes than six. I dare not criticise a physicist in his own field; but the figures I have quoted may suggest to you that an arithmetical generalisation, based on a corresponding volume of evidence, would be more than a little rash.

There are similar questions, of course, for squares, but the answers to these were found long ago by Euler and by Gauss, and belong to the classical mathematics. Suppose, for simplicity of statement, that the number in question is *prime*. Then, if it is of the form $4m + 1$, it is a sum of squares, and in one way only, while if it is of the form $4m + 3$ it is not so expressible; and this simple rule may readily be generalised so as to apply to numbers of any form. But there is no similar solution for our actual problem, nor, I need scarcely say, for the analogous problems for fourth, fifth, or higher powers. The smallest number known to be expressible in two ways by two biquadrates is

$$635318657 = 158^4 + 59^4 = 134^4 + 133^4;$$

and I do not believe that any number is known expressible in three. Nor, to my knowledge, has the bare existence of such a number yet been proved. When we come to fifth powers, nothing is known at all. The field for future research is unlimited and practically untrodden.

2. I pass to another question, again about cubes, but of a somewhat different kind. *Is every large number* (every number, that is to say, from a definite point onwards) *the sum of five cubes*? This is another exceptionally difficult problem. It is known that every number, without exception, is the sum of nine cubes; two numbers, 23 (which is $2 \cdot 2^3 + 7 \cdot 1^3$) and 239, actually require so many. It seems that there are just fifteen numbers, the largest being 454, which need eight, and 121 numbers, the largest being 8042, which need seven; and the evidence suggests forcibly that the six-cube numbers also ultimately disappear. In a lecture which I delivered on this subject at Oxford I stated, on the authority of Dr. Ruckle, that there were two numbers, in the immediate neighbourhood of 1,000,000, which could not be resolved into fewer cubes than six; but Dr. A. E. Western has refuted this assertion by resolving each of them into five, and is of opinion, I believe, that the six-cube numbers have disappeared entirely considerably before this point. It is conceivable that the five-cube numbers also disappear, but this, if it be so, is probably in depths where computation is helpless. The four-cube numbers must certainly persist,

for ever, for it is impossible that a number $9n + 4$ or $9n + 5$ should be the sum of three.

I need scarcely add that there is a similar problem for every higher power. For fourth powers the critical number is 16. There is no case, except the simple case of squares, in which the solution is in any sense complete. About the squares there is no mystery; every number is the sum of four squares, and there are infinitely many numbers which cannot be expressed by fewer.

3. I will next raise the question *whether the number* $2^{137} - 1$ *is prime*. I said that I would include one question which does not interest me particularly; and I should like to explain to you the kind of reasons which damp down my interest in this one. I do not know the answer, and I do not care greatly what it is.

The problem belongs to the theory of the so-called "perfect" numbers, which has exercised mathematicians since the times of the Greeks. A number is perfect if, like 6 or 28, it is the sum of all its divisors, unity included. Euclid proved that the number

$$2^m(2^{m+1} - 1)$$

is perfect if the second factor, is prime; and Euler, 2000 years later, that all *even* perfect numbers are of Euclid's form. It is still unknown whether a perfect number can be odd.

It would obviously be most interesting to know generally in what circumstances a number $2^n - 1$ is prime. It is plain that this can be so only if n itself is prime, as otherwise the number has obvious factors; and the 137 of my question happens to be the least value of n for which the answer is still in doubt. You may perhaps be surprised that a question apparently so fascinating should fail to arouse me more.

It was asserted by Mersenne in 1644 that the only values of n, up to 257, for which $2^n - 1$ is prime are

$$2, 3, 5, 7, 13, 17, 19, 31, 67, 127, 257;$$

and an enormous amount of labour has been expended on attempts to verify this assertion. There are no simple general tests by which the primality of a number chosen at random can be determined, and the amount of computation required in any particular case may be appalling. It has, however, been imagined that Mersenne perhaps knew something which later mathematicians have failed to rediscover. The idea is a little fantastic, but there is no doubt that, so long as the possibility remained, arithmeticians were justified in their determination to ascertain the facts at all costs. "The riddle as to, how Mersenne's numbers were discovered remains unsolved," wrote Mr. Rouse Ball in 1891. Mersenne, he observes, was a good mathematician, but not an Euler or a Gauss, and he inclines to attribute the discovery to the exceptional genius of Fermat, the only mathematician of the age whom any one could suspect of being hundreds of years ahead of his time.

These speculations appear extremely fanciful now, for the bubble has at last been pricked. It seems now that Mersenne's assertion, so far from hiding unplumbed depths of mathematical profundity, was a conjecture based on inadequate empirical evidence, and a somewhat unhappy one at that. It is now known that there are at least four numbers about which Mersenne is definitely wrong; he should have included at any rate 61, 89, and 107, and he should have left out 67. The mistake as regards 61 and 67 was discovered so long ago as 1886, but could be explained with some plausibility, so long as it stood alone, as a merely clerical error. But when Mr. R. E. Powers, in 1911 and 1914, proved that Mersenne was also wrong about 89 and 107, this line of defence collapsed, and it ceased to be possible to take Mersenne's assertion seriously.

The facts may be summed up as follows. Mersenne makes fifty-five assertions, for the fifty-five primes from 2 to 257. Of these assertions forty are true, four false, and eleven still doubtful. Not a bad result, you may think; but there is more to be said. Of the forty correct assertions many, half at least, are trivial, either because the numbers in question are comparatively small, or because they possess quite small and easily detected divisors. The test cases are those in which the numbers are prime, or Mersenne asserts that they are so; there are only four of these cases which are difficult and in which the truth is known; and in these Mersenne is wrong in every case but one.

It seems to me, then, that we must regard Mersenne's assertion as exploded; and for my part it interests me no longer. If he is wrong about 89 and 107, I do not care greatly whether he is wrong about 137 as well, and I should regard the computations necessary to decide as very largely wasted. There are so many much more profitable calculations which a computer could undertake.

I hope that you will not infer that I regard the problem of perfect numbers as uninteresting in itself; that would be very far from the truth. There are at least two intensely interesting problems. The first is the old problem, which so many mathematicians have failed to solve, whether a perfect number can be odd. The second is whether the number of perfect numbers is infinite or not. If we assume that all perfect numbers are even, we can state this problem in a still more arresting form. *Are there infinitely many primes of the form $2^n - 1$?* I find it difficult to imagine a problem more fascinating or more intricate than that. It is plain, though, that this is a question which computation can never decide, and it is very unlikely that it can ever give us any data of serious value. And the problem itself really belongs to a different chapter of the theory, to which I should like next to direct your attention.

4. *Are there infinitely many primes of the form $n^2 + 1$?* Let me first remind you of some well-known facts in regard to the distribution of primes.

There are infinitely many primes; their density decreases as the numbers increase, and tends to zero when the numbers tend to infinity. More accurately, the number of primes less than x is, to a first approximation,

$$\frac{x}{\log x}.$$

The chance that a large number n, selected at random, should be prime is, we may say, about, $\frac{1}{\log n}$. Still more precisely, the "logarithm-integral"

$$\operatorname{Li} x = \int_2^x \frac{dt}{\log t}$$

gives a very good approximation to the number of primes. This number differs from Li x by a function of x which oscillates continually, as Mr. Littlewood, in defiance of all empirical evidence to the contrary, has shown, between positive and negative values, and is sometimes large, of the order of magnitude $\sqrt{x}$ or thereabouts, but always small in comparison with the logarithm-integral itself.

Except for one lacuna, which I must pass over in silence now, this problem of the general distribution of primes, the first and central problem of the theory, is in all essentials solved. But a variety of most interesting problems remain as to the distribution of primes among numbers of special forms. The first and simplest of these is that of the arithmetical progressions: *How are the primes distributed among all possible arithmetical progressions $an + b$?* We may leave out of account the case in which a and b have a common factor; this case is trivial, since $an + b$ is then obviously not prime.

The first step towards a solution was made by Dirichlet, who proved for the first time, in 1837, that any such arithmetical progression contains an infinity of primes. It has since been shown that the primes are, to a first approximation at any rate, distributed evenly among all the arithmetical progressions. When we pursue the analysis further, differences appear; there are on the average, for example, more primes $4n + 3$ than, primes $4n + 1$, though it is not true, as the evidence of statistics has led some mathematicians to conclude too hastily, that there is always an excess to whatever point the enumeration is carried.

The problem of the arithmetical progressions, then, may also be regarded as solved; and the same is true of the problem of the primes of a given quadratic

15. The Theory of Numbers

form, say $am^2 + 2bmn + cn^2$, homogeneous in the two variables m and n. To take, for example, the simplest and most striking case, there is the natural and obvious number of primes $m^2 + n^2$. A prime is of this form, as I have mentioned already, if, and only if, it is of the form $4k + 1$. The quadratic problem reduces here to a particular case of the problem of the arithmetical progressions.

When we pass to cubic forms, or forms of higher degree, we come to the region of the unknown. This, however, is not the field of inquiry which I wish now to commend to your attention. The quadratic forms of which I have spoken are forms in two independent variables m and n; the form $n^2 + 1$ of my question is a non-homogeneous form in a single variable n, the simplest case of the general form $an^2 + 2bn + c$. It is clear that one may ask the same question for forms of any degree: are there, for example, infinitely many primes $n^3 + 2$ or $n^4 + 1$? I do not choose $n^3 + 1$, naturally, because of the obvious factor $n + 1$.

This problem is one in which computation can still play an important part. You will remember that I stated the same problem for perfect numbers. There a computer is helpless. For the numbers $2^n - 1$, which dominate the theory, increase with unmanageable rapidity, and the data collected by the computers appear, so far as one can judge, to be almost devoid of value. Here the data are ample, and, though the question is still unanswered, there is really strong statistical evidence for supposing a particular answer to be true. It seems that the answer is affirmative and that there is a definite approximate formula for the number of primes in question. This formula is

$$\frac{1}{2}\text{Li}\sqrt{x} \times \left(1 + \frac{1}{3}\right)\left(1 - \frac{1}{5}\right)\left(1 + \frac{1}{7}\right)\left(1 + \frac{1}{11}\right)\ldots,$$

where the product extends over all primes p, and the positive sign is chosen when p is of the form $4n + 3$. Dr. A. E. Western has submitted this formula to a most exhaustive numerical check. It so happens that Colonel Cunningham some years ago computed a table of primes $n^2 + 1$ up to the value 15,000 of n, a limit altogether beyond the range of the standard factor tables, and Cunningham's table has made practicable an unusually comprehensive test. The actual number of primes is 1199, while the number predicted is 1219. The error, less than 1 in 50, is much less than one could reasonably expect. The formula stands its test triumphantly, but I should be deluding you if I pretended to see any immediate prospect of an accurate proof.

5. The last problem I shall state to you is this: *Are there infinitely many prime-pairs p, $p + 2$?* One may put the problem more generally: *Does any*

group of primes, with assigned, and possible differences, recur indefinitely, and what is the law of its recurrence?

I must first explain what I mean by a "possible" group of primes. It is possible that p and $p+2$ should both be prime, like 3, 5, or 101, 103. It is not possible (unless p is 3) that p, $p+2$ and $p+4$ should *all* be prime, for one of them must be a multiple of 3: but p, $p+2$, $p+6$ or p, $p+4$, $p+6$ are possible triplets of primes. Similarly

$$p, p+2, p+6, p+8, p+12$$

can all be prime, so far as any elementary test of divisibility shows, and in fact 5, 7, 11, 13 and 17 satisfy the conditions. It is easy to define precisely what we understand by a "possible" group. We mean a group the differences in which, like 0, 2, 6, have at least one missing residue to every possible modulus. The "impossible" group 0, 2, 4 does not satisfy the condition, for the remainders after division by 3 are 0, 2, 1, a complete set of residues to modulus 3. There is no difficulty in specifying possible groups of any length we please.

We define in this manner, then, a "possible" group of primes, and we put the questions: Do all possible groups of primes actually occur, do they recur indefinitely often, and how often on the average do they recur? Here again it would seem that the answers are affirmative, that all possible groups occur, and continue to occur for ever, and with a frequency the law of which can be assigned. The order of magnitude of the number of prime-pairs, p, $p+2$, or p, $p+4$, or p, $p+6$, both members of which are less than a large number x, is, it appears,

$$\frac{x}{(\log x)^2}.$$

The order of magnitude of the corresponding number of triplets, of any possible type, is

$$\frac{x}{(\log x)^3}.$$

and so on generally. Further, we can assign the relative frequencies of pairs or triplets of different types; there are, for example, about twice as many pairs the difference of which is 6 as there are pairs with the difference 2. All these results have been tested by actual enumeration from the factor tables of the first million numbers; and a physicist would probably regard them as proved, though we of course know very well that they are not.

There is a great deal of mathematics the purport of which is quite impossible for any amateur to grasp, and which, however beautiful and important it may be, must always remain the possession of a narrow circle of experts, It

15. The Theory of Numbers

is the peculiarity of the theory of numbers that much of it could be published broadcast, and would win new readers for the *Daily Mail*. The positive integers do not lie, like the logical foundations of mathematics, in the scarcely visible distance, nor in the uncomfortably tangled foreground, like the immediate data of the physical world, but at a decent middle distance, where the outlines are clear and yet some element of mystery remains... There is no one so blind that he does not see them, and no one so sharp-sighted that his vision does not fail; they stand there a continual and inevitable challenge to the curiosity of every healthy mind. I have merely directed your attention for a moment to a few of the less immediately conspicuous features of the landscape, in the hope that I may sharpen your curiosity a little, and that some may feel tempted to walk a little nearer and take a closer view.

16
The Riemann Zeta-Function and Lattice Point Problems

E. C. Titchmarsh

(*Journal of the London Mathematical Society* 25 (1950), pp. 125–128)*

E. C. Titchmarsh wrote several books, among them *The Theory of Functions*, a highly regarded text. He was elected a Fellow of the Royal Society in 1931 and was the recipient of many honors over the course of his long career as Savilian Professor at Oxford.

---

* Reprinted with the kind permission of the publisher, the London Mathematical Society.

From the Editors
Both E. C. Titchmarsh and G. H. Hardy had long associations with the venerable London Mathematical Society (LMS) and its principal publications, this Journal *(established in 1926) and the older journal, the* Proceedings *(dating from 1865). Titchmarsh was president of the LMS between 1945 and 1947. Hardy had been President twice (1926–28 and 1939–41), and, further, Hardy had won its prestigious De Morgan Prize, named for the first president of the LMS (Augustus De Morgan). The Society named its fellowship for foreign students for Hardy.*

The theory of the Riemann zeta-function

$$\zeta(s) = \sum_{n=1}^{\infty} \frac{1}{n^s} = \prod_p \left(1 - \frac{1}{p^s}\right)^{-1}$$

as a function of a complex variable, and as a means of proving theorems in prime-number theory, was inaugurated by the brilliant but inconclusive analysis of Riemann (1860). In the hands of Hadamard and de la Vallée-Poussin, this led to the proof of the prime-number theorem (1896). A little later the modern theory of the function was started by H. Bohr, Landau and Littlewood. The great puzzle of the theory was the "Riemann hypothesis", that $\zeta(s)$ has all its complex zeros on the "critical line" $\Re(s) = \frac{1}{2}$.

This presented all workers in the field, as it still does, with a perpetual challenge. It was Hardy who first gave any sort of answer to it, with the discovery, in 1914, that $\zeta(s)$ has at any rate an infinity of zeros on the critical line. The main idea of the proof is as follows. The function

$$\Xi(t) = -\frac{1}{2}\left(t^2 + \frac{1}{4}\right)\pi^{-\frac{1}{4}-\frac{1}{2}it}\Gamma\left(\frac{1}{4} + \frac{1}{2}it\right)\zeta\left(\frac{1}{2} + it\right)$$

is real for real values of t, and has a real zero corresponding to each zero of $\zeta(s)$ on the critical line. Now it can be proved that

$$\lim_{a \to \frac{1}{4}x} \int_0^\infty \frac{\Xi(t)}{t^2 + \frac{1}{4}} t^{2n} \cosh at\, dt = \frac{(-1)^n \pi \cos\frac{1}{8}\pi}{2^{2n}}.$$

16. The Riemann Zeta-Function and Lattice Point Problems

If $\Xi(t)$ were ultimately of one sign, it would follow that

$$\int_0^\infty \frac{\Xi(t)}{t^2 + \frac{1}{4}} t^{2n} \cosh \frac{1}{4}\pi t \, dt = \frac{(-1)^n \pi \cos \frac{1}{8}\pi}{2^{2n}}$$

for all values of n. This, however, is impossible, since the left-hand side has the same sign for all sufficiently large values of n.

Several variants of the proof have since been given.

Later the method was, jointly with Littlewood, extended so as to give a much more precise result of the same kind. Let $N_0(T)$ denote the number of complex zeros of $\zeta(s)$ with real part $\frac{1}{2}$ and imaginary part between 0 and T. It was proved that

$$N_0(T) > AT,$$

where A is a positive constant. If $N(T)$ denotes the total number of zeros with imaginary part between 0 and T, it is known that

$$N(T) \sim \frac{1}{2\pi} T \log T.$$

The number of zeros accounted for by the Hardy-Littlewood theorem is therefore infinitesimal compared with the total number of zeros. It is only recently that their result has been surpassed by A. Selberg, with the proof that

$$N_0(T) > AT \log T.$$

Hardy used to say that anyone who had a really new idea about the zeta-function must surely prove the Riemann hypothesis, but Selberg's work seems to have disproved this.

The second main contribution of Hardy and Littlewood to the theory was the "approximate functional equation". If $s = \sigma + it$, we have

$$\zeta(s) = \sum_{n=1}^\infty \frac{1}{n^s} \quad (\sigma > 1),$$

and, by the functional equation,

$$\zeta(s) = 2^s \pi^{s-1} \sin \frac{1}{2} s\pi \, \Gamma(1-s) \sum_{n=1}^\infty \frac{1}{n^{1-s}} \quad (\sigma < 0).$$

If $0 \leq \sigma \leq 1$, neither formula holds, but it is possible to represent $\zeta(s)$ approximately by combining them. The result is

$$\zeta(s) = \sum_{n<x} \frac{1}{n^s} + 2^s \pi^{s-1} \sin \frac{1}{2} s\pi \, \Gamma(1-s) \sum_{n<v} \frac{1}{n^{1-s}} + O(x^{-\sigma})$$
$$+ O(y^{\sigma-1}|t|^{\frac{1}{2}-\sigma}),$$

where $2\pi xy = |t|$. It was found later that this discovery had been anticipated to a certain extent by Riemann himself. Riemann's unpublished papers were found to contain the main idea of such formulae, though the applications made by Hardy and Littlewood go far beyond anything in Riemann.

A later paper by Hardy and Littlewood contains a similar approximate functional equation for $\{\zeta(s)\}^2$.

The third main section of the work of Hardy and Littlewood is concerned with the proof of mean-value theorems for $\zeta(s)$. A general theorem on Dirichlet series shows that, if $\sigma > \frac{1}{2}$, then, as $T \to \infty$,

$$\int_1^T |\zeta(\sigma+it)|^2 dt \sim T\zeta(2\sigma).$$

The corresponding result for $\sigma = \frac{1}{2}$,

$$\int_1^T \left|\zeta\left(\frac{1}{2}+it\right)\right|^2 dt \sim T \log T,$$

is due to Hardy and Littlewood. Other important extensions are concerned with higher powers of $\zeta(s)$, the most striking result being

$$\int_1^T \left|\zeta\left(\frac{1}{2}+it\right)\right|^4 dt \sim \frac{1}{2\pi^2} T \log^4 T.$$

It was also proved that, if

$$\int_1^T \left|\zeta\left(\frac{1}{2}+it\right)\right|^{2k} dt = O(T^{1+\epsilon})$$

could be proved for all positive integers k, then the truth of Lindelöf's hypothesis, that

$$\zeta\left(\frac{1}{2}+it\right) = O(t^\epsilon),$$

would follow. But no further progress has been made in this direction.

16. The Riemann Zeta-Function and Lattice Point Problems

A direct attack on the problem of the order of $\zeta(s)$ in the critical strip also produced some striking results. Weyl had invented a method of obtaining inequalities for exponential sums of the form

$$\sum_n e^{2\pi i f(n)},$$

where $f(n)$ is a real function. Since

$$\sum n^{-s} = \sum n^{-\sigma-it} = \sum n^{-\sigma} e^{-it \log n},$$

such results can be applied to the estimation of the sums which occur in the theory of the zeta-function. A typical result is

$$\zeta\left(\frac{1}{2}+it\right) = O(t^{\frac{1}{8}+\epsilon}).$$

This is still a long way from Lindelöf's hypothesis, but only minute improvements on it have since been made.

The above theorems have important applications in the theory of divisor problems, i.e. in problems of approximations to sums of the form

$$\sum_{n<x} d_k(n),$$

where $d_k(n)$ is the number of decompositions of n into k factors.

Another subject to which Hardy made a fundamental contribution was that of the lattice-points in a circle. The number $R(x)$ of lattice-points in a circle of radius $\sqrt{x}$, i.e. of pairs of integers μ, ν, such that $\mu^2 + \nu^2 \le x$, is roughly equal to the area πx of the circle, but closer approximations to $R(x)$ are difficult to make. It had been proved by Sierpiński that, if

$$R(x) = \pi x + P(x),$$

then $P(x) = O(x^{\frac{1}{3}})$, but the true order of $P(x)$ was unknown.

Hardy obtained an exact formula for $R(x)$ as a series of Bessel functions. If x is not an integer this is

$$R(x) = \pi x + x^{\frac{1}{2}} \sum_{n=1}^{\infty} \frac{r(n)}{n^{\frac{1}{2}}} J_1\{2\pi \sqrt{(nx)}\},$$

where $r(n)$ is the number of solutions in integers of $\mu^2 + \nu^2 = n$. If x is an integer, $R(x)$ must be replaced by $R(x) - \frac{1}{2}r(x)$. This "exact formula" is very striking, but it is not of much use in the problem of the order of $P(x)$. If we could treat the series as a finite sum, the ordinary asymptotic formula for Bessel functions would give at once $P(x) = O(x^{\frac{1}{4}})$. It is tempting to suppose that at any rate $P(x) = O(x^{\frac{1}{4}+\epsilon})$, but nothing approaching this has ever been proved. What

Hardy did prove was that each of the inequalities $P(x) > Kx^{\frac{1}{4}}$, $P(x) < -Kx^{\frac{1}{4}}$, is satisfied, with some K, for some arbitrarily large values of x. The true order of $P(x)$ therefore lies some where between $x^{\frac{1}{4}}$ and $x^{\frac{1}{3}}$, and later research has done a little, but not much, to narrow this gap.

17
Four Hardy Gems

A book on G. H. Hardy should contain some real mathematics, for he certainly provided the world with more than his share. Many of his contributions are, however, of a highly technical nature and require a background far more extensive than what we envision for most readers of this volume.

We have thus included in this section four expository pieces that give a sense of Hardy's achievement without requiring unreasonable expertise. The proofs, presented in our own words, come with full mathematical detail. We believe that these results, although relatively elementary, are impressive and revealing examples of Hardy in action.

The four articles are:

I. **A Function.** This is adapted from "A Formula for the Prime Factors of Any Number," *Messenger of Mathematics* 35 (1906), pp. 145–146. There, Hardy provided an explicit function $\theta(n)$ that took any whole number as input and yielded the number's largest prime factor as output. It is quite remarkable. Unfortunately, the formula was so complicated it has no practical utility, but its mere existence was, in Hardy's words, "astonishing."

II. **An Inequality.** This result appeared in "A Note on Two Inequalities," *Journal of the London Mathematical Society* 11 (1936), pp. 167–170. Throughout his career, Hardy was a master of inequalities, and proving this one, which featured the number π in a most unexpected place, gave him a chance to shine.

III. **An Integral.** We found this clever evaluation as part of his solution to "Question 13979," *Educational Times* 71 (1899), pp, 61–62. Here Hardy determined the exact value of the improper integral $\int_0^\infty \frac{\ln x}{1+x^3} dx$.

IV. **An Application.** This is the so-called Hardy-Weinberg law of genetics, taken from his famous letter, "Mendelian Proportions in a Mixed Population," *Science* (N.S.) 28 (1908), pp. 49–50.

Of course, a full appreciation of Hardy's talents requires that his work be considered unabridged and undigested. For this, we refer the reader to his *Collected Papers*, published in seven volumes by Oxford University Press. These provide the true gauge of Hardy's mathematical genius.

A Function

It would be a number theorist's dream to find a function θ with the property that, if n is a whole number, then

$$\theta(n) = \text{the largest prime divisor of } n.$$

With such a function at hand, a mathematician could easily factor integers into primes by applying θ repeatedly. For instance, to factor $n = 150$, we would first find $\theta(150) = 5$, then $\theta(150/5) = \theta(30) = 5$, followed by $\theta(30/5) = \theta(6) = 3$, and then $\theta(6/3) = \theta(2) = 2$, whereupon the process ends. The prime factorization is thus

$$150 = 5^2 \cdot 3 \cdot 2.$$

All of this would be quite wonderful... *if* there were an elementary and useful formula for $\theta(n)$. That, alas, is asking too much.

But consider the opening passage of Hardy's remarkable paper from 1906:

A FORMULA FOR THE PRIME FACTORS OF ANY NUMBER.

By *G. H. Hardy*, Trinity College, Cambridge.

§1. My object in this note is to define a function $\theta(x)$ of x such that, if x is a positive integer n, $\theta(x)$ is the largest prime contained in x.

Here, Hardy provided an analytic expression for a function $\theta(x)$ that did the job. In what follows, we give his formula and accompanying proof and then illustrate it by revisiting our example of $n = 150$ above.

Theorem

$$\theta(x) = \lim_{r \to \infty} \lim_{m \to \infty} \lim_{k \to \infty} \sum_{i=0}^{m} \left(1 - \left[\cos \frac{\pi(i!)^r}{x}\right]^{2k}\right)$$

is the desired function.

Proof We begin, as did Hardy, by stipulating that n has prime factorization $n = a^{\alpha} b^{\beta} c^{\gamma} \ldots d^{\delta}$, where the primes $a, b, c \ldots, d$ are in descending order. Temporarily fix a value of r that equals or exceeds all the exponents in the factorization—i.e., fix $r \geq \max(\alpha, \beta, \gamma, \ldots, \delta)$. If $i \geq a$, (where a is the largest prime factor of n), it is clear that $\frac{(i!)^r}{n}$ is a whole number. By contrast, if $i < a$,

17. Four Hardy Gems

then $\frac{(i!)^r}{n}$ cannot be a whole number. In the former case, $\cos \frac{\pi(i!)^r}{n} = \pm 1$; in the latter, $-1 < \cos \frac{\pi(i!)^r}{n} < 1$. As a consequence, for any $r \geq \max(\alpha, \beta, \gamma, \ldots, \delta)$ and for each $i = 0, 1, \ldots, a-1$, we have

$$\lim_{k \to \infty} \left(1 - \left[\cos \frac{\pi(i!)^r}{n}\right]^{2k}\right) = 1 - 0 = 1.$$

On the other hand, if $i \geq a$, then

$$\lim_{k \to \infty} \left(1 - \left[\cos \frac{\pi(i!)^r}{n}\right]^{2k}\right) = 1 - [\pm 1]^{2k} = 0.$$

Thus, for any $r \geq \max(\alpha, \beta, \gamma, \ldots, \delta)$ and for any $m \geq a$, we see that

$$\lim_{k \to \infty} \sum_{i=0}^{m} \left(1 - \left[\cos \frac{\pi(i!)^r}{n}\right]^{2k}\right) = (1 + 1 \cdots + 1) + (0 + 0 + \cdots) = a.$$

As a consequence,

$$\theta(n) = \lim_{r \to \infty} \lim_{m \to \infty} \lim_{k \to \infty} \sum_{i=0}^{m} \left(1 - \left[\cos \frac{\pi(i!)^r}{n}\right]^{2k}\right) = a$$

$$= \text{the largest prime factor in } n. \qquad \text{Q.E.D.}$$

Here is a concrete example to illustrate Hardy's process in action:
Let $n = 150 = 5^2 \cdot 3^1 \cdot 2^1$, where the prime factors are in descending order. Because $\max(2, 1, 1) = 2$, we fix $r = 2$. We first want to consider $i \geq 5$, so, for the sake of specificity, we let $i = 7$. Then

$$\frac{(i!)^r}{n} = \frac{(7!)^2}{150} = \frac{7^2 \cdot 6^2 \cdot 5^2 \cdot 4^2 \cdot 3^2 \cdot 2^2 \cdot 1^2}{5^2 \cdot 3 \cdot 2},$$

which is a positive integer because of the wholesale cancellation of denominator into numerator. But if we now consider any $i < 5$—e.g., take $i = 3$—we have $\frac{(i!)^r}{n} = \frac{(3!)^2}{150} = \frac{3^2 \cdot 2^2 \cdot 1^2}{5^2 \cdot 3 \cdot 2}$, which most certainly is not a whole number. These conditions, along with the properties of the cosine, show that

$$\lim_{k \to \infty} \sum_{i=0}^{7} \left(1 - \left[\cos \frac{\pi(i!)^2}{150}\right]^{2k}\right) = (1+1+1+1+1) + (0+0+0) = 5,$$

and passage to the limit as $m \to \infty$ and as $r \to \infty$ does not affect the outcome. In short, $\theta(150) = 5$.

Hardy, it should be noted, was under no illusion about the *utility* of his formula. He wrote:

I need hardly say that this result does not pretend to be more than a curiosity, and that its value for any kind of application is *nil*.

Of course, a lack of mathematical applicability never troubled G. H. Hardy. He clearly was intrigued with this peculiar result, and so we give him the last word:

At the same time, it seems worthwhile to point out that it is possible to write down a closed analytical expression which properly represents a *function* of x in Euler's sense, and which does possess the at first sight astonishing property stated above.

An Integral

It is an understatement to say that G. H. Hardy was fascinated by integrals.

His collected works contain scores of articles on the subject, including an astonishing 69 papers in the journal *Messenger of Mathematics* that appeared under the heading "Notes on some points in the integral calculus." Many of his works, with daunting titles like "On absolutely convergent integrals of functions which are infinitely often infinite" (1902), were quite general in their focus. Others dealt with specific integrands such as $\int_0^\infty \frac{\sin x}{x} dx$, which we shall treat later in this volume.

Given Hardy's interest—perhaps "obsession" is a better word—we should consider at least one of his integrals here. The example below, a problem from 1899, will stand as representative of his work on this subject.

The challenge was to find the *exact* value of $\int_0^\infty \frac{\ln x}{1+x^3} dx$. The answer is by no means self-evident and might at first seem beyond the power of elementary calculus. But Hardy was up to the task. He evaluated this improper integral by means of a change of variable, a theorem of Euler, and the observation that, if $r > 0$, then

$$\int_0^1 x^r \ln x \, dx = -\frac{1}{(r+1)^2}.$$

This last is easily confirmed using integration by parts.

Theorem $\int_0^\infty \frac{\ln x}{1+x^3} dx = -\frac{2}{27}\pi^2$.

Proof Hardy began by writing $\int_0^\infty \frac{\ln x}{1+x^3} dx = \int_0^1 \frac{\ln x}{1+x^3} dx + \int_1^\infty \frac{\ln x}{1+x^3} dx$.

The substitution $x = 1/y$ transformed the right-hand integral into $-\int_0^1 \frac{y \ln y}{1+y^3} dy$ and so

$$\int_0^\infty \frac{\ln x}{1+x^3} dx = \int_0^1 \frac{\ln x}{1+x^3} dx - \int_0^1 \frac{y \ln y}{1+y^3} dy = \int_0^1 \frac{1-x}{1+x^3} \ln x \, dx.$$

17. Four Hardy Gems

Next, he replaced $\frac{1-x}{1+x^3}$ by the infinite series $1 - x - x^3 + x^4 + x^6 - x^7 - x^9 + \cdots$ and concluded that

$$\int_0^\infty \frac{\ln x}{1+x^3} dx = \int_0^1 (1 - x - x^3 + x^4 + x^6 - x^7 - x^9 + \cdots) \ln x \, dx$$

$$= -1 + \frac{1}{4} + \frac{1}{16} - \frac{1}{25} - \frac{1}{49} + \frac{1}{64} + \frac{1}{100} - \cdots \text{ by the observation above.}$$

So far, so good. It only remained to evaluate this infinite series. But here Hardy had assistance from Leonhard Euler, who in 1734 had famously summed the reciprocals of the squares to get

$$1 + \frac{1}{4} + \frac{1}{9} + \frac{1}{16} + \frac{1}{25} + \frac{1}{36} + \frac{1}{49} + \frac{1}{64} + \frac{1}{81} + \frac{1}{100} + \cdots = \frac{\pi^2}{6}.$$

It is then a straightforward calculation to evaluate the alternating series

$$1 - \frac{1}{4} + \frac{1}{9} - \frac{1}{16} + \frac{1}{25} - \frac{1}{36} + \frac{1}{49} - \cdots$$

$$= \left(1 + \frac{1}{4} + \frac{1}{9} + \frac{1}{16} + \frac{1}{25} + \frac{1}{36} + \frac{1}{49} + \cdots\right)$$

$$- \left(\frac{2}{4} + \frac{2}{16} + \frac{2}{36} + \frac{2}{64} + \cdots\right)$$

$$= \frac{\pi^2}{6} - \frac{1}{2}\left(1 + \frac{1}{4} + \frac{1}{9} + \frac{1}{16} + \frac{1}{25} + \frac{1}{36} + \cdots\right) = \frac{\pi^2}{6} - \frac{\pi^2}{12} = \frac{\pi^2}{12}.$$

Next we evaluate the following alternating series whose denominators are the squares of multiples of 3:

$$\frac{1}{9} - \frac{1}{36} + \frac{1}{81} - \frac{1}{144} + \frac{1}{225} - \cdots = \frac{1}{9}\left(1 - \frac{1}{4} + \frac{1}{9} - \frac{1}{16} + \frac{1}{25} - \cdots\right)$$

$$= \frac{1}{9}\left(\frac{\pi^2}{12}\right) = \frac{\pi^2}{108}.$$

Hardy then combined these last two series to get:

$$-1 + \frac{1}{4} + \frac{1}{16} - \frac{1}{25} - \frac{1}{49} + \frac{1}{64} + \frac{1}{100} - \cdots$$

$$= -\left(1 - \frac{1}{4} + \frac{1}{9} - \frac{1}{16} + \frac{1}{25} - \frac{1}{36} + \frac{1}{49} - \frac{1}{64} + \frac{1}{81} - \cdots\right)$$

$$+ \left(\frac{1}{9} - \frac{1}{36} + \frac{1}{81} - \frac{1}{144} + \frac{1}{225} - \cdots\right)$$

$$= -\frac{\pi^2}{12} + \frac{\pi^2}{108} = -\frac{2}{27}\pi^2.$$

In short, $\int_0^\infty \frac{\ln x}{1+x^3}dx = -1 + \frac{1}{4} + \frac{1}{16} - \frac{1}{25} - \frac{1}{49} + \frac{1}{64} + \frac{1}{100} - \cdots - \frac{2}{27}\pi^2$, as desired. Q.E.D.

Clearly, Hardy was an analyst after Euler's own heart.

An Inequality

In 1936, Hardy gave a proof of a result that he called "a curious inequality of Carlson," which had appeared in *Arkiv für Matematik*, Vol. 25 (1935). He stated it as follows:

Theorem *If $a_n > 0$ and all series converge, then $\left[\sum_{n=1}^\infty a_n\right]^4 < \pi^2 \left[\sum_{n=1}^\infty a_n^2\right] \cdot \left[\sum_{n=1}^\infty n^2 a_n^2\right]$.*

His argument, from *The Journal of the London Mathematical Society*, Vol. 11 (1936), pp. 167–170, is worthy of our consideration for two reasons.

First, it displays Hardy's characteristic agility in treating formulas of genuine sophistication. The result at first glance appears to require some heavy mathematical artillery; indeed, Fritz David Carlson's original derivation was of that sort. But Hardy found a brilliant shortcut to the same end.

Second, the result itself is striking because of the appearance of the constant π^2. The generality that accompanies its mild assumptions gives no clue as to where this unexpected number comes from.

Hardy's proof—which he described as "very simple"—required the arctangent integral, an integral/series comparison, and Cauchy's inequality. Only the last of these might be unfamiliar to a student of calculus, so we state it here in the form that Hardy needed it:

Cauchy's inequality: If $A_n > 0$ and $B_n > 0$, then $\left[\sum_{n=1}^\infty A_n B_n\right]^2 \le \left[\sum_{n=1}^\infty A_n^2\right] \cdot \left[\sum_{n=1}^\infty B_n^2\right]$, where all series are assumed to converge. A derivation can be found in any sufficiently thorough analysis book. See, for instance, section 2.4 of *Inequalties* by Hardy, Littlewood, and Pólya.

Armed with these weapons, Hardy began his proof. Setting $S = \sum_{n=1}^\infty a_n^2$ and $T = \sum_{n=1}^\infty n^2 a_n^2$, he proceeded as follows:

$$\left(\sum_{n=1}^\infty a_n\right)^2 = \left(\sum_{n=1}^\infty a_n \cdot \sqrt{T + Sn^2} \cdot \frac{1}{\sqrt{T + Sn^2}}\right)^2$$

$$\le \sum_{n=1}^\infty a_n^2(T + Sn^2) \cdot \sum_{n=1}^\infty \frac{1}{T + Sn^2} \quad \text{by Cauchy's inequality}$$

17. Four Hardy Gems

$$= \left(T\sum_{n=1}^{\infty} a_n^2 + S\sum_{n=1}^{\infty} n^2 a_n^2\right) \cdot \sum_{n=1}^{\infty} \frac{1}{T + Sn^2}$$

$$= (TS + ST) \cdot \sum_{n=1}^{\infty} \frac{1}{T + Sn^2}$$

$$< 2TS \int_0^{\infty} \frac{1}{T + Sx^2} dx \qquad \text{by the usual comparison of series and integrals.}$$

Hardy evaluated the improper integral as $\frac{1}{\sqrt{ST}}$ Arctan $\frac{\sqrt{S}x}{\sqrt{T}}\Big|_0^{\infty} = \frac{\pi}{2\sqrt{ST}}$, and so

$$\left(\sum_{n=1}^{\infty} a_n\right)^2 < 2ST \frac{\pi}{2\sqrt{ST}} = \pi\sqrt{S}\sqrt{T}.$$

Upon squaring both sides of the inequality, Hardy arrived at the desired result:

$$\left(\sum_{n=1}^{\infty} a_n\right)^4 < \pi^2 ST = \pi^2 \left[\sum_{n=1}^{\infty} a_n^2\right] \cdot \left[\sum_{n=1}^{\infty} n^2 a_n^2\right]. \qquad \text{Q.E.D.}$$

This was a triumph of economy and elegance... and a reminder of why G. H. Hardy was an undisputed master of the inequality.

An Application

In spite of Hardy's distaste for applied mathematics, he is properly credited with a significant application: the Hardy-Weinberg Law. This appeared in a brief letter to *Science* in 1908, at a time when the principles of genetics—both biological and mathematical—were very much in their infancy.

Suppose that a characteristic in a population—say, eye color or blood type—is expressed in genes of type A or a, where the former is dominant and the latter recessive. That is, if an offspring receives an A gene from either parent, then the offspring will have that characteristic (*e.g.*, brown eyes); only by receiving an a from both parents will the opposite characteristic (*e.g.*, blue eyes) appear.

Prior to Hardy, it had been speculated that, over generations, a dominant character would become so widespread as to eliminate its recessive counterpart. Superficially, this might seem plausible. But Hardy, using only simple algebra, established that

> ...there is not the slightest foundation for the idea that a dominant character should show a tendency to spread over a whole population, or that a recessive should tend to die out.

Here's how he did it:

Because each parent contributes one gene, an offspring's genetic make-up must be one of three types: AA, Aa, or aa. Hardy stipulated that the initial numbers of these were in the ratios of $p : 2q : r$. We shall slightly tweak his scenario to imagine a population of n individuals in which the probability of having an AA pair is p/n; the probability of an Aa pair is $2q/n$; and the probability of an aa pair is r/n. Of course this requires that

$$\frac{p}{n} + \frac{2q}{n} + \frac{r}{n} = 1 \quad \text{or, equivalently, that} \quad p + 2q + r = n.$$

As did Hardy, we assume that these probabilities are identical among females and males, and we assume as well that mating can be regarded as a random process. Under such conditions we ask, "What is the probability that a parent contributes an A gene to its offspring?"

This is easy. If the parent is itself an AA, then it is certain to contribute an A gene, for that is all it has to offer. If the parent is an Aa, then the probability is 1/2 that it will contribute an A. And if the parent is an aa, then there is no chance whatever of its contributing an A. Thus,

$$\text{Prob(parent contributes an } A \text{ gene)} = 1 \cdot \frac{p}{n} + \frac{1}{2} \cdot \left[\frac{2q}{n}\right] + 0 \cdot \frac{r}{n} = \frac{p+q}{n}.$$

Similarly,

$$\text{Prob(parent contributes an } a \text{ gene)} = 0 \cdot \frac{p}{n} + \frac{1}{2} \cdot \left[\frac{2q}{n}\right] + 1 \cdot \frac{r}{n} = \frac{q+r}{n}.$$

From this, Hardy found the probabilities of the three genetic types in the *next* generation by using the notion of independence (which he characterized as "... a little mathematics of the multiplication-table type"):

Prob (AA) = Prob(male parent contributes A *and* female parent contributes A)

$$= \text{Prob(male parent contributes A)} \times \text{Prob(female parent contributes } A)$$

$$= \frac{p+q}{n} \cdot \frac{p+q}{n} = \frac{(p+q)^2}{n^2};$$

Prob (Aa) = Prob(male parent contributes A *and* female parent contributes a)

$$+ \text{Prob(female parent contributes } A \text{ } and \text{ male parent contributes } a)$$

$$= \frac{p+q}{n} \cdot \frac{q+r}{n} + \frac{q+r}{n} \cdot \frac{p+q}{n} = \frac{2(p+q)(q+r)}{n^2};$$

Prob (aa) = Prob(male parent contributes a *and* female parent contributes a)

$$= \frac{q+r}{n} \cdot \frac{q+r}{n} = \frac{(q+r)^2}{n^2}.$$

17. Four Hardy Gems

He noted that ratios in the second generation need not match those of the initial population. For example, consider the simple case in which $p = 3$, $2q = 6$, and $r = 1$, so that

$$\text{Prob}(AA) = \frac{3}{10}, \quad \text{Prob}(Aa) = \frac{6}{10}, \quad \text{and} \quad \text{Prob}(aa) = \frac{1}{10}.$$

Thus, in Hardy's terminology, the ratios begin as $3 : 6 : 1$.

For the second generation, we find

$$\text{Prob}(AA) = \frac{(p+q)^2}{n^2} = \frac{36}{100} = \frac{9}{25},$$

$$\text{Prob}(Aa) = \frac{2(p+q)(q+r)}{n^2} = \frac{48}{100} = \frac{12}{25},$$

$$\text{and Prob}(aa) = \frac{(q+r)^2}{n^2} = \frac{16}{100} = \frac{4}{25}.$$

The second generation ratios have now changed to $9 : 12 : 4$.

But from here onward, all is stable. Hardy put it this way: "... whatever the values of p, q and r may be, the distribution will in any case continue unchanged after the second generation."

To see this, we apply our reasoning above as we move from the second to the third generation:

Prob(parent contributes an A gene)

$$= 1 \cdot \frac{(p+q)^2}{n^2} + \frac{1}{2} \cdot \left[\frac{2(p+q)(q+r)}{n^2} \right] + 0 \cdot \frac{(q+r)^2}{n^2}$$

$$= \frac{(p+q)^2 + (p+q)(q+r)}{n^2}$$

$$= \frac{(p+q)[p+q+q+r]}{n^2} = \frac{(p+q)n}{n^2}$$

$$= \frac{p+q}{n} \quad \text{because} \quad p + 2q + r = n.$$

This is the same probability of contributing an A gene in the previous generation. Similarly,

$$\text{Prob(parent contributes an } a \text{ gene)} = \frac{q+r}{n}, \text{ as before.}$$

Because these probabilities match those of the preceding generation, so must $\text{Prob}(AA)$, $\text{Prob}(Aa)$, and $\text{Prob}(aa)$. In short, the ratios have stabilized.

To illustrate, we revisit our numerical example above, where the ratios began as $3 : 6 : 1$ and then became $9 : 12 : 4$. As we move to the next generation, we see

$$\text{Prob}(A) = 1 \cdot \frac{9}{25} + \frac{1}{2} \cdot \left[\frac{12}{25}\right] + 0 \cdot \frac{4}{25} = \frac{3}{5} \quad \text{and} \quad \text{Prob}(a) = \frac{2}{5}, \text{ and so}$$

$$\text{Prob}(AA) = \left(\frac{3}{5}\right)^2 = \frac{9}{25}, \quad \text{Prob}(Aa) = 2\left(\frac{3}{5}\right)\left(\frac{2}{5}\right) = \frac{12}{25},$$

$$\text{and Prob}(aa) = \left(\frac{2}{5}\right)^2 = \frac{4}{25}.$$

The corresponding ratios remain at 9 : 12 : 4 in the third and in all succeeding generations. The dominant characteristic does not wipe out the recessive one.

For good reason, Hardy regarded this as a mere crumb from his mathematical table. Simple algebra was all he needed carry the argument along. Yet the result was sufficiently important to attach Hardy's name—like it or not!—to a mathematical *application*.

18

What Is Geometry?

(*Mathematical Gazette* 12 (1925), pp. 309–316)*
This was the Presidential Address to the Mathematical Association, 1925.

From the Editors

It was in 1870 that a mathematics master at a school in Birmingham proposed in Nature the establishment of an "Anti-Euclid Association" to move school mathematics away from its rigid devotion to Euclid. It was this organization that evolved to become the Mathematical Association (not to be confused with the Mathematical Association of America (MAA) that didn't come along until 1915). The logo of the British organization is Mα. In 1894 the English Mα started publishing The Mathematical Gazette, *which by now has a long distinguished history. The organizations and their journals serve somewhat different audiences. The English journal appeals not only to secondary school students and faculty but also to those who would be in the first years of college in the US. College faculty in America publish in the* Gazette, *as did G. H. Hardy on many occasions, as one can see from the contents of this collection. What we have here was Hardy's Presidential Address given in 1925.*

I have put the title of my address in the form of a definite question, to which I propose to return an equally definite answer. I wish to make it quite plain from the beginning that there will be nothing in the least degree original, still less anything paradoxical or sensational, in my answer, which will be the orthodox answer of the orthodox professional mathematician.

---

* Reprinted with the kind permission of the publisher of *The Mathematical Gazette:* The Mathematical Association.

I expect that you, as members of an Association which stands half-way between the ordinary mathematical teacher and the professional mathematician in the narrower sense, will probably agree with me that I am wiser to avoid topics of what is usually called a "pedagogical" character. I am sorry to be compelled to use the unpleasant word "pedagogical," and I am sure that you will believe me when I say that I do not use it in any contemptuous sense, and that I am enough of a pedagogue myself to realise the very genuine interest of many "pedagogical" questions. But I do not regard it as the business of a professional mathematician to concern himself primarily with such questions, and, even if I did, I should have very little to say about them. It has always seemed to me that in all subjects, and most of all in mathematics, questions concerning methods of teaching, whether this should come before that, and how the details of a particular chapter are best presented, however interesting they may be, are of secondary importance; and that in mathematics at all events there is one thing only of primary importance, that a teacher should make an honest attempt to understand the subject he teaches as well as he can, and should expound the truth to his pupils to the limits of their patience and capacity. In a word, I do not think it matters greatly what you teach, so long as yon are really certain what it is; and I feel that you might reasonably be impatient with me, whether you agreed with me or not, if I occupied your attention for an hour and had nothing more to say to you than that. It is obviously better that I should take some definite chapter of mathematical doctrine, a chapter which is at any rate of the most obvious and direct educational interest, and expound it to you as clearly as I can.

It is, however, quite likely that some of the more sophisticated of you, and particularly any genuine geometer who may be present, will criticise my choice of a subject in a manner which I might find a good deal more difficult to meet. You might object that it would be reasonable enough for me to try to expound the differential calculus, or the theory of numbers to you, because the view that I might find something of interest to say to you about it is not *prima facie* absurd; but that geometry is, after all, the business of geometers, and that I know, and you know, and I know that you know, that I am not one; and that it is useless for me to try to tell you what geometry is, because I simply do not know. And here I am afraid that we are confronted with a regrettable but quite definite cleavage of opinion. I do not claim to know any geometry, but I do claim to understand quite clearly what geometry is.

I think that this claim is in reality not quite so impertinent as it may seem. The question, "What is geometry?" is not, in the ordinary sense of the phrase, a geometrical question, and I certainly do not think it absurd to suppose that a logician, or even an analyst, may be better qualified to answer it than a geometer.

18. What Is Geometry?

There have been very bad geometers who could have answered it quite well, and very great geometers, such as Apollonius, Poncelet, or Darboux, who would probably have answered it extremely badly. It is a comfort, at any rate, to reflect that my answer can hardly be worse than theirs would in all probability have been.

I propose, then, to cast doubts of this sort aside, and to proceed to answer my question to the best of my ability. There are two things, I think, which become quite clear the moment we reflect about the question seriously. In the first place, there is not one geometry, but an infinite number of geometries, and the answer must to some extent be different for each of them. In the second place, the elementary geometry of schools and universities is not this or that geometry, but a most disorderly and heterogeneous collection of fragments from a dozen geometries or more. These are, or should be, platitudes, and I have no doubt that they are to some extent familiar to all of you; but it is a small minority of teachers of geometry that has envisaged such platitudes clearly and sharply, and it is probably desirable that I should expand them a little.

I begin with the second. It is obvious, first, that a great part of what is taught in schools and universities under the title of geometry is not geometry, or at any rate mathematical geometry, at all, but physics or perhaps philosophy. It is an attempt to set up some kind of ordered explanation of what has been humorously called the real world, the world of physics and sensation, of sight and hearing, heat and cold, earthquakes and eclipses; and earthquakes and eclipses are plainly not constituents of the world of mathematics.

It is dangerous to repeat truisms in public, and the particular truism which I have just stated to you is one which I have often expressed before, and which has sometimes been received in a manner very different from that which I had anticipated. But I am not speaking now to an audience of rude and simple physicists, or of philosophers dazed by centuries of Aristotelian tradition, but to one of mathematicians familiar with common mathematical ideas. I find it difficult to believe that any mathematician of the twentieth century is quite so unsophisticated as to suppose that geometry is primarily concerned with the phenomena of spatial perception, or the physical facts of the world of common sense. It is, however, perhaps unwise to take too much for granted, and I will therefore try to drive home my point by a simple illustration.

Imagine that I am giving an ordinary mathematical lecture at Oxford, let us suppose on elementary differential geometry, and that I write out the proof of a theorem on the blackboard. John Stuart Mill would have maintained that the theorem was at the best approximately true, and that the closeness of the approximation depended on the quality of the chalk; and, though Mill was a man for whom I feel in many ways a very genuine admiration, I can hardly believe that there is anybody quite so innocent as that to-day. I want, however, to push my illustration a stage further. Let us imagine now that a very violent dynamo, or an extremely heavy gravitating body, is suddenly introduced into the room. Einstein and Eddington tell us, and I have no doubt that they are right, that the whole geometrical fabric of the room is changed, and every detail of the pattern to which it conforms is distorted. Does common sense really tell us that my theorem is no longer true, or that the strength or weakness of the arguments by which I have established it has been in the very slightest degree affected? Yet that is the glaring and intolerable paradox to which anyone is committed who supports the old-fashioned view that geometry is "the science of space."

The simple view, then—the view which I will call for shortness the view of common sense, though there is uncommonly little common sense about it—the view that geometry is the science which tells us the facts about the space of physics and sensation, is one which will not stand a moment's critical examination; and this, of course, was plain enough before Einstein, though it is Einstein who, by enabling us to exhibit its paradoxes in so crude a form, has finally completed the demonstration. The philosophers, of course, have tried to restate the view of common sense in a more sophisticated form. Geometry, they have explained to us, tells us, not exactly the facts of physical or perceptual space, but certain general laws to which all spatial perception must conform. Philosophers have been singularly unhappy in their excursions into mathematics, and this is no exception. It is, as usual, an attempt to restrict the liberty of mathematicians, by proving that it is impossible for them to think except in some particular way; and the history of mathematics shows conclusively that mathematicians will never accept the tyranny of any philosopher. The moment a philosopher has demonstrated the impossibility of any mode of thought, some rebellious mathematician will employ it with unconquerable energy and conspicuous success. No sooner was the apodictic certainty of Euclid firmly established, than the non-Euclidean geometries were constructed; no sooner were the inherent contradictions of the infinite finally exposed, than Cantor erected a coherent theory. I do not think, then, that we need trouble ourselves with the views of the philosophers concerning geometry. They are, indeed, of much less interest than those of the man in the street, which do possess some interest, since there are valid reasons for supposing that others may share them.

It will be more profitable to leave the philosophers alone, and to consider what the mathematicians themselves have to say. We shall then have reasonable hope of making some substantial progress, since mathematicians, or those of them who are at all interested in the logic of mathematics, hold fairly definite views, and views which are in tolerable agreement, concerning this question of the relation of geometry to the external world. The views of the mathematicians

18. What Is Geometry? 239

are also much more modest than those which the philosophers have tried to impose upon them.

A geometry, like any other mathematical theory, is essentially a map or scheme. It is a *picture*, and a picture, naturally, of *something*; and as to what that something is opinions do and well may differ widely. Some will say that it is a picture of something in our minds, or evolved from them or constructed by them, while others, like myself, will be more disposed to say that it is a picture of some independent reality, outside them; and personally I do not think it matters very much which type of view you may prefer to adopt. What is much more important and much clearer is this, that there is one thing at any rate of which a geometry is *not* a picture, and that is the so-called real world. About this, I think that almost all modern mathematicians would agree.

This is only common mathematical orthodoxy, but it is an orthodoxy which outsiders very frequently misunderstand or misrepresent. I need hardly say that it does not mean that mathematicians regard the world of physical reality as uninteresting or unimportant. That would be on a par with the view that mathematicians are peculiarly absent-minded, always lose at bridge, and are habitually unfortunate in their investments. Still less does it mean that they regard as uninteresting or unimportant the contribution which mathematics can make to the study of the real world. The Ordnance Survey suggests to me that Waterloo Station, and Piccadilly Circus, and Hyde Park Corner lie roughly in a straight line. That is a geometrical statement about reality, and it enables me to catch my train at Paddington. Einstein is more daring, and issues his orders to the stars, and the stars halt in their courses to obey him. Einstein, and the Ordnance Survey, and even I, can all of us, armed with our mathematics, put forward suggestions concerning the structure of physical reality, and our suggestions will continually prove to be not merely interesting, but of the most direct and practical importance. We can point to this or that mathematical model, Euclidean or Lobatschewskian or Einsteinian geometry, and suggest that perhaps the structure of the universe resembles it, or can be correlated with it in one way or another; that that is a possibility at any rate which the physicists may find it worth their while to consider. We can offer these suggestions, but, when we have offered them, our function as mathematicians is discharged. We cannot, do not profess to, and do not wish to *prove* anything whatsoever. There is not, and cannot be, any question of a mathematician proving anything about the physical world; there is one way only in which we can possibly discern its structure, that is to say the laboratory method, the method of direct observation of the facts.

I will venture here on an illustration which I have used before. If one of you were to tell me that there are three dimensions in this room, but five in

Southampton Row, I should not believe him. I would not even suggest that we should adjourn our discussion and go outside to see. The assertion would, of course, be one of an exceedingly complicated character, and a very painstaking analysis might prove necessary before we were quite certain what it meant. However, I could attach a definite meaning to it. I should understand it to imply that, owing to particularities in the geography of London which had up to the present escaped my attention, the common three-dimensional model, sufficient for our purposes in here, becomes inadequate when we pass out into the street. And, however sceptical I might feel about such a theory, I should certainly not be so foolish as to advance mathematical arguments against it, for the all-sufficient reason that I am quite certain that there are none. I should be sceptical, not as a geometer but as a citizen of London, not because I am a mathematician, but in spite of it; and, indeed, I am sure that, if you appealed from me to the nearest policeman, you would find him not less but far more obstinately sceptical than me.

I must pass on, however, to what is really the proper subject-matter of my address. Geometries, I will ask you to agree provisionally, are *models*, and models of something which, whatever it may be in the last analysis, we may allow for our present purposes to be described as mathematical reality. The question which we have now to consider is that of the nature of these models, and the characteristics which distinguish one from another; and there is one great class of geometries for which the answer is immediate and easy, namely, that of the *analytical* geometries.

An analytical geometry, whether of one, two, three, four, or n dimensions, whether real or complex, projective or metrical, Euclidean or non-Euclidean, and it may, of course, be any of these, is a branch of analysis concerned with the properties of certain sets or classes of sets of numbers. I will take the simplest example, the two-dimensional Cartesian geometry which resembles very closely, though it is by no means the same as, the elementary "analytical geometry" taught in schools. I will call it, as I usually call it in lectures, *Common Cartesian Geometry*.

In Common Cartesian Geometry, a *point* is, by definition, a pair of real numbers (x, y), which we call its *coordinates*. A *line* is, again by definition, a certain class of points, viz. those which satisfy a linear relation $ax + by + c = 0$, where a, b, c are real numbers and a and b are not both zero. The relation itself is called the *equation* of the line. If the coordinates of a point satisfy the equation of a line, the line is said to *pass through* the point, and the point to *lie on* the line. And that is the end of Common Cartesian Geometry, in so far as it is projective, that is to say in so far as it does not use the so-called metrical notions of distance and angle, and in so far as it is concerned only with

18. What Is Geometry?

equations of the first degree. What remains is just algebraical deduction from the definitions.

Common Cartesian Geometry, as I have defined it, is a very simple and not a very interesting subject. It gains a great deal in interest, as you will readily imagine, when "metrical" concepts are introduced. We define the *distance* of two points (x_1, y_1) and (x_2, y_2) by the usual formula

$$d = \sqrt{\{(x_1 - x_2)^2 + (y_1 - y_2)^2\}},$$

and the *angle* between two lines by another common formula, which I need not repeat. We have still, however, only to explore the algebraical consequences of our definitions, and no new point of principle arises, so that I can illustrate what I want to say quite adequately from the projective and linear system. This system, trivial as it is, has certain features to which I wish to call your attention as characteristic of analytical geometries in general.

The first feature is this, that a point in Common Cartesian Geometry is *a definite thing*. This is so in all analytical geometries. Thus in any system of two-dimensional and homogeneous analytical geometry a point is a class of triads (x, y, z), those triads being classified together whose coordinates are proportional, and in the geometry of Einstein a point is a set of four numbers (x, y, z, t). This is a very obvious observation, but it is of fundamental importance, since it marks the most essential difference between analytical geometries and "pure" geometries, in which, as we shall see, a point is not a definite entity at all.

The next point which I ask you to observe is the absence of *axioms*. There are no axioms in any analytical geometry. An analytical geometry consists entirely of *definitions* and *theorems*; and this is only natural, since the object of axioms is, as we shall see, merely to limit our subject-matter, and in an analytical geometry our subject-matter is known.

It is most important to realise clearly that, in different geometrical systems, propositions verbally identical may occupy entirely different positions. What is an axiom in one system may be a definition in another, a true theorem in a third, and a false theorem in a fourth. You are accustomed, for example, to *proving* that the equation of a straight line is of the first degree, and I am not suggesting that the "proof" to which you are accustomed is meaningless, trivial, or false. You profess to be proving a theorem, and you are, in fact, genuinely proving *something*, though it might take us some time to ascertain exactly what it is. There is one thing, however, that is quite plain, and that is that the something which you are proving is not a theorem of analytical geometry, for your supposed theorem is, as a proposition of analytical geometry, not a theorem at all but the definition of a straight line.

Let us take another simple illustration, the "parallel postulate" of Euclid. *If L is a line, and P is a point which does not lie on L, then there is one and only one line through P which has no point in common with L.* This, in school geometry, is sometimes called an "axiom" and sometimes, I suppose, an "experimental fact." It cannot be either of these in analytical geometry, where there are neither axioms nor experimental facts, and it is obviously not a definition. It is, in fact, a theorem, which in Common Cartesian Geometry is true, though in other systems it may be false; and it is a theorem which any schoolboy can prove. It is the algebraical theorem that, given an equation $ax + by + c = 0$, and a pair of numbers, x_0, y_0, which do not satisfy this equation, then it is possible to find numbers A, B, C, such that

(i) $$Ax_0 + By_0 + C = 0$$

and (ii) the equations

$$ax + by + c = 0, \quad Ax + By + C = 0$$

are inconsistent with one another; and that the ratios $A : B : C$ are determined uniquely by these conditions.

These are the characteristics of Common Cartesian Geometry which it is most essential for us to observe at the moment. There are others which I should like to say something about if I had time. There is no infinite and no imaginary in this geometry; there are imaginaries, naturally, only in complex systems, and infinites in homogeneous systems. Further, the principle of duality is untrue. All these topics call for comment; and I should have liked particularly to say something on the subject of the geometrical infinite, since the tragical misunderstandings which have beset many writers of text-books of analytical geometry, and which have generated such appalling confusion in the minds of university students, are misunderstandings for which writers like myself of text-books on analysis have been largely though innocently responsible. The geometrical infinite, however, is a subject which would demand at least a lecture to itself. Apart from this, there is nothing in analytical geometry which presents any logical difficulty whatever, and I may pass to the slightly more delicate topic of pure geometry.

The nature of a system of pure geometry, such as the ordinary projective system, is most easily elucidated, I think, by contrast with analytical systems. The contrasts, which I have made by implication already, are sharp and striking, and when once they have been clearly observed the road to the understanding of the subject is open. I observed, first, that the points and lines of analytical geometry were *definite objects*, such as the pair of numbers (2, 3). Secondly, I observed that there were no *axioms* in an analytical geometry, which consists of

18. What Is Geometry?

definitions and theorems only; and that it is the definitions which differentiate one system of analytical geometry from another. The business of an analytical geometer is, in short, to investigate the properties of *particular systems of things*. The standpoint of a pure geometer is entirely different. He is not, except for incidental and subsidiary purposes, concerned with particular things at all. His function is always to consider *all things which possess certain properties*, and otherwise to be strictly indifferent to what they are. His "points" and "lines" are neither spatial objects, nor sets of numbers, nor this nor that system of entities, but *any* system of entities which are subject to a certain set of logical relations. The particular system of relations which he studies is that which is expressed by the *axioms* of his geometry. It is the axioms only which really matter; it is they which discriminate systems, and the definitions play an altogether subsidiary part.

Suppose, for example, if I may take a frivolous illustration, that a pure geometer and an analytical geometer were to go together to the Zoo. The analytical geometer might be interested in tigers, in their colour, their stripes, and in the fact that they eat meat. A point, he would say, is by definition a tiger, and the central theorems of my geometry are that "points are yellow," that "points are striped," and above all that "points eat meat." The pure geometer would reply that he was quite indifferent to tigers, except in so far as they possessed the properties of being yellow and striped; that *anything* yellow and striped was a point to him; that "points are yellow" and "points are striped" were the *axioms* of his geometry, and that all he wanted to know was whether "points eat meat" is a logical deduction from them.

You will, in fact, find, if you consult any standard work on pure geometry, such as Hilbert's *Grundlagen* or Veblen and Young's *Projective Geometry*, that a pure geometer begins somewhat as follows. We consider a system S of objects $A, B, C, \ldots$. We call these objects *points*, and their aggregate *space*; the *plane*, I may say, if I confine myself for simplicity to geometries of two dimensions. From the complete system S which constitutes space we pick out certain partial aggregates $L, M, N, \ldots$, which we call *lines*. If a point A belongs to the particular partial aggregate L, we say that A *lies on* L and that L *passes through* A. These are the *definitions*, and you will observe the quite subsidiary part they play. They are, in fact, purely verbal, and common to all systems; and they do not indicate or imply any special property whatever of the objects which they are said to define, which are indeed often called the *indefinables* of the geometry. The function of the definitions, in fact, is merely to point to the indefinables.

The serious business of the geometry begins when the axioms are introduced. We suppose next that our points and lines are subject to certain logical relations. These suppositions are assumptions, and we call them axioms. To construct a geometry is to state a system of axioms and to deduce all possible consequences from them.

Let us take an actual example. I select the following system of axioms:

AXIOM 1. *There are just three different points.*

AXIOM 2. *No line contains more than two points.*

AXIOM 3. *There is a line through any two points.*

These axioms are consistent with one another, for it is easy to construct a system of objects which satisfy them. We might, for example, take the numbers 1, 2, 3 as our points and the pairs of numbers 23, 31, 12 as our lines, in which case all our axioms are obviously satisfied. Further, the axioms are independent of one another. If the numbers 1, 2, 3 were still our points, but the pairs 23, 31 alone, and not the pair 12, were taken as lines, then the first and second axioms would be satisfied but not the third, and it naturally follows that Axiom 3 is incapable of deduction from the other two. You will have no difficulty in proving in a similar manner, if you care to do so, that each of the three axioms is logically independent of the others. I do not profess to have stated the axioms in the best form possible, but at any rate they are consistent and independent.

It is easy to deduce from our axioms:

THEOREM 1. *There are just three lines;*

and THEOREM 2. *There are just two lines through any point.*

The state of affairs in this geometry is, in short, that suggested by a figure consisting of three points on a blackboard and three lines joining them in pairs. With this, our geometry appears to be exhausted.

The geometry which I have constructed is not an interesting system, since it has no particular application and virtually no content. For our present purpose, however, that is an advantage, as it makes it possible for me to exhibit the system to you in its entirety. However little interest it may possess, it is a perfectly fair specimen of a pure geometry. All systems of pure geometry, projective geometry, metrical geometry Euclidean or non-Euclidean, are constructed in just this way. They are usually very much more complicated, for you must naturally be prepared to sacrifice simplicity to some extent if you wish to be interesting; but their differences from my trivial geometry are differences not at all of principle or of method, but merely of richness of content and variety of application.

18. What Is Geometry?

I have now given to you the substance of the orthodox answer to the question which I started by asking. I might expand it definitely in detail, but I should add nothing essentially new. Geometry is a collection of logical systems. The number of systems is infinite, and any of you can invent as many new systems as he pleases; I have myself, with the aid of a few pupils, constructed seven or eight in the course of an hour. There are two kinds of systems, analytical geometries and pure geometries. An analytical geometry attaches the usual geometrical vocabulary to more or less complicated systems of numbers, and investigates their properties by means of the ordinary machinery of algebra and analysis. A pure geometry, on the other hand, considers all possible fields of certain logical relations, and explores their connections without reference to the nature of the objects among which they hold.

I said when I started that I did not propose to offer any very definite suggestions about the teaching of mathematics; but I should like to conclude with a few words about some of the practical problems with which members of this Association are primarily concerned. It should be obvious to you by now, I think, that school geometry is, as I stated early in my address, not a well-defined subject, a rational exposition of a particular geometrical system, but a collection of miscellaneous scraps, a selection of airs from different pieces, strung together in the manner which experience shows to be the most enlivening. It would be very easy for me to illustrate my thesis by examining a few passages from current text-books of geometry. What is taught as projective geometry, for example, is not projective geometry, and makes very little pretence of being so, since it is based quite frankly on ratios of lengths and other obviously metrical concepts. Indeed, so far as I know, no English book on projective geometry proper exists, except Mathews' *Projective Geometry*, Dr. Whitehead's tract, and parts of Prof. Baker's treatise. On the other hand, a great deal of what is taught as analytical geometry is not analytical geometry, but an attempt to apply the methods of analytical geometry in other fields, partly to some rough kind of physical geometry supposed to be given intuitively, partly to some system of hybrid pure geometry of which some previous knowledge is assumed. But I must not enter into detail, since detail would mean criticism, and criticism of particular books and particular passages, which I have no time for, and am in any case anxious to avoid.

It is not my object now to offer criticisms of the present methods of geometrical teaching. There are a good many very obvious criticisms suggested by the doctrines which I have tried to explain to you, but I recognise that most of these criticisms would be to a very great extent unfair. It is obvious that the teaching of geometry must be based on what is at best a very illogical compromise, and I am prepared to believe that the compromise evolved by experience, and applied by people who know a good deal more about the practical necessity of compromise than I do, is in substance as reasonable a compromise as the difficulties of the problem permit. My object, so far, has been one not of criticism but of explanation.

I do propose, however, to conclude with one word of criticism, directed only to those of you whose pupils are comparatively able and comparatively mature. There is no doubt that the standard of teaching of analysis has improved out of all knowledge during the last twenty years. The elements of the calculus, even the elements of what foreign mathematicians call algebraical analysis, are taught in a manner with which I personally have comparatively little fault to find. The stupid old superstition that falsehood is always easy and attractive, the truth inevitably repulsive and dull, is almost dead, and it is no longer supposed that ignorance of analysis is in itself a proof either of superior intelligence, or high moral character, or profound geometrical or physical intuition. The teaching of higher geometry does not seem to me to have advanced in the same degree.

I think that it is time that teachers of geometry became a little more ambitious. Geometry in its highest developments may be, for all I know, a more difficult subject than analysis; it is not for me as an analyst to deny it. But what may be true enough of the theory of deformation of surfaces, or of algebraical curves in space, is not even plausible of the elements of higher geometry. Those stages of the subject are surely very much easier than the corresponding stages of analysis. There is something hard and prickly about the basic difficulties of analysis, definite stages on the road where definite types of mind seem to come to an inevitable halt. The difficulties of geometry seem to me a little softer and vaguer; knowledge and general intelligence will carry a student appreciably further on the way. And, if this is so, it seems to me regrettable that students are not given the opportunity, while still at school, of learning a good deal more about the real subject-matter out of which modern geometrical systems are built. It is probably easier, and certainly vastly more instructive, than a great deal of what they are actually taught. Anyone who can investigate properties of six or eight points on a conic is capable of understanding what projective geometry is. Anyone who has the faintest hope of a scholarship at Oxford or Cambridge could learn the nature of an axiom, and how a system of axioms may be shown to be consistent with, or independent of, one another. And anyone who can be taught to project two arbitrary points into the circular points at infinity could learn, what he certainly does not learn at present, to attach some sort of definite meaning to the process he performs. Small as my own knowledge of geometry is, and slight as are my qualifications for teaching it to anybody, I have not yet encountered the student who finds difficulty with such ideas when once they are put before him clearly. I am well aware of the very great services which the Association has rendered in the improvement of geometrical teaching. I think that it might well now concentrate its efforts on a general endeavour to widen the horizon of knowledge, recognising, as regards niceties of logic, sequence, and exposition, that the elementary geometry of schools is a fundamentally and inevitably illogical subject, about whose details agreement can never be reached.

19
The Case against the Mathematical Tripos

(*Mathematical Gazette* 13 (1926) 61–71)*
(Presidential Address to the Mathematical Association, 1926)

From the Editors

In this lengthy essay, G. H. Hardy took aim at the infamous Tripos examination. As we mentioned in our Introduction, this was a formidable ordeal for Cambridge students that consumed four or five days—eight days in 1854—and presented a set of mind-numbing problems often requiring clever tricks and facility with computations that were extraordinarily complex. It was required of all mathematically inclined Cambridge undergraduates, and the outcome could determine one's future in the world of British mathematics and beyond. Those with high scores (the Wranglers) were destined to receive fellowships, honors, and success. Unfortunately, among the students competing for "honours" the one getting the lowest score received the "wooden spoon" and his prospects were bleak.

Competitive examinations for undergraduates are not uncommon—in the United States and Canada, the prime example is the William Lowell Putnam Competition—and such examinations have long been given in Eastern Europe. But there is a difference. The Tripos examinations that Hardy abhorred featured highly technical problems drawn from a narrow band of mathematical topics, whereas these other examinations are not only entirely voluntary but cover a broader scope and reward ingenuity and mathematical creativity.

So why should we care about Hardy's argument against the Tripos? The dreaded test has long since disappeared (though Cambridge still has

---

* Reprinted with the kind permission of the publisher of *The Mathematical Gazette*: The Mathematical Association.

examinations in a variety of fields that carry the name Tripos, just as Oxford has its "Greats"). One thing that makes this 1926 lecture worth reading is that some of the issues raised by Hardy are still relevant. He started off with a strong assertion that the Tripos of the day should not be "reformed" but "abolished," at the same time making it clear that he still held to the belief that examinations are valuable: "Denunciation of examinations, like denunciation of lectures, is very popular now among education reformers... and most of what they say... is little better than nonsense." (Does that sound familiar?) It is relevant to point out that he was addressing the Mathematical Association, an organization that was, and remains, largely concerned with school mathematics as defined in Britain and overlaps with the early undergraduate curriculum in the U.S. It is therefore somewhat similar to the Mathematical Association of America in the U. S. and Canada.

In his argument Hardy alluded to the sorry state of English mathematics following Newton. He pointed out that Cayley was an outstanding mathematician, but one of very few from Britain, whereas France and Germany produced twenty or thirty such. "There has been no country, of first-rate status and high intellectual tradition, whose standard has been so low; and no first-rate subject, except music, in which England has occupied so consistently humiliating a position." (So much for Elgar and Vaughan Williams!) Part of the cause for this shortfall could well have been the absurd attention given to the Tripos at Cambridge.

Hardy eloquently made the case for pure mathematics: "It is hardly likely that anybody here will accuse me of any lack of devotion to the subject which has after all been the one great permanent happiness of my life. My devotion to mathematics is indeed of the most extravagant and fanatical kind; I believe in it and love it, and should be utterly miserable without it, and I have never doubted that, for any one who takes real pleasure in it and has a genuine talent for it, it is the finest intellectual discipline in the world." Then he added: "I believe also that a fair knowledge of mathematics is, even for those who have no pronounced mathematical talent, extremely useful and extremely stimulating and that it should be part of the ordinary intellectual capital of all intelligent men. I am prepared indeed to go further, since I believe that a very large proportion of students abandon mathematics merely because it is often very badly taught..."

Yet he went on to say that he was "not at all sure that, among all possible subjects which might be selected as special courses of study, for an intelligent young man of no particular talent, mathematics is not the worst." He was raising interesting questions here about the role of mathematics in the curriculum. And, as someone deeply suspicious of fountain pens and telephones, Hardy would

19. The Case against the Mathematical Tripos 251

probably not be comfortable with laptops on every student desk, or MOOCs, or flipped classrooms.

In this essay, Hardy was nothing if not outspoken. When evaluating one person's defense of the Tripos: "What... is the function of the Tripos?... Surely to examine and to make distinctions between young men," Hardy's response was: "It would indeed be difficult to compress a larger quantity of vicious educational doctrine into a smaller number of words."

Hardy himself took the Tripos a year earlier than his cohort and was only (only!) Fourth Wrangler. Others who placed higher in those years had much less illustrious careers. Hardy said: "If Einstein had taken the Mathematical Tripos, what would it matter what place he took? The world can recognize its Einsteins quickly enough when it gets the opportunity."

Whether one finds oneself in agreement with Hardy's position is probably less important than the enjoyment of reading his prose. Even after the Tripos was watered down—or, as he put it, "now that so many of its teeth have been successfully drawn"—Hardy remained unconvinced of its value and claimed "... that the system is vicious in principle, and that the vice is too radical for what is usually called reform. I do not want to reform the Tripos, but to destroy it."

What fun if G. H. Hardy were still around to comment on some of our current issues!

My address to-day is the result of an informal discussion which arose at our meeting last year after the reading of Mr. Bryon Heywood's paper. You may remember that Mr. Heywood put forward a number of suggestions, with whose general trend I found myself entirely in sympathy, for the improvement of the courses in higher pure mathematics in English universities. He did not criticise one university more than another; but Cambridge is admittedly the centre of English mathematics, so that it is almost inevitable that such suggestions should be considered from the Cambridge standpoint; and that, if my recollections are correct, is what actually happened in the discussion.

My own contribution to the discussion consisted merely in an expression of my feeling that the best thing that could happen to English mathematics, and to Cambridge mathematics in particular, would be that the Mathematical Tripos should be abolished. I stated this on the spur of the moment, but it is my considered opinion, and I propose to defend it at length to-day. And I am particularly anxious that you should understand quite clearly that I mean exactly what I say; that by "abolished" I mean "abolished", and not "reformed";

that if I were prepared to co-operate, as in fact I have co-operated in the past, in "reforming" the Tripos, it would be because I could see no chance of any more revolutionary change; and that my "reforms" would be directed deliberately towards destroying the traditions of the examination and so preparing the way for its extinction.

There are, however, certain possible grounds of misunderstanding which I wish to remove before I attempt to justify my view in detail. The first of these is unimportant and personal, but probably I shall be wise if I refer to it and deal with it explicitly. Our proceedings here do not as a rule attract a great deal of attention, but they are occasionally noticed in the press; and the writer of a well-known column in an evening paper, who was inspired last year to comment on these particular remarks of my own, observed that it was unnecessary to take such iconoclastic proposals seriously, since Cambridge mathematicians were very unlikely to be disturbed by the criticisms of an Oxford man. Perhaps, then, I had better begin by stating that the Mathematical Tripos is an institution of which I have an extensive and intimate knowledge. It is true that I have not taken any part in it during the last six years; but I was a candidate in both parts of it, I took my degree on it, I have examined in it repeatedly under both the old regulations and the new, and, when the old order of merit was abolished in 1910, I was a secretary of the committee which forced this and other changes through a reluctant Senate. I am not then a mere jealous outsider, itching to destroy an institution which I cannot comprehend, but a critic perfectly competent to express an opinion on a subject which I happen to know unusually well.

The second possible misapprehension which I am anxious to remove is decidedly more important. It is possible that some of you may have come here expecting me to deliver a general denunciation of examinations; and if so I am afraid that I shall disappoint you. Denunciation of examinations, like denunciation of lectures, is very popular now among educational reformers, and I wish to say at once that most of what they say, on the one topic and on the other, appears to me to be little better than nonsense. I judge such denunciations, naturally, as a mathematician; and it has always seemed to me that mathematics among all subjects is, up to a point, the subject most obviously adapted to teaching by lecture and to test by examination. If I wish to teach twenty pupils, for example, the exponential theorem, the product theorem for the sine, or any of the standard theorems of analysis or geometry, it seems to me that by far the best, the simplest, and the most economical course is to assemble them in a lecture-room and explain to them collectively the essentials of the proofs. It seems to me also that, if I wish afterwards to be certain that they have understood me, the obviously sensible way of finding out is to ask them to reproduce the substance of what I said, or to apply the theorems which

19. The Case against the Mathematical Tripos

I proved to simple examples. In short, up to a point, I believe in formal lectures, and I believe also in formal examinations.

There are in fact certain traditional purposes of examinations, purposes for which they always have been used, and for which they seem to me to be the obvious and the appropriate instrument. There are certain qualities of mind which it is often necessary to test, and which can be tested by examination much more simply and more effectively than in any other way. If a teacher wishes to test his pupils' industry, for example, or their capacity to understand something he has told them, something perhaps of no high order of difficulty, but difficult enough to require some little real intelligence and patience for its appreciation, it seems to me that his most reasonable course is to subject them to some sort of examination. Examinations have been used in this manner, from time immemorial, in every civilised country; there are, in England, quite a number of large, elaborately organised, and, so far as I can judge, quite sensibly conducted examinations of this type; and with such examinations I have no sort of quarrel.

There are, however, in England, and, so far as I know, in no other country in the world, a number of examinations, of which the Mathematical Tripos at Cambridge, and Greats at Oxford, have been the outstanding examples, which are of quite another type, and which fulfil, or purport to fulfil, quite different and very much more ambitious ends. These examinations originate in Oxford and Cambridge, and are found in their full development there only, though they have been copied to a certain extent by our modern universities. They are described as "honours" examinations, and pride themselves particularly on their traditions and their "standards." To these examinations are subjected a heterogeneous mass of students of entirely disparate attainments, and the examination professes to sort out the candidates and to label them according to the grade of their abilities. Thus in the old mathematical Tripos there were three classes, each arranged in order of merit, while in the new there are three classes and two degrees of marks of special distinction. It is evident that such an examination is not content with fulfilling the ends which I have admitted that an examination can fulfil so well; it is not, and prides itself that it is not, merely a useful test of industry, intelligence, and comprehension. It purports to appraise, and it must be admitted that to some extent, though very imperfectly, it does appraise, higher gifts than these. A "b^*" in the Tripos, or a first in Greats, is taken to be, and in a measure is, an indication of a man quite outside the common run. It is these examinations and these only, these examinations with reputations and standards and traditions, which seem to me mistaken in their principle and useless or damaging in their effect, and which I would destroy if I had the power. An examination can do little harm, so long as its standard is low.

I suppose that it would be generally agreed that Cambridge mathematics, during the last hundred years, has been dominated by the Mathematical Tripos in a way in which no first-rate subject in any other first-rate university has ever been dominated by an examination. It would be easy for me, were the fact disputed, to justify my assertion by a detailed account of the history of the Tripos, but this is unnecessary, since you can find an excellent account, written by a man who was very much more in sympathy with the Tripos than I am, in Mr. Rouse Ball's *History of Mathematics in Cambridge*. I must, however, call your attention to certain rather melancholy reflections which the history of Cambridge mathematics suggests. You will understand that when I speak of mathematics I mean primarily pure mathematics, not that I think that anything which I say about pure mathematics is not to a great extent true of applied mathematics also, but merely because I do not want to criticise where my competence as a critic is doubtful.

Mathematics at Cambridge challenges criticism by the highest standards. England is a first-rate country, and there is no particular reason for supposing that the English have less natural talent for mathematics than any other race; and if there is any first-rate mathematics in England, it is in Cambridge that it may be expected to be found. We are therefore entitled to judge Cambridge mathematics by the standards that would be appropriate in Paris or Göttingen or Berlin. If we apply these standards, what are the results? I will state them, not perhaps exactly as they would have occurred to me spontaneously—though the verdict is one which, in its essentials, I find myself unable to dispute—but as they were stated to me by an outspoken foreign friend.

In the first place, about Newton there is no question; it is granted that he stands with Archimedes or with Gauss. Since Newton, England has produced no mathematician of the very highest rank. There have been English mathematicians, for example Cayley, who stood well in the front rank of the mathematicians of their time, but their number has been quite extraordinarily small; where France or Germany produces twenty or thirty, England produces two or three. There has been no country, of first-rate status and high intellectual tradition, whose standard has been so low; and no first-rate subject, except music, in which England has occupied so consistently humiliating a position. And what have been the peculiar characteristics of such English mathematics as there has been? Occasional flashes of insight, isolated achievements sufficient to show that the ability is really there, but, for the most part, amateurism, ignorance, incompetence, and triviality. It is indeed a rather cruel judgment, but it is one which any competent critic, surveying the evidence dispassionately, will find it uncommonly difficult to dispute.

19. The Case against the Mathematical Tripos

I hope that you will understand that I do not necessarily endorse my friend's judgment in every particular. He was a mathematician whose competence nobody could question, and whom nobody could accuse of any prejudice against England, Englishmen, or English mathematicians; but he was also, of course, a man developing a thesis, and he may have exaggerated a little in the enthusiasm of the moment or from curiosity to see how I should reply. Let us assume that it is an exaggerated judgment, or one rhetorically expressed. It is, at any rate, not a *ridiculous* judgment, and it is serious enough that such a condemnation, from any competent critic, should not be ridiculous. It is inevitable that we should ask whether, if such a judgment can really embody any sort of approximation to the truth, some share of the responsibility must not be laid on the Mathematical Tripos and the grip which it has admittedly exerted on English mathematics.

I am anxious not to fall into exaggeration in my turn and use extravagant language about the damage which the Tripos may have done, and it would no doubt be an extravagance to suggest that the most ruthless of examinations could destroy a whole side of the intellectual life of a nation. On the other hand it is really rather difficult to exaggerate the hold which the Tripos has exercised on Cambridge mathematical life, and the most cursory survey of the history of Cambridge mathematics makes one thing quite clear: the reputation of the Tripos, and the reputation of Cambridge mathematics stand in correlation with one another, and the correlation is large and negative. As one has developed, so has the other declined. As, through the early and middle nineteenth century, the traditions of the Tripos strengthened, and its importance in the eyes of the public grew greater and greater, so did the external reputation of Cambridge as a centre of mathematical learning steadily decay. When, in the years perhaps between 1880 and 1890, the Tripos stood, in difficulty, complexity, and notoriety, at the zenith of its reputation, English mathematics was somewhere near its lowest ebb. If, during the last forty years, there has been an obvious revival, the fortunes of the Tripos have experienced an equally obvious decline.

Perhaps you will excuse me if I interpolate here a few words concerning my own experience of the Tripos, which may be useful as a definite illustration of part of what I have said. I took the first part of the Tripos in 1898, and the second in 1900: you must remember that it was then the first part which produced wranglers and caught the public eye.

I am inclined to think that the Tripos had already passed its zenith in 1898. There had already been one unsuccessful attempt to abolish the order of merit, a reform not carried finally till 1910. When the first signs of decline might have been detected I cannot say, but the changes in the Smith's Prize examination, and the examination for Trinity Fellowships, must have been partly responsible, and these had been determined by dissertation for a considerable time. At any rate it was beginning to be recognised, by the younger dons in the larger colleges, and to some extent by undergraduates themselves, that the difference of a few places in the order of merit was without importance for a man's career. This, however, is comparatively unimportant, since it is less the examination itself than its effect on teaching in the university that I wish to speak of now.

The teaching at Cambridge when I was an undergraduate was, of course, quite good of its kind. There were certain definite problems which we were taught to solve; we could learn, for example, to calculate the potential of a nearly spherical gravitating body by the method of spherical harmonics, or to find the geodesics on a surface of revolution. I do not wish to suggest that the two years which I spent over the orthodox course of instruction—my second two years were occupied in a different way—were altogether wasted. It remains true that, when I look back on those two years of intensive study, when I consider what I knew well, what I knew slightly, and of what I had never heard, and when I compare my mathematical attainments then with those of a continental student of similar abilities and age, or even with those of a Cambridge undergraduate of to-day, it seems to me almost incredible that anyone not destitute of ability or enthusiasm should have found it possible to take so much trouble and to learn no more. For I was indeed ignorant of the rudiments of my profession. I can remember two things only that I had learnt. Mr. Herman of Trinity had taught me the elements of differential geometry, treated from the kinematical point of view; this was my most substantial acquisition, and I am grateful for it still. I had also picked up a few facts about analysis, towards the end of those two years, from Prof. Love. I owe, however, to Prof. Love something much more valuable than anything he taught me directly, for it was he who introduced me to Jordan's *Cours d'analyse*, the bible of my early years; and I shall never forget the astonishment with which I read that remarkable work, to which so many mathematicians of my generation owe their mathematical education, and learnt for the first time as I read it what mathematics really meant.

It has often been said that Tripos mathematics was a collection of elaborate futilities, and the accusation is broadly true. My own opinion is that this is the inevitable result, in a mathematical examination, of high standards and traditions. The examiner is not allowed to content himself with testing the competence and the knowledge of the candidates; his instructions are to provide a test of more than that, of initiative, imagination, and even of some sort of originality. And as there is only one test of originality in mathematics, namely the accomplishment of original work, and as it is useless to ask a youth of twenty-two to perform original research under examination conditions, the examination necessarily degenerates into a kind of game, and instruction for it into initiation

19. The Case against the Mathematical Tripos

into a series of stunts and tricks. It was in any case certainly true, at the time of which I am speaking, that an undergraduate might study mathematics diligently throughout the whole of his career, and attain the very highest honours in the examination, without having acquired, and indeed without having encountered, any knowledge at all of any of the ideas which dominate modern mathematical thought. His ignorance of analysis would have been practically complete. About geometry I speak with less confidence, but I am sure that such knowledge as he possessed would have been exceedingly one-sided, and that there would have been whole fields of geometrical knowledge, and those perhaps the most fruitful and fascinating of all, of which he would have known absolutely nothing. A mathematical physicist, I may be told, would on the contrary have received an appropriate and an excellent education. It is possible; it would no doubt be very impertinent for me to deny it. Yet I do remember Mr. Bertrand Russell telling me that he studied electricity at Trinity for three years, and that at the end of them he had never heard of Maxwell's equations; and I have also been told by friends whom I believe to be competent that Maxwell's equations are really rather important in physics. And when I think of this I begin to wonder whether the teaching of applied mathematics was really quite so perfect as I have sometimes been led to suppose.

I remember asking another friend, who was Senior Wrangler some years later, and has since earned a very high reputation by research of the most up-to-date and highbrow kind, how the Tripos impressed him in his undergraduate days, and his reply was approximately as follows. He had learnt a little about modern mathematics while he was still at school, and he understood perfectly while he was an undergraduate, as I certainly did not, that the mathematics he was studying was not quite the real thing. But, he continued, he regarded himself as playing a game. It was not exactly the game he would have chosen, but it was the game which the regulations prescribed, and it seemed to him that, if you were going to play the game at all, you might as well accept the situation and play it with all your force. He believed—and remember, if you think him arrogant, that his judgment was entirely correct—that he could play that game at least as well as any of his rivals. He therefore decided deliberately to postpone his mathematical education, and to devote two years to the acquisition of a complete mastery of all the Tripos techniques, resuming his studies later with the Senior Wranglership to his credit and, he hoped, without serious prejudice to his career. I can only add—lost as I am in hopeless admiration of a young man so firmly master of his fate—that every detail of these precocious calculations has been abundantly justified by the event.

I feel, however, that I am laying myself open at this point to a challenge which I shall certainly have to meet sooner or later, and which I may as well deal with now. It will be said—I know from sad experience that such things are always said—that I am applying entirely wrong criteria to what is after all an examination for undergraduates. I shall be told that I am assuming that the principal object of the Cambridge curriculum is to increase learning and to encourage original discovery, and that this is false; that learning and research are admirable things, but that a great university must not allow itself to be overshadowed by them; and that, in short, a German professor of mathematics, however universal his reputation and profound his erudition, is not necessarily the noblest work of God. Indeed, at this point I seem to hear the voice of my opponent grow a little louder, as he points out to me that I am entirely misconceiving the function of an English university, that the universities of England are not at all intended as machines for the generation of an infinite sequence of professors, but as schools for the development of intellect and character, as training grounds of teachers, civil servants, statesmen, captains of industry, and proconsuls, in short as nurseries where every young Englishman may learn to add his quotum to the fulfilment of the destinies of an imperial race. I wish very heartily, I confess, that I was not going to be told all this, but I know very well that it is coming, for have I not heard it all a hundred times already, and did we not hear it all in 1910, from all the Justices who had been wranglers in their day?

Perhaps, however, I shall not be wasting your time entirely if I occupy a few minutes in an attempt to examine this indictment as dispassionately as I can. I find it very difficult to believe that most of the quite considerable body of quite intelligent people who continue to use this kind of language at the present day, and to turn it to the defence of our present university education, can have considered at all coolly some of the implications of what they say. On the other hand I recognise that it is a good deal easier to laugh at these people than to refute them, and that, if I were to attempt a reasoned reply to their contention, considered as a general principle to guide us in the construction of an educational system, then I should have a long and tiresome argument before me.

Fortunately, this is unnecessary. We are not now discussing educational systems generally, but the merits of a particular examination. We have not to undertake a general defence of mathematics and the position which is at present allowed to it in education, or to repel the very formidable onslaught which might be directed against it by Philistinism pure and simple. You and I and the Justices are after all agreed in wanting to see some sort of education in higher mathematics, and differ only in the kind of mathematical education which we prefer. The question is merely whether it is possible to defend the

19. The Case against the Mathematical Tripos 259

Mathematical Tripos on these lines, and we can appeal here, I think, to the method of *reductio ad absurdum*.

I have already put forward one test of a mathematical education, namely that it should produce mathematicians, as "mathematician" is understood by the leading mathematicians of the world; and this test, whatever its defects may be, has one merit at any rate, namely that it is clear and sharp and easily applied. It is also a test to which I suppose that everybody would agree in attaching *some* degree of importance, since it must be extraordinarily difficult for any English mathematician to maintain that it is of no importance whatever whether English mathematics be good or bad. The question therefore is not of the validity of the test, but only of its relative importance.

Now there is one obvious difference between my test, which I will call for shortness the professional test, and the slightly more orotund test which I have tried to state in general terms. My test has certainly this advantage, that I am testing a mathematical education as a means to one of the ends which a mathematical education may reasonably be expected to secure, and which it is hardly possible to secure in any other way. When, on the other hand, we attempt to test a training in higher mathematics, the highest such training the country offers, by its effects generally on the intelligence and character of those who submit to it, we are at once confronted with a question which is obviously more fundamental, whether intelligence and character of the type at which we aim are really developed very effectively by a training in higher mathematics. And as we are all mathematicians here, we need not indulge in humbug about it. We know quite well that the answer is No.

It is hardly likely that anybody here will accuse me of any lack of devotion to the subject which has after all been the one great permanent happiness of my life. My devotion to mathematics is indeed of the most extravagant and fanatical kind; I believe in it, and love it, and should be utterly miserable without it, and I have never doubted that, for any one who takes real pleasure in it and has a genuine talent for it, it is the finest intellectual discipline in the world. I believe also that a fair knowledge of mathematics is, even for those who have no pronounced mathematical talent, extremely useful and extremely stimulating, and that it should be part of the ordinary intellectual capital of all intelligent men. I am prepared indeed to go further, since I believe that a very large proportion of students abandon mathematics merely because it is often very badly taught, and might push their mathematical studies a good deal further than they do at present with very great profit to themselves. But I do not believe for a moment, and I do not believe that the majority of competent mathematicians believe, that the intensive study of higher mathematics, whether it be understood as it would be in a foreign university, or whether it be understood as it has in the past been understood in the Mathematical Tripos, forms a good basis of a general education. I am not at all sure that, among all possible subjects which might be selected as special courses of study, for an intelligent young man of no particular talent, mathematics is not the worst. Indeed, I think that this is being gradually recognised both by teachers of mathematics and by students themselves, and that it is for this reason that the Mathematical Tripos is more and more becoming, and rightly becoming, the special preserve of professional mathematicians. And if this be so, then surely it is quite obviously futile to judge the Tripos by anything but a professional standard.

It seems to me, then, that the opponents of the professional standard are committed from the beginning to a very paradoxical position, and yet it seems— such is the attraction of a paradox—that they are actually dissatisfied with its already sufficiently serious difficulties and determined to surround it by still more fantastic entrenchments. For they generally go on to maintain that mathematics may indeed be made the finest of intellectual disciplines, but only if it is taught in a manner which ignores or rejects every development of recent years. It will teach you to think, so long as you are not allowed to think quite correctly; it will widen your interests and stimulate your imagination, so long as you are carefully confined to problems in which mathematicians have lost interest for fifty years. In a word, the mathematics of the amateur is all right, and that precisely because it is so much more than a little wrong, but if we once allow mathematics to be dominated by the professionals, that is to say by the men who live in the subject and are familiar with its vital developments, then its energy will be sapped and its educational efficacy destroyed. And of all insane paradoxes, surely, this is one of the most portentous.

I have told you already that I am not much of a believer in the general educational efficacy of a specialised mathematical training. I do not believe that it is possible to build a character or an empire on a foundation of mathematical theories; but surely it must be still more impossible to build either on a foundation of Tripos problems. If I were compelled to undertake so crazy an enterprise, I would select the true theorems rather than the false, the fundamental facts of mathematics rather than its trivial excrescences, the problems which are alive to-day rather than those which perished in the mid-Victorian era.

I would suggest to you, then, that, when you have next to listen to the mathematical reactionary who laments the good old days, if you doubt your competence to judge for yourself the merits of his complaints, you should apply to what he says Hume's test of the greater improbability. It does not seem very likely that the modern experts are all wrong, but it is quite possible. It is also

19. The Case against the Mathematical Tripos

possible that the times have really left a conservatively-minded mathematician a little bit behind; that his lectures and his text-books have run out of date; that there is a good deal in modern mathematics which he finds it too great an effort to master; and that it gives him a good deal less trouble to abuse the modern tendencies than to repair the gaps in his own mathematical equipment. This also is, of course, extremely improbable; but you must ask yourself which is the greater improbability of the two.

I do not propose to waste further time on the discussion of this question: in what more I have to say about the Tripos I shall adopt a frankly professional view. I shall judge the Tripos by its real or apparent influence on English mathematics. I have already told you that in my judgment this influence has in the past been bad, that the Tripos has done negligible good and by no means negligible harm, and that, so far from being the great glory of Cambridge mathematics, it has gone a very long way towards strangling its development. There are further questions to consider. We may ask in the first place, if it be granted that what I have said about the past is roughly true, how far have things improved? Is it not true already that the Tripos means a great deal less, and English mathematics appreciably more, than forty years ago, and is it not extremely likely that, even if there be no further radical changes, this process will continue? Then, if we are not content to answer this question by a simple affirmative and leave it there, we may ask what really are the fundamental faults of an examination on the Tripos model, and whether it is not possible to make less drastic suggestions for its improvement.

I began my address with what was to a certain extent a defence of examinations. I said that under certain conditions I believed in examinations, that is to say in examinations of a sufficiently lowly type, which do not profess to be more than a reasonable test of certain rather humdrum qualities. The phrases which I used were vague, and I ought no doubt to attempt to define my own standard a little more precisely. This is naturally not quite easy, but I will risk some sort of definition. I should say, roughly, that the qualities which I have in mind—reasonable industry, reasonable intelligence, reasonable grasp—would be about sufficient to carry a candidate, in any of the orthodox Oxford or Cambridge examinations, into a decent second class. Beyond that, I do not believe in recognising differences of ability by examination.

I said this here last year, and I was at once challenged. I was asked, whatever could you do, if you could not tell the quality of a man by looking at his examination record? I wonder whether my questioner realised that these elaborate honours examinations, so far from being one of the fundamental necessities of modern civilisation, are a phenomenon almost entirely individual to Oxford and Cambridge, copied in a half-hearted fashion by other English universities, and, beyond that, having hardly a parallel in the world? Does Germany suffer from intellectual stagnation, because there are no honours examinations in her universities? Germany does not think in terms of firsts and seconds; we think in terms of them, so far as we do so think, and perhaps the practice is to some extent abating, merely because we have heard so much about them that they have become to us like bitter ale or eggs and bacon, and we have forgotten that we could get on quite happily without them.

I remember, if you will excuse my referring once more to the forgotten controversies of 1910, a curious saying of, I think, Mr. Justice Romer. Mr. Justice Romer circulated a flysheet to the Senate, deploring, of course, the proposal to abolish the Senior Wrangler. "What", he asked judicially, "is the function of the Tripos?", and he replied "Surely to examine and to make distinctions between young men". It would indeed be difficult to compress a larger quantity of vicious educational doctrine into a smaller number of words. The exactly opposite doctrine, that no distinctions should be made by examination except such as practical necessities may make imperative, is surely somewhere a little nearer to the truth.

Let us then consider, with the view of meeting the objection which was raised to what I said last year, whether the kind of distinctions made by the Mathematical Tripos are, in fact, of any particular practical utility. The evidence of ability provided by the Tripos is as follows. A candidate may obtain a first, second, or a third. He may obtain a mark—the "b" mark—of adequate knowledge of some special subject, or a higher mark—the "b^*" mark—of special distinction in that subject. The test case for us, and the only one I have time to consider now, is the highest mark. When a candidate has attained this mark, what has he gained?

In the first place; he has gained the natural feeling of satisfaction which everyone experiences when he is adjudged to have performed a definite task at least as well as anybody else. He will feel with pleasure and pride that the world is awarding honest work, and these entirely creditable feelings may spur him on to further effort. Has he gained anything of more tangible or permanent value?

A man who can attain the highest honours in the Tripos is generally a good enough mathematician to hope for a permanent academic career. How far will his "b^*" assist him along this career? Will anyone give him a position, a fellowship or a lectureship, on the strength of it? If he thinks that, he will be very quickly disillusioned.

It is possible that there are positions, in the junior grades of the teaching staffs of certain universities, which are sometimes filled on the strength of an examination record. I have never come across such a post myself, but it is probable enough that they exist. Academic positions are usually bestowed, not on examination record, but from personal knowledge or on the strength of private recommendations from competent people. I have taken part myself in many such appointments. When applications are invited, the testimonials submitted by candidates contain statements of their academic qualifications, and often of their performance in examinations, and it would be an exaggeration to say that such records are never referred to. There are usually a fair proportion of the candidates whose qualifications seem obviously below the standard expected, and a glance at their examination records often provides useful evidence in confirmation of this view. I do not remember any case of any other kind in which such a record has played any part in the decision, or has been referred to in the discussion by any member of the board of electors.

I suppose that this is generally understood, and that candidates for such positions are not usually under any delusion about the attention paid to the records which they submit. It may, however, be urged that an examination record of high distinction might often determine the decision if the post were of a less purely academic kind, for example if it were a mastership in one of the big public schools. It may be so, but I must confess that—at any rate in the particular case which we are considering—I am uncommonly sceptical about it. In the first place, people who obtain "b^*"s have usually scientific ambitions, and the last thing they want is a mastership in the most historic of public schools. It is not possible now for the richest or the most aristocratic school to obtain a really distinguished mathematician, even if it wants one, which of course in general it does not. Finally, even if the demand existed and the supply were there, the headmasters of the great public schools do not, so far as my experience has shown me, select their assistants in this way, but proceed much more in the spirit of a board of electors, though naturally in a more capricious and autocratic way.

My conclusion, then, is that the highest certificate of merit offered by the Tripos might just as well be scrapped, for all the influence it exerts on the careers of those who obtain it. I suppose, in fact, that the universities, and most of the other bodies in whose hands educational patronage is vested, have come in practice to very much the same conclusion as my own, that examinations are an admirable test of competence and industry, but ineffective and erratic as a test of any higher gift. The Government stands alone, so far as I know, in attaching a definite money value to an examination class, and even the Government stops short of rewarding the only mark which could plausibly pretend to be a mark of real distinction.

If such distinctions are in effect futile, why should we waste our time and our energies in making them, even if we were certain that they do no harm? If Einstein had taken the Mathematical Tripos, what would it matter what place he took? The world can recognise its Einsteins quickly enough when it gets the opportunity. If Einstein sits for an examination, let him have his degree, assuming that he can satisfy the examiners. What is the object of taking all these pains to make to-day, uncertainly and half-heartedly, distinctions which, if they have any foundation in reality, the world will make in its own much sharper fashion to-morrow?

I have left to the last the defence of the Tripos which I find myself most difficult to meet. It is a defence difficult to overcome, because it proceeds on what a chess-player would call close lines. This defence, which I have often heard from mathematicians whose judgment I value, and which I wish to treat with all respect, is simply this: that the examination has already been considerably relaxed, and that the effects of its relaxation can already be traced in a corresponding strengthening of English mathematics; that it may be indefensible in principle, but that the spirit of emulation which it fosters may conceivably do some slight positive good; and that, now that so many of its teeth have been successfully drawn, it is not very obvious that it does any very serious harm. This is undeniably the case for the Tripos in its strongest and sanest form.

I should admit that, up to a point, the defence is sound. I would go so far as to admit that the system now does little harm to men of what I may define roughly as fellowship standard. The truth is that the principles for which I am contending have been so far recognised that a man of this degree of ability need not really disturb himself very seriously about the examination. Such a man may pursue a course of serious mathematical study with every confidence that, unless he is wilfully neglectful, he can obtain without any intensive effort all such honours as the Tripos can bestow. The test, in short, does him no harm, because for him it has lost its meaning. This I admit, and, of course, I recognise that it is a very large admission, since it destroys a good deal of the case which could be urged so irresistibly against the Tripos thirty years ago on strictly professional grounds. It is no longer true that Cambridge is notably behind the times, or that its courses compare particularly unfavourably, at any rate in the subjects about which I am best qualified to judge, with those at any but the very best of continental universities. This, I think, Mr. Heywood did not recognise sufficiently; it was the only point in his address from which I particularly dissented. It is no longer true that the development of a decent school of English mathematics is being steadily throttled by the vices of its principal examination.

19. The Case against the Mathematical Tripos

We must recognise this and rejoice that things have moved so far, and if they have moved just because the glamour of the Tripos has faded, we shall only rejoice the more. We need not rush to the conclusion that the whole case against the Tripos has been destroyed. We have to think of its effects, not only on students of the highest class, but also upon teachers of mathematics in the university, and upon students a little less gifted than those of whom I have spoken hitherto. I am afraid that it is still true that mathematical teaching is hampered very seriously by the examinations, both in Cambridge and in Oxford, where the system is different in detail, but in essentials the same.

In the first place, it is still true that a large proportion of students, either wilfully, because they exaggerate the importance of the examination, or from ignorance, because they have never heard of anything better, or (and I am afraid that this is the most common explanation) because they are driven to it by tutors who have to justify themselves in the eyes of college authorities greedy for firsts, for one or other of these reasons allow their mathematical education to be stunted by absorption in examination technique. They spend hour after hour, which ought to be devoted to lectures or reading, in working through examination papers, or the collections of problems in which English text-books are so rich, exhausting themselves and their tutors in the struggle to turn a comfortable second into a marginal first. It is possible that the effect of all this mistaken exertion is more directly damaging to the tutors than to the students themselves; but a pupil cannot draw much inspiration from a tutor who is always tired, and there is hardly a tutor in Oxford, and not very many in Cambridge, who has not about twice as much teaching as any active mathematician should be asked to undertake. A professor at Oxford or Cambridge is very much his own master, but even a professor may be handicapped very seriously in his teaching by the recollection of the syllabus of the schools. It is often very hard to ask your pupils to go on listening to you when you know that what you tell them will gain them no credit in an examination to which they attach enormous importance and over which you have practically no control. I should always like to ignore the examination completely, and often summon up the courage to do so for a while, only to be pulled up short a few weeks later by the thought that after all it is hardly fair.

Whatever, then, may be said about the improvement of Cambridge teaching, and however much the dominion of the Tripos may have abated of its rigours of thirty years ago, I adhere to the view which I expressed to you last year, that the system is vicious in principle, and that the vice is too radical for what is usually called reform. I do not want to reform the Tripos, but to destroy it. And if you ask me whether the Tripos is a peculiar case, or whether what I have said applies to all other high-grade honours examinations, I can only answer that, so far as I can see, it does. The Tripos is the worst case. It is the oldest examination and the most famous, and generally the most strongly entrenched; and mathematics is a subject in which it is particularly easy to examine ferociously, so that the evils of the system stand out here in the clearest light. But, of course, the greater part of what I have said about the Tripos could be applied with almost equal force to Greats.

I wish, then, to abolish the Tripos, and as I know perfectly well that neither I nor anyone else will succeed in doing so, since the practical difficulties would be so serious, and the force of tradition is so terribly strong, I may reasonably be asked what seem to me the best practical steps for a more moderate reformer to take. You will probably have inferred from my remarks that I am not prepared with any very illuminating suggestions. I could, of course, suggest many changes of detail, both in the schedules and in the conduct of the examination; but my suggestions would be comparatively unimportant; and I should not be prepared to expend my energies in pressing them, since they would all be inspired by the same ideal, and that an ideal whose realisation will no doubt remain hopeless for many years. Indeed I am afraid that my advice to reformers might sound like a series of stupid jokes. I should advise them to let down the standard at every opportunity; to give first classes to almost every candidate who applied; to crowd the syllabus with advanced subjects, until it was humanly impossible to show reasonable knowledge of them under the conditions of the examination. In this way, in the course of years, they might succeed in corrupting the value of the prizes which they have to offer, and in all probability time would do the rest.

20

The Mathematician on Cricket

C. P. Snow

(*Saturday Book*, edited by Leonard Russell, Hutchinson, London, 1948, pp. 65–73)*

From the Editors

Hardy was widely known to be an ardent fan of the game of cricket, following scores and commentaries as the cricket season moved each year from continent to continent, and when there were no cricket scores he had to make do with American baseball or even football. Here we read the details from the popular novelist and sometimes cricketer, C. P. Snow, who wrote the Introduction to Hardy's A Mathematician's Apology.

Above all, G. H. Hardy was a man of genius. He was one of the great pure mathematicians of the world: in his own life-time he altered the whole course of pure mathematics in this country. He was also a man whose intelligence was so brilliant, concentrated, and clear that by his side anyone else's seemed a little muddy, a little pedestrian and confused. No one ever spoke to him for five minutes without feeling that, whatever genius means, here was one born with it. And no one ever spoke to him for five minutes—not even serious-minded Central European mathematicians—without hearing a remark about the game of cricket. Others may have gained as much delight from cricket as he did; no one can possibly have gained more. His creative mathematics was, as he wrote

---

* Reprinted with permission of Curtis Brown Group Ltd, London.

himself in *A Mathematician's Apology*, the one great sustained happiness of his life; but cricket, from his childhood until he died in 1947 at the age of seventy, was his continual refreshment.

It was to that fact that I owed my friendship with him. I remember vividly the first time we met. He had just returned to Cambridge to occupy the Sadleirian chair, and was dining as a guest in Christ's. He was then in his early fifties, and his hair was already grey, above skin so deeply sunburned that it stayed a kind of Red Indian bronze. His face was beautiful—with high cheek bones, thin nose, spiritual and austere but capable of dissolving into convulsions of internal, gamin-like amusement. He had opaque brown eyes, bright as a bird's—a kind of eye not uncommon among those with a gift for conceptual thought. I thought that night that Cambridge was a town where the streets were full of unusual and distinguished faces, but even there Hardy's could not help but stand out.

As we sat round the combination-room table after dinner, someone said that Hardy wanted to talk to me about cricket. I had only been elected a year, but the pastimes of even the very young fellows were soon detected, in that intimate society. I was taken to sit by him—never introduced, for Hardy, shy and self-conscious in all formal actions, had a dread of introductions. He just put his head down, as if it were in a butt of acknowledgment, and without any preamble whatever began:

'You're supposed to know something about cricket, aren't you?'

Yes, I said, I knew something.

Immediately he proceeded to put me through a moderately stiff viva. Did I play? What did I do? I half-guessed that he had a horror of persons who devotedly learned their Wisden's backwards but who, on the field, could not distinguish between an off-spinner and short-leg. I explained, in some technical detail, what I did with the ball. He appeared to find the reply partially reassuring, and went on to more tactical questions. Whom should I have chosen as captain for the last test a year before (in 1930)? If the selectors had decided that Snow was the man to save us, what would have been my strategy, and tactics? ('You are allowed to act, if you are sufficiently modest, as non-playing captain.') And so on, oblivious to the rest of the table. He was quite absorbed. The only way to measure someone's knowledge, in Hardy's view, was to examine him. If he had bluffed and then wilted under the questions, that was his look-out. First things came first, in that brilliant and concentrated mind. It was necessary to discover whether I should be tolerable as a cricket companion. Nothing else mattered. In the end he smiled with immense charm, with child-like openness, and said

20. The Mathematician on Cricket

that Fenner's might be bearable after all, with the prospect of some reasonable conversation.

Except on special occasions, he still did mathematics in the morning, even in the cricket season, and did not arrive at Fenner's until after lunch. He used to walk round the cinderpath with a long, loping, heavy-footed stride (he was a slight spare man, physically active), head down, hair, tie, sweaters, papers, all flowing, a figure that caught everyone's eye. ('There goes a Greek poet, I'll be bound,' once said some cheerful farmer as Hardy passed the scoreboard.) He made for his favourite place, at the Wollaston Road end, opposite the pavilion, where he could catch every ray of sun—for he was impatient when any moment of sunshine went by and he was prevented from basking in it. In order to deceive the sun into shining, he brought with him, even on a fine May afternoon, what he called his 'anti-God battery.' This consisted of three or four sweaters, an umbrella belonging to his sister, and a large envelope containing mathematical manuscripts, such as a Ph.D. dissertation, a paper which he was refereeing for the Royal Society, or some Tripos answers. He would then explain, if possible to some clergyman, that God, believing that Hardy expected the weather would change and give him a chance to work, counter-suggestibly arranged that the sky should remain cloudless.

There he sat. To complete his pleasure in a long afternoon watching cricket, he liked the sun to be shining and a companion to join in the fun. But he was never bored by any cricket in any circumstances; he was fond of saying that no one of any vitality, intellectual or other, should know what it was like to be bored. As for being bored at cricket, that was manifestly impossible. He had watched the game since, as a child, he had gone to the Oval in the great days of Surrey cricket, with Tom Richardson, Lockwood, Abel in their prime: as a schoolboy at Winchester, an undergraduate at Trinity, with W. G. in his Indian summer and Ranjitsinjhi coming on the scene: through Edwardian afternoons, when Hardy was already recognized as one of the mathematicians of the age: in the Parks at Oxford after the first war, the serenest time of his whole life: and now in the 'thirties at Fenner's, his delight in the game as strong as ever.

His first interest was technical. He was secretly irritated when people assumed that he knew every record in cricket history; in fact, his book-knowledge was considerable but in no way remarkable; it was greater than mine, but less than that of several acquaintances. He had been a creative person all his life, without the taste for that kind of recondite scholarship. He was far more occupied with the backswing of the man at that moment batting, or the way in which one could make a legbreak dip, or the difference between the hooking mechanism of Bradman and Sutcliffe. His interest was primarily a games player's, and not a scholar's. He himself had an unusually fine eye, and when well into the sixties could still offdrive or make an old-fashioned square leg sweep with astonishing certainty. Asked who lived the most enviable of lives, he would have said, quite simply, a creative mathematician—for he knew that no one could have led a life more creatively satisfying than his own. His second choice would not have been a scientist (for science he had surprisingly little sympathy): I think he would have said instead a first-rate creative writer. And I am sure his third choice would have been a great batsman.

His second interest was in tactics, in that whole area of small decisions—about bowling changes, field-placings, batting orders, and the like—in which cricket is so rich. This was an interest which in exciting matches sometimes occupied him entirely. He used to say that P. G. H. Fender's famous passage about the last day of the Fourth Test in the 1928–9 series was the finest expression of sheer intellectual agony in any language.

The point was this: there was nothing in the game, Australia needed about 100 to win with 5 wickets to fall, and J. C. White was the only effective bowler on the English side. White was bowling, of course, slow left arm, pitching on the leg and middle and going away a bit. Chapman had set a field with a fine leg to stop the glance for a single, but with no short leg to stop the push. The Australians kept making these safe singles. Fender's agony grew. Why could

20. The Mathematician on Cricket

not the captain see that, if one cannot block both places, one should in all sanity block the safe push and leave the glance open, when it is the riskier of the two strokes against a bowler as accurate as White? Fender was, naturally, worried about the result as the singles kept creeping up: but chiefly he was dismayed by anyone missing such a pure intellectual point. At last lunch-time came. It is not stated in the book, but one imagines a desperate piece of lucid exposition from Fender. After lunch there was a short leg instead of a fine leg. Fender settled down in intellectual-content; in the description, it comes almost as an anti-climax when England win by 11 runs.

With Hardy, it would have been just the same. I have sat by him, and seen him distressed for half-an-hour over some similar tactical blindness. How was it possible for a sentient human being to miss such a clear, simple, beautiful point?

Everything Hardy did was light with grace, order, a sense of style. To those competent to respond, I have been told, his mathematics gave extreme aesthetic delight; he wrote, in his own clear and unadorned fashion, some of the most perfect English of our time (of which samples can be read in *A Mathematician's Apology* or the preface to *Ramanujan*). Even his handwriting was beautiful; the Cambridge University Press had the inspiration to print a facsimile on the dust-jacket of the *Apology*. It was natural that he should find much formal beauty in cricket, which is itself a game of grace and order. But he found beauty after his own style, not anyone else's. He was deeply repelled by all the 'literary' treatment of the game; he did not want to hear about white flannels on the green turf, there was nothing he less wanted to hear; he felt that Mr. Neville Cardus, despite gifts which Hardy was too fair-minded to deny, had been an overwhelmingly bad influence on the cricket writing of the last twenty years. Fender, analytical, informed, alive with intellectual vitality and a nagging intellectual integrity, was by a long way Hardy's favourite cricket writer. Fender has an involved and parenthetical style, which bears a faint family resemblance to Proust and Henry James. Hardy commented that all three were trying to say genuinely difficult things, but that Fender, like Proust and unlike Henry James, had within his chosen field an instinct for the essential.

Technique, tactics, formal beauty—those were the deepest attractions of cricket for Hardy. But there were two others, which to many who have sat within earshot must have seemed more obvious. Of these minor attractions, one was his relish for the human comedy. He would have been the first to disclaim that he possessed deep insight into any particular human being. That was a novelist's gift; he did not pretend to compete. But he was the most intelligent of men, he had lived with his eyes open and read much, and he had obtained a good generalized sense of human nature—robust, indulgent, satirical, and utterly

free from moral vanity. He was spiritually candid as few men are, and he had a mocking horror of pretentiousness, self-righteous indignation, and the whole stately pantechnicon of the hypocritical virtues. Now cricket, the most beautiful of games, is also the most hypocritical: one ought to prefer to make 0 and see one's side win than make 100 and see it lose (J. B. Hobbs, like Hardy a man of innocent candour, remarks mildly that he never managed to feel so). Such statements were designed to inspire Hardy's sense of fun, and in reply he used to expound, with the utmost solemnity, a counter-balancing series of maxims.

'Cricket is the only game where you are playing against eleven of the other side and ten of your own.'

'If you are nervous when you go in first, nothing restores your confidence so much as seeing the other man get out.'

'After a pogrom, the Freshman's Match is the best place to see human nature in the raw.'

No match was perfect unless it produced its share of the human comedy. He liked to have his personal sympathies engaged; and, if he did not know anyone in whom to invest either sympathy or antipathy, he proceeded to invent them. In any match at Fenner's, he decided on his favourites and non-favourites: the favourites had to be the underprivileged, young men from obscure schools, Indians, the unlucky and diffident. He wished for their success and, alternatively, for the downfall of their opposites—the heartily confident, the overpraised heroes from the famous schools, the self-important, those designed by nature to boom their way to success and moral certitude ('the large-bottomed,' as he called them: the attribute, in this context, was psychological).

20. The Mathematician on Cricket

So each match had its minor crises. 'The next epic event,' Hardy would say, 'is for Iftikar Ali to get into double figures.' And the greatest of the minor joys of cricket for him was the infinite opportunity for intellectual play. It happens that unlike any kind of football or the racket games, cricket is a succession of discrete events: each ball is a separate mark in the score-book: this peculiarity makes it much easier to describe, talk about, comment on, remember in detail. (Incidentally, it is this peculiarity which makes its great climaxes so intense). It could not have been better suited to the play of Hardy's mind.

'Cricket is a game of numbers,' said Hardy cheerfully to those who wanted to compare an innings to a musical composition, and asked them what was the maximum number of times the same integer could appear on the score-board at one instant in the innings. (The integer is 1; one batsman must have retired hurt: the score is 111 for 1, batsmen 1 and 11 each 11, bowlers 1 and 11, last player 11, caught by 11). As his professional life had been devoted to the theory of numbers, he could see something interesting in any score-board at any time.

Any newspaper in the cricket season had the same interest. Numbers, the structure of a score, the personal fate of clergymen, against whom he carried on an ironical private war: few things gave him more mischievous glee than to read that a clerical batsman had been run out. His great triumph in that direction, however, was a little different. It happened in a Gentlemen v. Players match at Lord's. It was early in the morning's play, and the sun was shining over the pavilion. One of the batsmen, facing the Nursery end, complained that he was unsighted by a reflection from somewhere unknown. The umpires, puzzled, padded round by the sight-screen. Motor-cars? No. Windows? None on that side of the ground. At last, with justifiable triumph, an umpire traced the reflection down—it came from a large pectoral cross resting on the middle of an enormous clergyman. Politely the umpire asked him to take it off. Close by, Hardy was doubled up in Mephistophelian delight. That lunch time, he had no time to eat; he was writing post-cards (such post-cards as came to many out of the blue) to each of his clerical friends.

But in his war against clergymen, victory was not all on one side. On a quiet and lovely May evening at Fenner's, the chimes of six o'clock fell across the ground. 'It's rather unfortunate,' said Hardy with his usual candour, 'that some of the happiest hours of my life should have been spent within sound of a Roman Catholic Church.'

Sometimes, not often, his ball-by-ball interest flagged. Then he promptly demanded that we should pick teams—teams whose names began with HA (first wicket pair Hadrian and Hayward (T.)), LU, MO—the combinations of any of our friends round us, the all-time teams of Trinity, Christ's (first wicket pair Milton and Darwin, which takes a lot of beating), teams of humbugs, clubmen, bogus poets, bores.... Or he ordered: 'Mark that man,' and someone had to be marked out of 100 in the categories Hardy had long since invented and defined: STARK, BLEAK, 'a stark man is not necessarily bleak: but all bleak men without exception want to be considered stark,' DIM, OLD BRANDY, and SPIN. There were other categories which cannot be printed; of the five above, STARK, BLEAK, and DIM are self-explanatory, SPIN meant a subtlety and delicacy of nature that Hardy loved ('X may not be a great man, but he does spin the ball just a little all the time,') and OLD BRANDY was derived from a mythical character who said that he never drank anything but very old brandy.

So, by elaboration, Old Brandy came to mean a taste that was eccentric, esoteric, but just within the confines of reason. To say that one would rather watch Woodfull than any other batsman, would be a typical 'old brandy' remark. But one had to keep one's head in all these games with Hardy. Claiming Proust's novels as the best in the world's literature came within the permissible limits of old brandy: but a young man who did the same for *Finnegans Wake* was dismissed as an ass. 'Young men ought to be conceited, but they oughtn't to be imbecile,' Hardy grumbled afterwards.

Walking home after the close of play, he would maintain the same flow of spirits. At half-past six on a summer evening, Parker's Piece was crammed with boys' matches, square leg in one game dangerously near to cover in the next.

'I'll bet you sixpence that we see three wickets fall. Another sixpence that one chance is missed.' That bet meant that we had to keep a steady walking pace; as a rule, he wanted to stop and study any conceivable kind of game. He was too shy to offer to umpire, but if the boys invited him he settled down to an hour's entertainment.

Safely across the Piece, he would have some new concept, such as persuading me to stay in some dingy hotel. 'How much should I have to pay you to spend a night there? No, a pound is excessive.' Then the last excitement of the day's cricket, as he bought a local paper by the side of what was then the New Theatre. Stop press news and Surrey first: I can still hear his *cri de coeur*, some time in the middle 'thirties, when Lancashire were playing Surrey, and most of the present test team unknown. 'Washbrook 196 not out. Washbrook! Who the hell is *Washbrook?*'

The summer days passed. After one of the short Fenner's seasons, there was the University match; arranging to meet him in London was not always simple, for he had a profound suspicion of any mechanical contrivance such as a telephone. I do not think he ever used a fountainpen: while in his rooms at Trinity or his flat in St. George's Square, he used to say, in a disapproving and slightly sinister tone: 'If you *fancy yourself* at the telephone, there happens to be one in the next room.' His idea of communication was, if possible, to call in on foot: as a second line, to write a postcard.

Yet, punctually, he arrived at Lord's. There he was at his most sparkling, year after year. Surrounded by friends, men and women, he was quite released from shyness; he was the centre of all our attention, which he by no means disliked; and one could often hear the party's laughter from a quarter of the way round the ground. Having been a professor at both Oxford and Cambridge, he reserved the right to sympathize with either side; but in fact, except when his beloved friend John Lomas was batting for Oxford, his heart stayed more faithful to Cambridge than he liked to admit.

In those years, I used to go abroad soon after the University match, and to villages round the Mediterranean and Adriatic arrived post-cards in a beautiful hand, often mysteriously covered with figures. Those post-cards marked Hardy's August progress, Oxford, the Oval, Folkestone. Sometimes they contained nothing but a single sentence. 'How does N. F. Armstrong of your county hit the ball so hard without moving his feet, arms, or even apparently his bat?' 'Bradman is a whole class above any batsman who has ever lived: if Archimedes, Newton, and Gauss remain in the Hobbs class, I have to admit the possibility of a class above them, which I find difficult to imagine.' 'The half-mile from St. George's Square to the Oval is my old brandy nomination for the most distinguished walk in the world.'

In 1934 he sent me a large exercise book. On the left-hand pages he had written an over-by-over account of the fifth Test; on the right hand, he had spread himself in disquisitions on the players, cricket in general, human nature, and life. Maddeningly, I lost the book in a move during the war, or it would give a better picture of him than any second-hand account. He promised that, when he had finished completely with creative mathematics, he would do something in the same form, but on a more ambitious scale, as his last non-mathematical testament. It was to be called *A Day at the Oval*. It would have been an eccentric minor classic; but it was never written.

Almost up to the end I hoped that he would do it. His creative power left him, much later than with most mathematicians, but still too early: how harsh a deprivation it was for him, anyone can read in his *Apology*, which, for all its high spirits, is a book of intolerable sadness. His heart was failing: he took it stoically, but he had always been active, his enjoyments had been those of a young man until he was sixty, and he found it bitter to grow old. All this happened during the war: he hated war, not as we all do, but with a personal and desperate loathing. The world had gone dark for him, and, because of the war, there was not much cricket for him to watch, which would have been an amelioration, which would, at least for occasional afternoons, have made him gay again.

So he never wrote *A Day at the Oval*. In his last illness, in the summer and autumn of 1947, he thought of it again, but he could not make the effort. Yet cricket, during those last months of his life, was his chief, almost his only, comfort and interest. His sister read to him every scrap of cricket news that she could find. Until the end of the English season, there was plenty of material, but after that she had to fill in with World Series Baseball before the Indians started their tour in Australia. That was his final interest. I had left Cambridge some years before, but during those months I went to talk cricket with him as often as I could get away; in each visit, he liked to spend a few minutes discussing death, and then hear everything I could tell him about the latest cricket gossip. Edrich was a particular favourite of his, and he showed all the old delight when I brought the news that Edrich had, right on the post, passed Tom Hayward's record. The last conversation I had with Hardy was four or five days before he died: it was about Vinoo Mankad: was he, or was he not, an all-rounder of the Rhodes or Faulkner class?

It was in that same week that he told his sister: 'If I knew I was going to die to-day, I think I should still want to hear the cricket scores.'

Each evening that week before she left him, she read a chapter from a history of Cambridge cricket. One such chapter was the last thing he heard, for he died suddenly, in the morning.

21

Cricket for the Rest of Us

John Stillwell
February 13, 2015

Some mathematicians may be hazy about Hardy's mathematical work, but I suspect that many more will be baffled by his interest in cricket. That means you, mathematicians in the US and other countries where cricket is not played! So, in an effort to bridge the gap and to enable those unfamiliar with cricket to understand the previous article by C. P. Snow, I will attempt to explain the game of cricket in the language of baseball. As will be seen from the article, Hardy himself appreciated both games, because he wanted to know the World Series scores when no cricket news was available.

I am not fluent in baseball but hopefully my knowledge will suffice to roughly compare and contrast the two games. For a start, they are both games in which a ball is hit by a bat and runs are scored. We extend the analogy a bit further in the following table of roughly equivalent concepts.

Baseball	Cricket
bat	bat
ball	ball
run	run
out	out
inning	innings
batter	batsman
pitcher	bowler
catcher	wicketkeeper
fielder	fieldsman
ball (pitch outside the strike zone)	wide
home run	four or six
base	wicket
base hit	scoring stroke
diamond	pitch

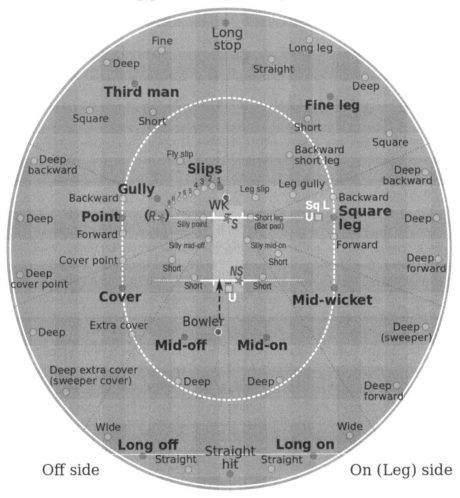

21. Cricket for the Rest of Us

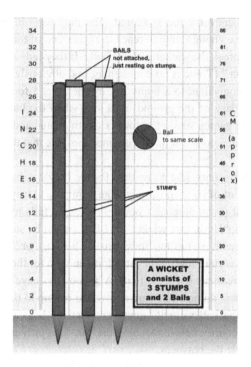

Of course, some of these equivalences are quite rough. For example, it is much easier to strike the ball with a cricket bat, which has a flat face about 4 inches wide. Because of this, and the fact that it is not compulsory to run, hundreds of runs can be scored. Instead of a diamond with four bases there is a 22-yard strip of turf called the "pitch" with a "wicket" at each end. The wicket consists of three cylindrical wooden stumps, resembling a little fence 28 inches high and 9 inches wide. In effect, there are only two bases, and the bases are loaded right up to the end of the innings, when just one batsman remains not out.[1] The accompanying diagrams from Wikimedia Commons show the layout of a cricket ground, with the many possible fielding positions, and what a wicket looks like.

There is a batsman at each end of the pitch: one faces the bowler, who delivers the ball from the opposite end, and the other stands ready to take his

---

[1] I guess the term "Innings" is supposed to mean the time period when the batsmen of one team are "in"; that is, not out. But don't ask me why the word ends in "s" when it is singular. In cricket each team has a "first innings" and a "second innings", not the nine innings that baseball has.

partner's place whenever a run is scored. The batsmen run only when they judge that they have enough time to reach the other end of the pitch before the fielding team can break the wicket with the ball.[2] If they judge badly, one of the batsmen is "run out." (There is no double play, and obviously no triple play.)

The nearest equivalent to a home run in cricket is a four (when the ball is hit to the boundary of the ground) or a six (when the ball is hit over the boundary). As the names suggest, a four is worth 4 runs and a six is worth 6 runs, but the batsmen do not actually run once they see that the ball will reach the boundary.

This no doubt sounds like a raw deal for the bowlers, but there are ways to get a batsman out other than run out. The commonest ways a batsman can get out are: caught (as in baseball), bowled (when he fails to prevent the ball hitting his wicket), "leg before wicket" or LBW (when he stops the ball from hitting his wicket with part of his body, usually his leg), or "stumped" (when he strays beyond a certain line on the pitch long enough for the wicketkeeper to take the ball and break the wicket with it). Also, bowlers have certain freedoms that baseball pitchers do not. They are allowed to run up to the delivery point, they can deliver the ball virtually anywhere that the bat can reach, and the ball is allowed to bounce before reaching the bat. This provides many ways to intimidate or puzzle the batsman.

A fast bowler can deliver the ball at over 90 mph—close to that of a baseball fastball—the only restraint being that the ball must be delivered with a straight arm (that is, "bowled" rather than thrown), bouncing it unpredictably towards the head or body.[3] If the batsman is hit, unlike in baseball, he gets no compensation, just a short break in play to regain his composure. In Hardy's day, batsmen did not even wear helmets. The slow bowler's weapon is spin, often cleverly disguised, so that the ball turns dramatically and unexpectedly as it bounces off the turf. On occasion a spin bowler can turn the ball a yard, getting round the batsman's body behind his back before hitting the stumps.

Because the cricket pitch is symmetric, and placed in the center of the ground, bowlers can (and do) operate from either end. After a bowler has bowled six deliveries from one end, another bowler bowls six from the other. These alternating sets of six are called "overs. " The alternation evens out wear on the pitch, and also compensates for asymmetric conditions such as wind or slope of the ground. (Amazingly, at the traditional home of cricket, the Lord's

---

[2] The wicket is deemed to be broken when at least one of the bails falls to the ground.
[3] One of Australia's renowned fast bowlers, Jeff Thomson, made a virtue of being unpredictable. He said: "If I don't know where the ball is going, I figure the batsman won't know either."

21. Cricket for the Rest of Us

ground in London, the pitch has an appreciable slope from one side to the other. Naturally, it is easier to turn the ball from the "high" side to the "low" side.)

The role of the wicketkeeper is similar to that of the catcher in baseball, but it is complicated by the nature of batting and bowling. There are no "foul balls" in cricket, so the wicketkeeper's job is to stop all balls that travel behind the wicket whether they have been hit or not (because runs can be scored even on a swing-and-a-miss if the wicketkeeper fails to collect the ball). Alone among the members of the fielding team, the wicketkeeper wears gloves. All other fieldsmen take the ball barehanded.

A cricket match can last for up to five days, with six hours of play on each day. This is perhaps the most incomprehensible aspect of the game to Americans, but to my mind it is not unlike waiting for a World Series to play out. At one time there was in fact no time limit on the so-called "test" matches—the top-level matches between teams of different nations. The 5-day time limit was introduced after a test match in South Africa in 1939 had to be abandoned after nine days of play, in order for the England team to catch the boat home.

Because of the time limit, many test matches since 1939 have ended in a "draw," when time runs out before the last innings is completed. This can be boring, but it can also produce a nail-biting finish, as a team hopelessly behind on the scoreboard fights to keep its last batsmen in. Some of the most thrilling test cricket of all time has occurred in heroic efforts to produce a draw—for example, with batsmen stopping fast balls with their chest to avoid hitting up catches.

The many nuances of the game call for great tactical skill on the part of team captains. The captain of the fielding side has to be aware of the strengths and weaknesses of opposing batsmen, and to choose his bowlers and place his fieldsmen accordingly.[4] As C. P. Snow tells us, Hardy was fascinated by the tactical side of the game, and annoyed when people assumed he would know all the records and averages, just because he was a specialist in the theory of numbers.

Nevertheless, I am sure that Hardy was well aware of cricket statistics and that he defined the "Hobbs class" and the "Bradman class" on that basis. Cricket, like baseball, is a game best appreciated against a background of past great performances, so the more one knows (or is informed, by commentators with a database at their fingertips) the more one enjoys a game. Even a dull game

---

[4] The different fielding positions have bizarre and amusing names such as long leg, short leg, square leg, slips, cover, silly mid off, and the like. They do not all have a rational explanation, but the diagram of fielding positions should help.

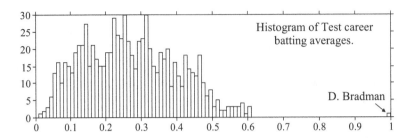

can become interesting if it breaks the record for the number of consecutive balls without a run being scored.

That brings me to the statistics behind the concepts of the "Hobbs class" and the "Bradman class." Jack Hobbs was an English batsman in the early 20th century, and one of the best of that or any other era. Perhaps the Hobbs class is similar to the class of .400 hitters in baseball, with the likes of Ty Cobb, Joe Jackson, and Ted Williams. Don Bradman was an Australian batsman whose career began in the late 1920s, as the career of Hobbs was coming to a close and Hardy had already formed his opinion that the "Hobbs class" was the highest level of batting. In the 1930s and 1940s, Bradman outclassed Hobbs to an extent that has no analogue in baseball. To get an idea of Bradman's ability, try to imagine a .600 hitter, because Bradman's batting average of 99.94 was more than 50% better than any other batsman.[5] That is why Hardy had to conceive the "Bradman class."

"It's not cricket"

The image of cricket conveyed by Snow—a thoughtful and civilized game, ideal for a quiet afternoon in an English university town—is part of its mystique. So too is the tradition of fair play and sportsmanship, which led to the expression "It's not cricket" being used throughout the cricket-playing world for any breach of fair and civil behavior. However, there have always been tensions in the game, leading to outbreaks of incivility. Some of these stem from the cultural clash between the traditional rivals England and Australia; some from a class divide

---

[5] This comparison is inexact, because the cricket batting average is defined as (total number of runs)/(number of times out), hence it has no upper bound in principle. However, the accompanying histogram should make it clear why Bradman was in a class of his own.

in English cricket itself, between "gentlemen" (who could afford to play cricket as a hobby) and "players" (who played cricket for a living).[6]

After one particularly torrid session of play, the Australian captain Bill Woodfull[7] famously said: "There are two teams out there. One is playing cricket and the other is not."

Both of these tensions came to a head in the 1930s, in the notorious "bodyline" series played in Australia. In this series, England was led by the aristocratic Douglas Jardine—a "gentleman," naturally—and his main weapon was the fast bowler, and "player," Harold Larwood. Jardine's strategy was to direct Larwood to frequently bounce fast balls towards the batsman's body, and to set the field to catch the likely deflections off the bat.

The series reverted to "playing cricket" in an unexpected way when Larwood broke a bone in his foot while running in to bowl. Jardine ordered him to complete his over anyway. Unable to run, Larwood was reduced to bowling harmless slow balls to the batsman, who happened to be Woodfull. But Woodfull was so incensed by Jardine's behavior that he refused to take advantage—simply blocking each ball until Larwood completed the over.

Is Woodfull's action still comprehensible to sports fans today? I don't know, but in Hardy's day *that* was cricket.

---

[6] The Gentlemen v. Players match, mentioned in Snow's article, was a highlight of the English cricket season from 1806 to 1962. A total of 274 of these matches were played, with Players winning 125 and Gentlemen 68. The matches were abolished, in 1963, when increasing professionalism in the game made the concept of "gentleman" virtually meaningless as batsmen tried to defend themselves. The strategy was very successful, as batsmen were often out if they managed to deflect the ball, and injured if they did not.

[7] Woodfull was a mathematics teacher who later became headmaster of Melbourne High School. I was a student there at the time, so I have his autograph—in my school report book.

22
A Mathematical Theorem about Golf
(*Mathematical Gazette* 29 (1945), pp. 226–227)*

From the Editors
As we saw earlier Hardy broke his rule about being exclusively a pure mathematician only once—the development in a brief letter in Science *of what came to be known as the Hardy-Weinberg Law in genetics. But perhaps we have here another example—a rather frivolous one—an analysis of a problem in the game of golf.*

It is usually held that steadiness, as against brilliancy, will tell more by strokes than by holes. It should follow that, if A is the steadier of two nearly equal players, B the more brilliant, and they are equal over a series of medal rounds, then B should have the advantage in a series of matches by holes. This seems to be the commonly accepted doctrine, but the evidence is inconclusive, since matches by holes between any given pair of leading golfers are comparatively rare. It may therefore be worth while to point out that a little mathematical analysis points, for what it is worth, to the opposite conclusion.

I construct a mathematical model as follows. Suppose that the course consists of 18 holes, all par fours (an immaterial simplification). Suppose that A is a completely mechanical player who does every hole in four (so that the conditions are similar to those of a bogey competition); and that B has equal chances x per stroke of making a *supershot* which gains a stroke or a *subshot* which costs one. It is plain that x cannot exceed $\frac{1}{2}$, and will be fairly small in any case at all corresponding to reality; that B will average par; and that the

---
* Reprinted with the kind permission of the publisher of *The Mathematical Gazette:* The Mathematical Association.

players will be equal over a series of medal rounds. According to the apparently accepted doctrine, B should win a long series of matches by holes.

To see the working of the model, suppose first that we may neglect terms of order x^2, *i.e.* consequences of B's playing more than one abnormal shot at the same hole. Then, to do a three, B must produce a supershot at one of his first *three* strokes, while he will take five if he makes a subshot at one of his first *four*. He will thus have a net expectation $4x - 3x$ or x of loss on the hole, and should lose the match, contrary to common expectation.

This approximation is too rough, but there is no difficulty in calculating the chances exactly. I find that B's chance of winning a hole is $3x - 9x^2 + 10x^3$, and his chance of losing $4x - 18x^2 + 40x^3 - 35x^4$, so that there is a balance

$$f(x) = x - 9x^2 + 30x^3 - 35x^4$$

against him. Mr. A. M. Binnie has plotted $f(x)$ for me. It increases to a maximum, about .037, for x about .09; and then decreases, vanishing for x about .37 and falling to $-\frac{3}{16}$ for $x = \frac{1}{2}$. There are two inflexions, for x about .16 and .27, but these merely flatten part of the curve and do not affect its appearance seriously. The model has, of course, lost all its plausibility by the time that x is near $\frac{1}{2}$; a player with $x = \frac{1}{2}$ would be a "completely erratic" player, *all* of whose strokes are abnormal, and the conclusion that such a player would win, paradoxical as it may seem, need not disturb us. But for smaller x, and in particular for x about $\frac{1}{10}$, the model seems fairly reasonable.

If experience points the other way—and I cannot deny it, since I am no golfer—what is the explanation? I asked Mr. Bernard Darwin, who should be as good a judge as one could find, and he put his finger at once on a likely flaw in the model. To play a "subshot" is to give yourself an *opportunity* of a "supershot" which a more mechanical player would miss: if you get into a bunker you have an opportunity of recovering without loss, and one which you are naturally keyed up to take. Thus the less mechanical player's chance of a supershot is to some extent automatically increased. How far this may resolve the paradox, if it is one, I cannot say, and changes in the model make it unpleasantly complex.

I add one word about possible testing of the model. There are few data about matches between, say, Hagen and R. T. Jones. But the model, as I said, is more like that of a bogey competition, and there must be abundant information about them. If a player averages bogey, does he win or lose when he plays against bogey by holes? There should be plenty of data for testing that.

23

Mathematics in War-Time

(*Eureka* 1–3 (1940), pp. 5–8)*

From the Editors

The Cambridge University Mathematics Society of undergraduates was known as the Archimedeans and its journal was called, appropriately, Eureka. *These reflections on the role of mathematicians during wartime were prescient and they appeared later that same year in* A Mathematician's Apology.

The editor asked me at the beginning of term to write an article for *Eureka*, and I felt that I ought to accept the invitation; but all the subjects which he suggested seemed to me at the time quite impossible. "My views about the Tripos"—I have never really been much interested in the Tripos since I was an undergraduate, and I am less interested in it now than ever before. "My reminiscences of Cambridge"—surely I have not yet come to that. Or, as he put it, "something more topical, something about mathematics and the war"—and that seemed to me the most impossible subject of all. I seemed to have nothing at all to say about the functions of mathematics in war, except that they filled me with intellectual contempt and moral disgust.

I have changed my mind on second thoughts, and I select the subject which seemed to me originally the worst. Mathematics, even my sort of mathematics, has its "uses" in war-time, and I suppose that I ought to have something to say about them; and if my opinions are incoherent or controversial, then perhaps so much the better, since other mathematicians may be led to reply.

---

* Reprinted with the kind permission of the Editor of *Eureka*, Jasper Bird.

I had better say at once that by "mathematics" I mean *real* mathematics, the mathematics of Fermat and Euler and Gauss and Abel, and not the stuff which passes for mathematics in an engineering laboratory. I am not thinking only of "pure" mathematics (though that is naturally my first concern); I count Maxwell and Einstein and Eddington and Dirac among "real" mathematicians. I am including the whole body of mathematical knowledge which has permanent aesthetic value, as for example the best Greek mathematics has, the mathematics which is eternal because the best of it may, like the best literature, continue to cause intense emotional satisfaction to thousands of people after thousands of years. But I am not concerned with ballistics or aerodynamics, or any of the other mathematics which has been specially devised for war. That (whatever one may think of its purposes) is repulsively ugly and intolerably dull; even Littlewood could not make ballistics respectable, and if he could not, who can?

Let us try then for a moment to dismiss these sinister byproducts of mathematics and to fix our attention on the real thing. We have to consider whether real mathematics serves any purposes of importance in war, and whether any purposes which it serves are good or bad. Ought we to be glad or sorry, proud or ashamed, in war-time, that we are mathematicians?

It is plain at any rate that the real mathematics (apart from the elements) has no *direct* utility in war. No one has yet found any war-like purpose to be served by the theory of numbers or relativity or quantum mechanics, and it seems very unlikely that anybody will do so for many years. And of that I am glad, but in saying so I may possibly encourage a misconception.

It is sometimes suggested that pure mathematicians glory in the "uselessness" of their subject, and make it a boast that it has no "practical" applications.* The imputation is usually based on an incautious saying attributed to Gauss,† which has always seemed to me to have been rather crudely misinterpreted. If the theory of numbers could be employed for any practical and honourable purpose, if it could be turned directly to the furtherance of human happiness or the relief of human suffering (as for example physiology and even chemistry can), then surely neither Gauss nor any other mathematician would have been

---

* I have been accused of taking this view myself. I once stated in a lecture, which was afterwards printed, that "a science is said to be useful if its development tends to accentuate the existing inequalities in the distribution of wealth, or more directly promotes the destruction of human life"; and this sentence, written in 1915, was quoted in the *Observer* only a few months ago. It was, of course, a conscious rhetorical flourish (though one perhaps excusable at the time when it was written).

† To the effect that, if mathematics is the queen of the sciences, then the theory of numbers is, because of its supreme "uselessness", the queen of mathematics. I cannot find an accurate quotation.

so foolish as to decry or regret such applications. But if on the other hand the applications of science have made, on the whole, at least as much for evil as for good—and this is a view which must always be taken seriously, and most of all in time of war—then both Gauss and lesser mathematicians are justified in rejoicing that there is one science at any rate whose very remoteness from ordinary human activities should keep it gentle and clean.

It would be pleasant to think that this was the end of the matter, but we cannot get away from the mathematics of the workshops so easily. Indirectly, we are responsible for its existence. The gunnery experts and aeroplane designers could not do their job without quite a lot of mathematical training, and the best mathematical training is training in real mathematics. In this indirect way even the best mathematics becomes important in war-time, and mathematicians are wanted for all sorts of purposes. Most of these purposes are ignoble and dreary—what could be more soul-destroying than the numerical solution of differential equations?—but the men chosen for them must be mathematicians and not laboratory hacks, if only because they are better trained and have the better brains. So mathematics is going to be really important now, whether we like it or regret it; and it is not so obvious as it might seem at first even that we ought to regret it, since that depends upon our general view of the effect of science on war.

There are two sharply contrasted views about modern "scientific" war. The first and the most obvious is that the effect of science on war is merely to magnify its horror, both by increasing the sufferings of the minority who have to fight and by extending them to other classes. This is the orthodox view, and it is plain that, if this view is just, then the only possible defence lies in the necessity for retaliation. But there is a very different view which is also quite tenable. It can be maintained that modern warfare is *less* horrible than the warfare of pre-scientific times, so far at any rate as combatants are concerned; that bombs are probably more merciful than bayonets; that lachrymatory gas and mustard-gas are perhaps the most humane weapons yet devised by military science; and that the "orthodox" view rests solely on loose-thinking sentimentalism. This is the case presented with so much force by Haldane in *Callinicus*.* It may also be urged that the equalisation of risks which science was expected to bring would be in the long run salutary; that a civilian's life is not worth more than a soldier's, or a woman's than a man's; that anything is better than the concentration of savagery on one particular class; and that, in short, the sooner war comes "all out" the better. And if this be the right view, then scientists

---

* J. B. S. Haldane, *Callinicus; a defence of chemical warfare* (Kegan. Paul, 1924).

in general and mathematicians in particular may have a little less cause to be ashamed of their profession.

It is very difficult to strike a balance between these extreme opinions, and I will not try to do so. I will end by putting to myself, as I think every mathematician ought to, what is perhaps an easier question. Are there *any* senses in which we can say, with any real confidence, that mathematics "does good" in war? I think I can see two (though I cannot pretend that I extract a great deal of comfort from them).

In the first place it is very probable that mathematics will save the lives of a certain number of young mathematicians, since their technical skill will be applied to "useful" purposes and will keep them from the front. "Conservation of ability" is one of the official slogans; "ability" means, in practice, mathematical, physical, or chemical ability; and if a few mathematicians are "conserved" then that is at any rate something gained. It may be a bit hard on the classics and historians and philosophers, whose chances of death are that little much increased; but nobody is going to worry about the "humanities" now. It is better that some should be saved, even if they are not necessarily the most worthy.

Secondly, an older man may (if he is not *too* old) find in mathematics an incomparable anodyne. For mathematics is, of all the arts and sciences, the most austere and the most remote, and a mathematician should be of all men the one who can most easily take refuge where, as Bertrand Russell says, "one at least of our nobler impulses can best escape from the dreary exile of the actual world." But he must not be too old—it is a pity that it should be necessary to make this very serious reservation. Mathematics is not a contemplative but a creative subject; no one can draw much consolation from it when he has lost the power or the desire to create; and that is apt to happen to a mathematician rather soon. It is a pity, but in that case he does not matter a great deal anyhow, and it would be silly to bother about him.

24

Mathematics

(*The Oxford Magazine*, 48 (1930), pp. 819–821)

From the Editors

When Hardy moved from Cambridge to Oxford he found that mathematics at Oxford did not command the same level of prestige that it did at Cambridge, though he admitted that Cambridge lagged behind the great centers on the Continent, Paris and Göttingen. In this piece Hardy advocated the establishment of a Mathematical Institute at Oxford. Unfortunately the new Institute did not appear until 1953, well after Hardy's death.

Mathematics is one of the traditional subjects of study in Oxford as elsewhere, but it probably attracts less notice in Oxford than in any other considerable university. The first thing which a Cambridge man notices when he migrates to Oxford is that mathematics and physics, the 'ranking' subjects in his old University, are overshadowed in Oxford not merely by the literary schools but even by other sciences. The friend who conveyed to me the invitation to write this article remarked to me (I believe quite innocently) that 'the Editor wondered whether it would be worth while to include in this series an article on the School of Mathematics.' I take the invitation as a compliment in the circumstances, but I cannot help reflecting that a similar question would seem a very odd one in Paris or Göttingen or Harvard or Princeton. I have encountered a number of curious instances of the ordinary Oxford estimate of mathematics. One is in Jowett's delightfully naïve memoir of Henry Smith, printed in Smith's *Collected Mathematical Papers*. Smith, says Jowett, 'was not the author of any considerable work'; he 'lived and died almost unknown to the world at large.' I wonder how many Germans knew the work of Smith, for one who had ever heard of Benjamin Jowett.

Another example was in an amusing little undergraduate book called *Isis*, which I picked up last Term in Blackwell's shop. The thesis of the book was, briefly, 'to Hell with Science.' Stamp out these insufferable scientists, and these uncouth and illiterate scholars from the secondary schools, and Oxford may be once again a home for a gentleman and a Hellenist. Remembering that mathematics ranks as a science in some universities, but as an art in others, and hoping against hope that we mathematicians might escape condemnation, I turned hastily over the pages. I am sorry to say that what I found was more humiliating than the worst that I had dreaded. Mathematics was not mentioned; the author did not know that such a subject existed.

What are the reasons for this ignorance and indifference? Mathematics is an ancient, and one would have supposed quite a gentlemanly, study. If Oxford desires above all things to be Hellenistic, it could hardly find a more Hellenistic activity; Pythagoras and Archimedes, after all, were Greeks, and Archimedes could have given Aristotle a good fifteen. Why then should Oxford tolerate its mathematicians but be so totally uninterested in what they do?

It is usually supposed that mathematics has languished in Oxford because the School here has been overshadowed by the Cambridge School. Cambridge has had its Tripos, the most difficult, the most notorious, and the best advertised examination in the world, and all the best mathematicians in England have flocked to Cambridge to distinguish themselves in this ordeal, so that Oxford has been starved of ability. This is the common explanation, and no doubt it contains a certain amount of truth. Oxford has produced a good many more distinguished mathematicians than is generally realised, but the majority of English mathematicians are Cambridge men, and they almost monopolise the more important chairs; and this is naturally discouraging to an Oxford student who wishes to make a name and a career for himself as a mathematician.

I do not think that it ought to be very difficult for the Oxford School of Mathematics to eradicate this rather humiliating inferiority complex. We may begin by reflecting (and it is easy for me as a Cambridge man to say so) that English opinion has overrated very grossly the status of Cambridge mathematics; if we have lagged behind them, they have lagged still further behind the great continental centres, and we have only to beat an opponent who has been beaten very badly a great many times already.

We can then remember that there is no subject in which it would be easier to develop a fine school than in mathematics, if once we had decided that we really wanted to do so. Harvard has created a school of the first rank in thirty years, Princeton in twenty, Hamburg in ten. 'The circumstances of the University,' says Jowett, 'hardly admitted' of Smith's 'raising up a school of mathematical pupils.' It is possible, but, if so, it was the fault of Jowett and people like him. There is no good reason at all why we should not do what these other universities have done. Mathematics, indeed, has here two advantages which no other subject combines; it is indispensable and it is comparatively inexpensive. It is indispensable because every other science (however little it may like it) finds itself compelled, sooner or later, to use mathematical methods; and it is inexpensive because it does not require either an enormous library or a costly laboratory equipment. No mathematician in the world needs very much more than brains and leisure.

In any case, we need not distress ourselves about Cambridge and its Tripos. If we cannot hope to rival those faded glories, so much the better; it is not possible to found a school of learning on an examination. I must not succumb to the temptation of airing views on 'honours schools' which most of my colleagues might reject. Still, the mathematical world at large has never cared two pins about the Tripos, Cambridge mathematicians themselves have long since ceased to attach any exaggerated importance to it, and, if we have here no examination of equal status and fame, our freedom from such shackles should be an inestimable advantage to us in the development of our School.

If then Oxford wishes to have a first-rate School of Mathematics, it can have one; the difficulty for us is to persuade the University (or rather the Colleges) that such a School is an asset worth possessing. It is a question, primarily, of buying the men; mathematicians are reasonably cheap, but they cannot be had for nothing, and at present there is a very serious lack of openings for rising men. It is here that we find the really serious contrast with Cambridge. The University itself is liberal enough; the number of professorships here and at Cambridge is the same; but we find a distressing disparity as soon as we look at college appointments. In this respect mathematics falls between two stools. It is not a 'modern' subject, with a centre of its own independent of college life, and it is not one of the favoured subjects on which scholarships and fellowships are showered. At Cambridge, Trinity has four mathematical lecturers, St. John's three, and every college one at least. Here Exeter, Lincoln, Magdalen, Oriel, Pembroke, Trinity, University, Wadham, and Worcester have not a mathematician between them; in none of those great centres of philosophy is there a man who can hope to understand Hilbert or Russell, Einstein or Schrödinger.

This is the great handicap with which Oxford mathematics has to contend, and I imagine that it can be overcome only gradually and indirectly. We can raise the prestige of Oxford mathematics in Oxford only by raising it first outside, and it is plain that we can do this only by making our school primarily a school of advanced study and research. There is no word which excites such bitter feeling in a 'teaching university' as the word 'research,' and I do not wish to suggest that the advancement of knowledge is the only worthy outlet for a don's activities. The fact remains that it is quite impossible to erect a school of any science, and in particular a School of Mathematics, on any other foundation. The world, when asked to admire the Oxford School, will inquire 'what it has done,' and will be quite uninterested unless it can show that it has done something substantial. It is already practically impossible for a mathematician who has not 'added to knowledge' even to start on an academical career.

It is fortunate that we can claim that our School already stands this test quite well. The current periodicals certainly show no lack of substantial contributions by Oxford men. The new *Quarterly Journal*, captured from Cambridge by the enterprise of the Clarendon Press, should put our activities still more clearly in evidence and add materially to our prestige abroad. We have now a journal, and we demand next a home. Why should Oxford, almost alone among great Universities, have no centre of any kind for its mathematical work? Why should a professor have to grovel before a College Bursar when he wants a room and a decent blackboard?

We are, therefore, asking for a Mathematical Institute as a centre for Oxford mathematical life. It is an obviously reasonable, and a modest, demand, remarkable only because it has been delayed so long, and sooner or later it must be met. Whether it is met sooner or later, the proposal is in itself a most encouraging sign, as evidence of a new spirit of enterprise and assertiveness on the part of Oxford mathematicians, and proof that they are beginning to rebel against the subjection, unworthy of a splendid subject, in which they have acquiesced too long.

25

Asymptotic Formulae in Combinatory Analysis (excerpts)

(with S. Ramanujan)

(*Proceedings of the London Mathematical Society* (2) 17 (1918), 75–115)*

From the Editors

This is a selection from an article, jointly written by Hardy and Ramanujan, about p(n), the number of unrestricted partitions of a whole number n. In it, they provided a short history of the function, going back to Euler, and then derived for the first time its asymptotic properties. Our excerpt ends with the statement of their key result: $\ln p(n) \sim C\sqrt{n}$, *where* $C = 2\pi/\sqrt{6}$.

1. Introduction and summary of results

1.1. The present paper is the outcome of an attempt to apply to the principal problems of the theory of partitions the methods, depending upon the theory of analytic functions, which have proved so fruitful in the theory of the distribution of primes and allied branches of the analytic theory of numbers.[†]

The most interesting functions of the theory of partitions appear as the coefficients in the power-series which represent certain elliptic modular functions. Thus $p(n)$, the number of unrestricted partitions of n, is the coefficient of

---

* Reprinted with the kind permission of the publisher, the London Mathematical Society.
[†] A short abstract of the contents of part of this paper appeared under the title "Une formule asymptotique pour le nombre des partitions de *n*," in the *Comptes Rendus*, January 2nd, 1917.

x^n in the expansion of the function

(1.11) $$f(x) = 1 + \sum_{1}^{\infty} p(n)x^n = \frac{1}{(1-x)(1-x^2)(1-x^3)\cdots}.$$ [*]

If we write

(1.12) $$x = q^2 = e^{2\pi i \tau},$$

where the imaginary part of τ is positive, we see that $f(x)$ is substantially the reciprocal of the modular function called by Tannery and Molk[†] $h(\tau)$; that, in fact,

(1.13) $$h(\tau) = q^{\frac{1}{12}} q_0 = q^{\frac{1}{12}} \prod_{1}^{\infty} (1 - q^{2n}) = \frac{x^{\frac{1}{24}}}{f(x)}.$$

The theory of partitions has, from the time of Euler onwards, been developed from an almost exclusively algebraical point of view. It consists of an assemblage of formal identities—many of them, it need hardly be said, of an exceedingly ingenious and beautiful character. Of *asymptotic* formulæ, one may fairly say, there are none[‡]. So true is this, in fact, that we have been unable

---

[*] P. A. MacMahon, *Combinatory Analysis*, Vol. II, 1916, p. 1.
[†] J. Tannery and J. Molk, *Functions elliptiques*, Vol. II, 1896, pp. 31 *et seq*. We shall follow the notation of this work whenever we have to quote formulæ from the theory of elliptic functions.
[‡] We should mention one exception to this statement, to which our attention was called by Major MacMahon. The number of partitions of *n into parts none of which exceed r* is the coefficient $p_r(n)$ in the series
$$1 + \sum_{1}^{\infty} p_r(n) x^n = \frac{1}{(1-x)(1-x^2)\cdots(1-x^r)}.$$
This function has been studied in much detail, for various special values of r, by Cayley, Sylvester, and Glaisher: we may refer in particular to J. J. Sylvester, "On a discovery in the theory of partitions," *Quarterly Journal*, Vol. I, 1857, pp. 81–85, and "On the partition of numbers," *ibid.*, pp. 141–152 (Sylvester's *Works*, Vol. II, pp. 86–89 and 90–99); J. W. L. Glaisher, "On the number of partitions of a number into a given number of parts," *Quarterly Journal*, Vol. XL, 1909, pp. 57–143; "Formulæ for partitions into given elements, derived from Sylvester's Theorem," *ibid.*, pp. 275–348; "Formulæ for the number of partitions of a number into the elements 1, 2, 3, ..., n up to $n = 9$," *ibid.*, Vol. XLI, 1910, pp. 94–112: and further references will be found in MacMahon, *loc. cit.*, pp. 59–71, and E. Netto, *Lehrbuch der Combinatorik*, 1901, pp. 146–158. Thus, for example, the coefficient of x^n in
$$\frac{1}{(1-x)(1-x^2)(1-x^3)}$$
is
$$p_3(n) = \frac{1}{12}(n+3)^2 - \frac{7}{72} + \frac{1}{8}(-1)^n + \frac{2}{9}\cos\frac{2n\pi}{3};$$

25. Asymptotic Formulae in Combinatory Analysis (excerpts)

to discover in the literature of the subject any allusion whatever to the question of the order of magnitude of $p(n)$.

1.2. The function $p(n)$ may, of course, be expressed in the form of an integral

(1.21) $$p(n) = \frac{1}{2\pi i} \int_\Gamma \frac{f(x)}{x^{n+1}} dx,$$

by means of Cauchy's theorem, the path Γ enclosing the origin and lying entirely inside the unit circle. The idea which dominates this paper is that of obtaining asymptotic formulæ for $p(n)$ by a detailed study of the integral (1.21). This idea is an extremely obvious one; it is the idea which has dominated nine-tenths of modern research in the analytic theory of numbers: and it may seem very strange that it should never have been applied to this particular problem before. Of this there are no doubt two explanations. The first is that the theory of partitions has received its most important developments, since its foundation by Euler, at the hands of a series of mathematicians whose interests have lain primarily in algebra. The second and more fundamental reason is to be found in the extreme complexity of the behaviour of the generating function $f(x)$ near a point of the unit circle.

It is instructive to contrast this problem with the corresponding problems which arise for the arithmetical functions $\pi(n), \vartheta(n), \psi(n), \mu(n), d(n), \ldots$ which have their genesis in Riemann's Zeta-function and the functions allied to it. In the latter problems we are dealing with functions defined by Dirichlet's series. The study of such functions presents difficulties far more fundamental than any which confront us in the theory of the modular functions. These difficulties, however, relate to the distribution of the zeros of the functions and their general behaviour at infinity: no difficulties whatever are occasioned by the crude singularities of the functions in the finite part of the plane. The single finite singularity of $\zeta(s)$, for example, the pole at $s = 1$, is a singularity of

---

as is easily found by separating the function into partial fractions. This function may also be expressed in the forms

$$\frac{1}{12}(n+3)^2 + \left(\frac{1}{2}\cos\frac{1}{2}\pi n\right)^2 - \left(\frac{2}{3}\sin\frac{1}{3}\pi n\right)^2,$$

$$1 + \left[\frac{1}{12}n(n+6)\right], \quad \left\{\frac{1}{12}(n+3)^2\right\},$$

where $[n]$ and $\{n\}$ denote the greatest integer contained in n and the integer nearest to n. These formulæ do, of course, furnish incidentally asymptotic formulæ for the functions in question. But they are, from this point of view, of a very trivial character: the interest which they possess is algebraical.

the simplest possible character. It is this pole which gives rise to the *dominant* terms in the asymptotic formulæ for the arithmetical functions associated with $\zeta(s)$. To prove such a formula rigorously is often exceedingly difficult; to determine precisely the order of the error which it involves is in many cases a problem which still defies the utmost resources of analysis. But to write down the dominant terms involves, as a rule, no difficulty more formidable than that of deforming a path of integration over a pole of the subject of integration and calculating the corresponding residue.

In the theory of partitions, on the other hand, we are dealing with functions which do not exist at all outside the unit circle. Every point of the circle is an essential singularity of the function, and no part of the contour of integration can be deformed in such a manner as to make its contribution obviously negligible. Every element of the contour requires special study; and there is no obvious method of writing down a "dominant term."

The difficulties of the problem appear then, at first sight, to be very serious. We possess, however, in the formulae of the theory of the linear transformation of the elliptic functions, an extremely powerful analytical weapon by means of which we can study the behaviour of $f(x)$ near any assigned point of the unit circle*. It is to an appropriate use of these formulae that the accuracy of our final results, an accuracy which will, we think, be found to be quite startling, is due.

1.3. It is very important, in dealing with such a problem as this, to distinguish clearly the various stages to which we can progress by arguments of a progressively "deeper" and less elementary character. The earlier results are naturally (so far as the particular problem is concerned) superseded by the later. But the more elementary methods are likely to be applicable to other problems in which the more subtle analysis is impracticable.

We have attacked this particular problem by a considerable number of different methods, and cannot profess to have reached any very precise conclusions as to the possibilities of each. A detailed comparison of the results to which they lead would moreover expand this paper to a quite unreasonable length. But we have thought it worth while to include a short account of two of them. The first is quite elementary; it depends only on Euler's identity

$$\frac{1}{(1-x)(1-x^2)(1-x^3)\cdots} = 1 + \frac{x}{(1-x)^2} + \frac{x^4}{(1-x)^2(1-x^2)^2} + \cdots$$
(1.31)

---

* See G. H. Hardy and J. E. Littlewood, "Some problems of Diophantine approximation (II: The trigonometrical series associated with the elliptic Theta-functions)," *Acta Mathematica*, Vol. XXXVII, 1914, pp. 193–238, for applications of the formulae to different but not unrelated problems.

25. Asymptotic Formulae in Combinatory Analysis (excerpts)

—an identity capable of wide generalisation—and on elementary algebraical reasoning. By these means we shew, in section 2, that

(1.32) $$e^{A\sqrt{n}} < p(n) < e^{B\sqrt{n}},$$

where A and B are positive constants, for all sufficiently large values of n.

It follows that

(1.33) $$A\sqrt{n} < \log p(n) < B\sqrt{n};$$

and the next question which arises is the question whether a constant C exists such that

(1.34) $$\log p(n) \sim C\sqrt{n}.$$

We prove that this is so in section 3. Our proof is still, in a sense, "elementary." It does not appeal to the theory of analytic functions, depending only on a general arithmetic theorem concerning infinite series; but this theorem is of the difficult and delicate type which Messrs Hardy and Littlewood have called "Tauberian." The actual theorem required was proved by us in a paper recently printed in these *Proceedings*. It shews that

(1.35) $$C = \frac{2\pi}{\sqrt{6}};$$

in other words that

(1.36) $$p(n) = \exp\left\{\pi\sqrt{\left(\frac{2n}{3}\right)}(1+\epsilon)\right\},$$

where ϵ is small when n is large. This method is one of very wide application.

26

A New Solution of Waring's Problem (excerpts)

(with J. E Littlewood)

(*Quarterly Journal of Mathematics* 48 (1920), 272–293)*

From the Editors

In this paper, Hardy and Littlewood presented a new attack on Waring's problem about decomposing whole numbers into sums of like powers. We include here only the introductory pages, up to the statement of their main result: $G(k) \leq 2^{k-1}k + 1$.

We should clarify the notation. In 1770, Lagrange proved that any whole number can be written as the sum of four or fewer perfect squares. We cannot get by with three *or fewer perfect squares because there are no three squares that sum to 7.*

This remarkable theorem can be expressed by introducing the function $g(k)$ to indicate the smallest number of kth powers sufficient to generate any whole number. For $k = 2$ (i.e., for "squares"), mathematicians write Lagrange's theorem as $g(2) = 4$. Likewise it is known that any whole number can be written as the sum of nine or fewer cubes, but there are numbers (e.g., 23) that cannot be written as the sum of just eight cubes. Hence, $g(3) = 9$.

However, as Hardy and Littlewood noted, it is of equal if not of more interest to consider the function $G(k)$—the smallest number of kth powers sufficient to yield any whole number after a certain point. *Thus, although it*

---

* Reprinted with the permission of the Oxford University Press.

is known that $g(4) = 19$, *it has been proved that* $G(4) = 16$. *In other words, although it requires* 19 *or fewer fourth powers to sum to any whole number, we can get by with* 16 *or fewer after a certain point. Those extra three fourth powers are necessary to handle the smaller numbers, but for sufficiently large numbers, sixteen will do the trick.*

1. It was asserted by Waring* in 1782 that every number is the sum of at most four squares, nine positive cubes, nineteen fourth powers, and, in general, $g(k)$ positive k-th powers, where $g(k)$ is a number depending upon k alone. Waring advanced no argument of any kind in support of his assertion; and there is no reason to suppose that it rested on any basis more substantial than the examination of a number of particular cases.

If the proposition 'every number is the sum of at most m positive k-th powers' is true for any particular value of m, it is true *a fortiori* for any larger value. The number $g(k)$ is, by definition, the smallest value of m for which the proposition is true. The problem suggested by Waring then falls naturally into two parts. The first is the proof of the existence of $g(k)$, the second the determination of its actual value as a function of k. It is the first of these two problems that has generally been described as *Waring's Problem*. This problem was solved by Hilbert† in 1909, the existence of $g(k)$ having been proved before only when k is 2, 3, 4, 5, 6, 7, 8, or 10. The second problem is still unsolved, except when k is 2 or 3.

Hilbert based his proof, in the first instance, on considerations drawn from the integral calculus. The proof falls into two parts. In the first he considers the properties of a volume integral in space of five dimensions‡, from which he deduces the existence of certain algebraical identities leading to an induction from k to $2k$. As the theorem is known to be true when $k = 2$, it follows that it is true when k is any power of 2. This completes the first part of the proof: the second, in which the conclusion is extended to an arbitrary k, is purely algebraical.

---

* E. Waring, *Meditationes Algebraicae*, ed. 3, 1782, pp. 349–350.
† D. Hilbert, 'Beweis für die Darstellbarkeit der ganzen Zahlen durch eine feste Anzahl *n*ter Potenzen (Waringsche, Problem)', *Göttinger Nachrichten*, 1909, pp. 17–36, and *Math. Annalen*, vol. lxvii, 1909, pp. 281–300.
‡ Twenty-five in the first version of his memoir.

26. A New Solution of Waring's Problem (excerpts)

Hilbert's proof was reconsidered and simplified by Hausdorff*, Stridsberg†, and Remak‡, who succeeded ultimately in eliminating all reference to the integral calculus, and indeed all reasoning of a transcendental character. The proof, as left by Remak, is by no means easy; but it is purely algebraical and, in the technical sense, entirely elementary§.

2. In this note we propose to give a short account of another solution of Waring's Problem which we have discovered recently and which proceeds on entirely different lines. This solution is not, in any sense of the word, elementary. It is based throughout on Cauchy's Theorem and the ordinary machinery of the theory of analytic functions, and has, from beginning to end, no point of contact with Hilbert's solution. It might seem that a highly transcendental proof of a theorem which has already been proved, and that in an entirely elementary manner, is unnecessary. This view, we think, would rest upon a misapprehension. It seems to us most desirable and important that Waring's Problem, and all similar problems of Combinatory Analysis, should be brought into relation with the transcendental side of the Analytic Theory of Numbers. Further, the method which we follow, and which we describe shortly as *the method of Farey dissection*, is in many ways the most natural and, in spite of its considerable technical difficulties, the most straightforward method for the discussion of any problem of this character; and it is a method of great power and wide scope, applicable to almost any problem concerning the decomposition of integers into parts of a particular kind‖, and to many against which it is difficult to suggest any other obvious method of attack.

---

* F. Hausdorff, 'Zur Hilbertschen Lösung des Waringschen Problems', *Math. Annalen*, vol, lxvii., 1909, pp. 301–305.

† E. Stridsberg, 'Sur la démonstration de M. Hilbert du théorème de Waring', *Math. Annalen*, vol. lxxii, 1912, pp. 145–152. This paper gives references to two earlier notes by the author in the *Arkiv för Matematik*, vol. vi, 1910 and 1911.

‡ R. Remak, 'Bemerkung zu Herrn Stridsbergs Beweis dea Waringschen Theorems', *Math. Annalen*, vol. lxxii, 1912, pp. 153–156.

§ That is to say, it does not depend upon arguments which involve limiting processes.

‖ See G. H. Hardy and S. Ramanujan, 'Une formule asymptotique pour le nombre des partitions de n', *Comptes Rendus*, 2 Jan. 1917; 'Asymptotic formulæ in Combinatory Analysis', *Proc. London Math. Soc.*, ser. 2, vol. xvii, 1918, pp. 75–115; 'On the co-efficients in the expansions of certain modular functions', *Proc. Royal Soc.* (A), vol. xcv, 1918, pp. 144–155; S. Ramanujan, 'On certain trigonometrical sums and their applications in the analytic theory of numbers', *Trans. Camb. Phil. Soc.*, vol. xxii, 1918, pp. 259–276: G. H. Hardy, 'On the expression of a number as the sum of any number of squares, and in particular of five or seven', *Proc. National Acad. of Sciences*

Moreover, when our method is applied to this particular problem, it yields a great deal more than can be obtained by more elementary methods. In particular it enables us to assign a definite upper bound, in the form of a function of k, not indeed for $g(k)$ but for another number $G(k)$ which seems in some ways more fundamental. This number $G(k)$ is the least number m of which it is true that every number *from a certain point onwards* is the sum of at most m positive k-th powers. It is obvious that $G(k) \leq g(k)$, and it would seem that in general $G(k) < g(k)$; and that $G(k)$ is more fundamental than $g(k)$ because its value is less likely to be determined, in any particular case, by a number of arithmetical coincidences. The value of $G(k)$ is not known for any value of k save 2; nor has any general upper bound for $G(k)$ yet been determined. Our method enables us to prove that

(2.1) $$G(k) \leq 2^{k-1}k + 1.$$

Finally our method yields not only a proof that representations of n in the form required exist, but also asymptotic formulæ for the number of representations.

(Washington), vol. iv, 1918, pp. 189–193: G. H. Hardy and J. E. Littlewood, 'Note on Messrs. Shah and Wilson's paper entitled *An empirical formula connected with Goldbach's Theorem'*, Proc. Camb. Phil. Soc., vol. xix (1919), pp. 245–254.

27

Some Notes on Certain Theorems in Higher Trigonometry

(*Mathematical Gazette* 3 (1906), 284–288)*

From the Editors

Hardy celebrated "Higher Trigonometry," which he more or less defined by its four indispensible results: DeMoivre's theorem, the "exponential theorem," Euler's identity—i.e., $\exp(ix) = \cos x + i \sin x$—and the factorization of $\sin x$. In this paper, he focused especially on Euler's identity and on how it could best be taught to "schoolboys."

I admire the ingenuity of the method by which Prof. Nanson establishes the expansions of $\sin x$ and $\cos x$ as power series in x. But I must confess that his attempt to 'simplify' the 'accepted accurate' proofs of these expansions seems to me a fundamentally mistaken one. Nor can I admit that the result is at all satisfactory 'from the elementary didactic point of view.' His proof is to my mind essentially an artificial *verification*, and altogether *unnatural*, in the sense that it has no place in any natural and logical way of developing the ideas which lead to the result. I speak with diffidence as one who has had far less experience of teaching than Prof. Nanson. But I am convinced that my criticism is a fair one: and I think that many of the proofs usually given, in English text-books, of many of the theorems of 'Higher Trigonometry,' are open to fair criticism on similar grounds, when (as is too seldom the case) they are not to be condemned for 'hopeless complexity' or utter lack of force.

---

* Reprinted with the kind permission of the publisher of *The Mathematical Gazette*: The Mathematical Association.

And if this is so, so it will and must be so long as this subject continues to occupy its present place in the mathematical course. At present, if one may judge by results, boys are taught these theorems quite prematurely and before they can have had the opportunity of acquiring anything like a sufficient grounding of general mathematical knowledge. The great majority of boys (I am speaking, of course, only of those whose abilities are up to something approaching a scholarship standard) never really grasp the meaning of them, and one of the most fascinating and instructive regions of mathematics is, from an educational point of view, practically wasted.

How are we to define what we mean by 'Higher Trigonometry'? It is, I suppose, the theory of the exponential, logarithmic, and circular functions, and of the series and products associated with them, for real and complex values of the variable. It is true that a certain part of this theory is usually lopped off and labelled as a part of 'Higher Algebra,' but the most conservatively-minded of teachers would probably admit the absurdity of this. At any rate this is what I mean by 'Higher Trigonometry,' and there seem to me to be four theorems or groups of theorems which may be said roughly to mark the stages of our progress in the subject. These are:—

(A) De Moivre's theorem for a real and rational index.

(B) The exponential theorem, that is to say the theorem that if x is real and rational one of the values of $(1 + \frac{1}{1!} + \frac{1}{2!} + \cdots)^x$ is $1 + \frac{x}{1!} + \frac{x^2}{2!} + \cdots$, or that one of the values of e^x is exp. x.

(C) The theorem that if x is real exp. $(ix) = \cos x + i \sin x$.

(D) The factor-theorem for $\sin x$.

It is with theorem (C) that I am particularly concerned at present. But there are one or two remarks which I should like to make about the other theorems.

(A), I think, presents no difficulty, as soon as the elements of the Argand diagram are mastered. *Irrational* powers should at this stage be severely and *explicitly* left alone. Any further extensions of the theorem should in any case be regarded as corollaries of (C); and so indeed *may* (A) itself; but the more direct treatment is better put first. (B) is a direct corollary from the multiplication theorem for exp. x. This last theorem should of course be proved for *all* values of x: it naturally requires some elementary notions with regard to convergence, and how far these should be insisted upon is a matter for judgment in particular cases. The theorem

$$\lim_{n \to \infty} \left(1 + \frac{x}{n}\right)^n = \exp. x$$

(with whatever degree of rigour or generality it may be stated) is instructive and important, but essentially accessory and not fundamental, and to base the exponential theorem on it, although possible, is logically quite wrong. The logarithmic expansion is more difficult and may well be postponed a while. Finally (D) is *really* difficult, and the question of what to do about it is a real *crux* in the teaching of trigonometry. There is no direct and rigid proof which is at all possible for boys, and I cannot think that anyone who has had to give a proof in lectures on the theory of functions will imagine that any indirect method of procedure is likely to lead to better results. One has only to read the account of the matter in Harkness' and Morley's excellent *Introduction to the Theory of Analytic Functions* to understand that we have here to deal with difficulties which we cannot evade. Fortunately a rigid proof is in no way necessary for profitable employment of the result, and the result is easily made *plausible* from either point of view or both.

So much for (A), (B), and (D). With regard to (C) there is much more to be said, as the theorem may be approached from so many points of view.

(i) In the first place we may postpone the theorem until a knowledge of Taylor's theorem has been acquired. There is then no difficulty at all. But although I am heartily in favour of pushing the Calculus, both Differential and Integral, as far forward in mathematical teaching as can possibly be done, I hardly think this theorem need wait so long, and we certainly lose something if it does. Moreover, if we adopt this method, the theorem (C) will naturally appear as a *consequence* of the expansion-theorems for $\sin x$ and $\cos x$. This, to my mind, is an inversion of the natural logical order, and in so far an argument against this method of procedure.*

(ii) The same argument also tells against the second possible method, which is that adopted in Chrystal's *Algebra*. The principle of this method is to give an accurate proof, independent of the notions of the Calculus, of the expansions, and to deduce (C) from them. Every deference is due to so eminent an authority as Prof. Chrystal; but I cannot help thinking that, from the elementary 'didactic' point of view, this method combines every possible disadvantage.

(iii) Thirdly, we may presuppose a little more knowledge of the foundations and a little less of the superstructure of the Calculus, and we may serve up the old line of argument,

$$y = \cos x + i \sin x, \quad \frac{dy}{dx} = -\sin x + i \cos x, \quad \frac{dy}{y} = i\,dx,$$

which may be made to lead to the desired result in a perfectly simple and satisfactory way. This is in substance the way followed by Jordan in his *Cours d'Analyse*, and is the way I should generally choose myself. The only point

---

* A modification of this method has, I find, been suggested by Prof. Bromwich (*Math. Gaz.*, vol. iii, p. 85). There is a good deal to be said for the line of proof which he adopts, but I cannot regard it as the best.

about it is that it essentially involves the ideas of y varying along a circle in the Argand diagram in such a way that dy/y is purely imaginary, and of the logarithm as a many-valued inverse of the exponential function of a complex variable.* These ideas are simple and fundamental, and should be acquired much earlier than is usually the case; and there is nothing essentially new, to anyone who understands the meaning of an integral, in the notion of $\int \frac{dy}{y}$ when y varies along any contour. And I am sure that anyone who has not yet acquired these ideas will derive more profit from the attempt to do so than from puzzling himself with theorems which are in reality applications of them: for this is the real *genesis* of (C).

(iv) Fourthly, we may follow Stolz and Gmeiner's method in their *Einleitung in die Funktionentheorie*. They presuppose less than is necessary if the preceding method is adopted.

Suppose
$$\exp.(ix) = u + iv.$$

Then
$$u^2 + v^2 = \exp.(ix) \cdot \exp.(-ix) = 1,$$

and the modulus of $u + iv$ is unity. Hence
$$u + iv = \cos\phi + i\sin\phi,$$

where ϕ is *some* real angle: we may suppose $0 \leqq \phi < 2\pi$. In particular
$$\exp.(i) = \cos\phi + i\sin\phi,$$

where $0 < \phi < 2\pi$, as $\exp.(i)$ is evidently not equal to unity. By De Moivre's theorem
$$\exp.(\xi i) = \cos\xi(\phi + 2k\pi) + i\sin\xi(\phi + 2k\pi),$$

where k is an integer. If $\phi + 2k\pi = \psi$, and we equate imaginary parts, we obtain
$$\xi - \frac{\xi^3}{3!} + \cdots = \sin\xi\psi,$$

and since
$$\lim_{\xi \to 0} \frac{\sin\xi\psi}{\xi} = \psi$$

it follows that $\psi = 1$.

---

* It has to be assumed that the exponential series can be differentiated term by term. I should never scruple, at this stage, to make assumptions such as this.

27. Some Notes on Certain Theorems in Higher Trigonometry

This method, which I had not seen before the publication of Stolz and Gmeiner's book, is very simple and there is a great deal to be said in its favour.

(v) Fifthly, we may establish the two equations

$$\lim_{n \to \infty} \left(1 + \frac{x+iy}{n}\right)^n = \exp.(x+iy),$$

$$\lim_{n \to \infty} \left(1 + \frac{x+iy}{n}\right)^n = e^x(\cos y + i \sin y),$$

directly and independently. This method (in conjunction with (ii)) is followed in Hobson's *Trigonometry*. As I have indicated already, I cannot consider it a good one, and it is probably more difficult than any of the others.

It will be observed that (iii) (iv) and (v) agree with one another in making the expansion-theorems for $\cos x$ and $\sin x$ corollaries of (C), whereas in (i) and (ii) this interdependence is inverted. For this reason the three latter methods appear to me to be logically preferable, and for my part I should discard (ii) and (v) and adopt (iii) or (iv), and if possible, *both*, though of course not losing sight of (i) for such pupils as are familiar with Taylor's theorem. The simplest method is doubtless (iv), but (iii) seems to me by far the most instructive. And if the objection is raised that the range of ideas required for (iii) is impossible for schoolboys, I should reply first, that, if this is so, then Higher Trigonometry, except in so far as it is concerned with *finite* series and products only, had better be postponed; and secondly, that I do not believe that it is so, and that a wider range of general ideas and less practice in detailed application of them is precisely what is wanted in elementary teaching, in Analysis even more than in Geometry.

28

The Integral $\int_0^\infty \frac{\sin x}{x} dx$ and Further Remarks on the Integral $\int_0^\infty \frac{\sin x}{x} dx$

(*Mathematical Gazette* 5 (1909), 98–103 and 8 (1916), 301–303)*

From the Editors

In these two papers, Hardy addressed various ways to establish that $\int_0^\infty \frac{\sin x}{x} dx = \frac{1}{2}\pi$. But he did more: he assigned to the different derivations a set of "marks" to indicate the simplicity of the method. According to him, the argument with the fewest marks was the best. Besides containing some clever mathematics, these articles feature incomparable Hardy-isms like, "As the upper limit is infinite we ought to assign 40 marks for this...; but as the two operations of differentiation involve inversions of the same character I reduce this to 30, adding 6, however, for the repeated integration by parts." Nobody writes like this anymore.

No one can possibly welcome more cordially than I do the widening of the school curriculum in mathematics that has taken place during the last twenty years. Even I can remember the days when 'Conics' and 'Mechanics' were the privilege of a select few, and only a prodigy was initiated into the mysteries of the Differential (much less the Integral) Calculus. All this is changed now, and it is fairly safe (at least that is my experience) to assume that a boy who has won a scholarship has learnt something about the Integral Calculus at school.

I do not believe for a moment that the curriculum has been made *harder*. These happy results are not due to any striking development in the schoolboy

---

* Reprinted with the kind permission of the publisher of *The Mathematical Gazette*: The Mathematical Association.

himself, either in intelligence, or in capacity or willingness for work. They are merely the result of improvement in the methods of teaching and a wider range of knowledge in the teacher, which enables him to discriminate, in a way his predecessor could not, between what is really difficult and what is not. No one now supposes that it is easy to resolve $\sin x$ into factors and hard to differentiate it!—and even in my time it was the former of these two problems to which one was set first.

Still there are difficulties, even in the Integral Calculus; it is not every schoolboy who can do right everything that Todhunter and Williamson did wrong; and, when I find my pupils handling definite integrals with the assurance of Dr. Hobson or Mr. Bromwich, I must confess I sometimes wonder whether even the school curriculum cannot be made too wide. Seriously, I do not think that schoolboys ought to be taught anything about definite integrals. They really are difficult, almost as difficult as the factors of $\sin x$: and it is to illustrate this point that I have written this note.

I have taken the particular integral

$$(1) \qquad \int_0^\infty \frac{\sin x}{x} dx = \frac{1}{2}\pi$$

as my text for two reasons. One is that it is about as simple an example as one can find of the definite integral *proper*—that is to say the integral whose value can be expressed in finite terms, although the indefinite integral of the subject of integration cannot be so determined. The other is that two interesting notes by Mr. Berry and Prof. Nanson have been published recently,* in which they discuss a considerable number of different ways of evaluating this particular integral. I propose now to apply a system of *marking* to the different proofs of the equation (1) to which they allude, with a view to rendering our estimate of their relative difficulty more precise.

Practically all methods of evaluating any definite integral depend ultimately upon the inversion of two or more operations of procedure to a limit†—*e.g.* upon the integration of an infinite series or a differentiation under the integral sign: that is to say they involve 'double-limit problems.'‡ The only exceptions to this statement of which I am aware (apart, of course, from cases in which the

---

* A. Berry, *Messenger of Mathematics*, vol. 37, p. 61; E. J. Nanson, *ibid.* p. 113.
† Whether this is true of a proof by 'contour integration' depends logically on what proof of Cauchy's Theorem has been used (see my remarks under 3 below).
‡ For an explanation of the general nature of a 'double-limit problem' see my *Course of Pure Mathematics* (App. 2, p. 420); for examples of the working out of such problems see Mr. Bromwich's *Infinite Series* (*passim*, but especially App. 3, p. 414 *et seq.*).

28. The Integral $\int_0^\infty \frac{\sin x}{x} dx$ and Further Remarks on the Integral

indefinite integral can be found) are certain cases in which the integral can be calculated directly from its definition as a sum, as can the integrals

$$\int_0^{-\frac{1}{2}\pi} \log \sin\theta \, d\theta, \quad \int_0^\pi \log(1 - 2p\cos\theta + p^2) d\theta.$$

Such inversions, then, constitute what we may call the standard difficulty of the problem, and I shall base my system of marking primarily upon them. *For every such inversion involved in the proof I shall assign 10 marks.* Besides these marks I shall add, in a more capricious way, marks for artificiality, complexity, etc. The proof obtaining least marks is to be regarded as the simplest and best.

1. *Mr. Berry's first proof.** This is that expressed by the series of equations

$$\int_0^\infty \frac{\sin x}{x} dx = \int_0^\infty \lim_{a \to 0}\left(e^{-ax}\frac{\sin x}{x}\right) dx = \lim_{a \to 0} \int_0^\infty e^{-ax}\frac{\sin x}{x} dx$$

$$= \lim_{a \to 0} \int_0^\infty e^{-ax} dx \int_0^1 \cos tx \, dt = \lim_{a \to 0} \int_0^1 dt \int_0^\infty e^{-ax}\cos tx \, dx$$

$$= \lim_{a \to 0}\int_0^1 \frac{a\,dt}{a^2 + t^2} = \lim_{a \to 0} \arctan\left(\frac{1}{a}\right) = \frac{1}{2}\pi.$$

This is a difficult proof, theoretically. Each of the first two lines involves, on the face of it, an inversion of limits. In reality each involves more: for integration, with an infinite limit, is in itself an operation of *repeated* procedure to a limit; and we ought really to write

$$\int_0^\infty \lim_{a \to 0} = \lim_{x \to \infty} \int_0^x \lim_{a \to 0} = \lim_{x \to \infty} \lim_{a \to 0}\int_0^x = \lim_{a \to 0} \lim_{x \to \infty}\int_0^x = \lim_{a \to 0}\int_0^\infty.$$

Thus *two* inversions are involved in the first line; and so also in the second. This involves 40 marks. On the other hand this proof is, apart from theoretical difficulties, simple and natural: I do not think it is necessary to add more marks. Thus our total is 40.

2. *Mr. Berry's second proof.* This is expressed by the equations

$$\int_0^\infty \frac{\sin x}{x} dx = \frac{1}{2}\int_{-\infty}^\infty \frac{\sin x}{x} dx = \frac{1}{2}\sum_{i=-\infty}^\infty \int_{i\pi}^{(i+1)\pi}\frac{\sin x}{x} dx$$

$$= \frac{1}{2}\sum_{-\infty}^\infty (-1)^i \int_0^\pi \frac{\sin x}{x - i\pi} dx = \frac{1}{2}\int_0^\pi \sin x \sum_{-\infty}^\infty \frac{(-1)^i}{x - i\pi} dx$$

$$= \frac{1}{2}\int_0^\pi \sin x \csc x \, dx = \frac{1}{2}\pi.$$

---

* I mean, of course, the first proof mentioned by Mr. Berry—the proof itself is classical.

It is obvious that this has a great advantage over the first proof in that only one inversion is involved. On the other hand the uniform convergence of the series to be integrated is, as Mr. Berry remarks, 'not quite obvious.' Moreover the fact that the cosecant series contains the terms

$$\frac{1}{x} - \frac{1}{x-\pi},$$

which become infinite at the limits, although it does not really add to the difficulty of the proof, does involve a slight amount of additional care in its statement. I am inclined to assign 15 marks, therefore, instead of 10 only, for this inversion. For the first three steps in the proof I assign 3 marks each; for the use of the cosecant series 4. Thus the total mark may be estimated at $9 + 15 + 4 = 28$.

3. *Mr. Berry's third proof (by contour integration).* It is naturally a little harder to estimate the difficulties of a proof which depends upon the theory of analytic functions. It seems to me not unreasonable to assign 10 marks for the use of Cauchy's theorem in its simplest form, when we remember that the proof of the theorem which depends upon the formula*

$$\iint \left(\frac{\partial q}{\partial x} - \frac{\partial p}{\partial y}\right) dx\,dy = \int (p\,dx + q\,dy)$$

involves the reduction of a double integral partly by integration with respect to x and partly by integration with respect to y. The actual theoretical difficulty of this proof of Cauchy's theorem I should be disposed to estimate at about 20; but half that amount seems sufficient for the mere use of a well known standard theorem.†

The particular problem of contour integration with which we are concerned is, however, by no means a really simple one. The range of integration is infinite, the pole occurs *on* the contour; we have to use the fact that

$$\lim_{R\to\infty} \lim_{\epsilon\to 0} \left(\int_{-R}^{-\epsilon} + \int_{\epsilon}^{R}\right) \frac{e^{ix}}{x} dx = \int_{-\infty}^{\infty} \frac{\sin x}{x} dx.$$

For this I assign 8 marks; for the evaluation of the limit of the integral round the 'small semicircle' another 8. There remains the proof that the integral round the 'large semicircle' tends to zero. As Mr. Berry points out, the ordinary proof of this is really rather difficult; I estimate its difficulty at 16. Mr. Berry has suggested a simplification of this part of the argument; the difficulty of

---

* Forsyth, *Theory of Functions*, p. 23.
† Goursat's proof is better and more general, and does not involve my inversion of limit-operations, but its difficulties are far too delicate for beginners.

his proof I estimate at 8. Thus we have $10+8+8+16 = 42$ marks for the ordinary proof, and $10+8+8+8 = 34$ for Mr. Berry's modification of it.

4. *Prof. Nanson's proof.* This proof, I must confess, does not appeal to me. Starting from the integral

$$u = \int_0^\infty \frac{a \cos cx}{a^2 + x^2} dx,$$

where a and c are positive, we show, by a repeated integration by parts, and a repeated differentiation under the integral sign, that u satisfies the equation

$$\frac{d^2 u}{da^2} = c^2 u.$$

As the upper limit is infinite we ought to assign 40 marks for this, following the principles we adopted in marking the first proof; but as the two operations of differentiation involve inversions of the same character I reduce this to 30, adding 6, however, for the repeated integration by parts.

The transformation $x = ay$ shows that u is a function of ac, so that $u = Ae^{ac} + Be^{-ac}$, where A and B are independent of both a or c. This step in the proof I assess at 4.

That $A = 0$ is proved by making c tend to ∞ and observing that

$$|u| < \int_0^\infty \frac{a dx}{a^2 + x^2} = \frac{1}{2}\pi.$$

For this I assign 6 marks.

That $c = \frac{1}{2}\pi$ is proved by making $c = 0$. Here we assume the continuity of the integral, and this should involve 20 marks. But as the proof is simple I reduce this to 12. Thus by the time that we have proved that

$$u = \frac{1}{2}\pi e^{-ca},$$

we have incurred $30+6+4+6+12 = 58$ marks.

We have then

$$\int_0^\infty \frac{a \sin mx \, dx}{x(a^2+x^2)} = \int_0^\infty \frac{a dx}{a^2+x^2} \int_0^m \cos cx \, dc$$

$$= \int_0^m u \, dc = \frac{\pi}{2a}(1 - e^{-am}).$$

This involves 20 marks. We then obtain

$$\int_0^\infty \frac{x \sin mx}{a^2 + x^2} dx = \frac{1}{2}\pi e^{-am}$$

by two differentiations with respect to m. This should strictly involve 40, and I cannot reduce the number to less than 30, as the second differentiation is none too easy to justify. The result then follows by multiplying the last formula but one by a and adding it to the second.

I do not think that the $58 + 20 + 30 = 108$ marks which we have assigned are more than the difficulties of the proof deserve. But of course it is hardly fair to contrast this heavy mark with the 40, 28, 42, 36 which we have obtained for the others; for Prof. Nanson has evaluated the integrals

$$\int_0^\infty \frac{\sin mx}{x(a^2+x^2)}dx, \int_0^\infty \frac{\cos mx}{a^2+x^2}dx, \int_0^\infty \frac{x\sin mx}{a^2+x^2}dx$$

as well as the integral (1).

5. *Mr. Michell's proof.** This depends on the equations

$$\int_0^\infty \frac{\sin x}{x}dx = \lim_{h\to 0, H\to\infty} \int_h^H \frac{\sin x}{x}dx$$

$$= -\lim_{h\to 0, H\to\infty} \int_h^H dx \int_0^{\frac{1}{2}\pi} \frac{d}{dx}\{e^{-x\sin\theta}\cos(x\cos\theta)\}d\theta$$

$$= -\lim_{h\to 0, H\to\infty} \int_0^{\frac{1}{2}\pi} d\theta \int_h^H \frac{d}{dx}\{e^{-x\sin\theta}\cos(x\cos\theta)\}dx$$

$$= -\lim_{h\to 0, H\to\infty} \int_0^{\frac{1}{2}\pi} \{e^{-H\sin\theta}\cos(H\cos\theta) - e^{-h\sin\theta}\cos(h\cos\theta)\}d\theta$$

$$= -(0 - \frac{1}{2}\pi) = \frac{1}{2}\pi.$$

I have myself for several years used a proof, in teaching, which is in principle substantially the same as the above, though slightly more simple in details and arrangement, viz.:

$$\int_0^\infty \frac{\sin x}{x}dx = \frac{1}{2i}\int_0^\infty dx \int_0^\pi e^{i(t+xe^{it})}dt$$

$$= \frac{1}{2i}\int_0^\pi e^{it}dt \int_0^\infty e^{ixe^{it}}dx$$

$$= \frac{1}{2}\int_0^\pi dt = \frac{1}{2}\pi.$$

The successive steps of the proof can of course be stated in a form free from i by merely taking the real part of the integrand. To justify the inversion

---

* This proof is also referred to by Prof. Nanson. The only proofs which are 'classical' are 1, 2, 3, viz. the three discussed by Mr. Berry.

28. The Integral $\int_0^\infty \frac{\sin x}{x} dx$ and Further Remarks on the Integral 317

we have to observe that

$$\int_0^\pi e^{it} dt \int_0^X e^{ixe^{it}} dx = \int_0^X dx \int_0^\pi e^{it+ixe^{it}} dt$$

for any positive value of X, in virtue of the continuity of the subject of integration, and that

$$\left| \int_0^\pi e^{it} dt \int_X^\infty e^{ixe^{it}} dx \right| = \left| \int_0^\pi e^{iXe^{it}} dt \right| < \int_0^\pi e^{-X \sin t} dt$$

$$< 2 \int_0^{\frac{1}{2}\pi} e^{-X \sin t} dt < 2 \int_0^{\frac{1}{2}\pi} e^{-2Xt/\pi} dt$$

$$= \pi(1 - e^{-X})/X,$$

the closing piece of argument being essentially the same as that at the end of Prof. Nanson's paper. Hence

$$\int_0^\pi \int_0^\infty = \lim_{X \to \infty} \int_0^\pi \int_0^X = \lim_{X \to \infty} \int_0^X \int_0^\pi = \int_0^\infty \int_0^\pi .$$

This proof involves an inversion of a repeated integral, one limit being infinite. This involves 20 marks. To this I add 15 on account of the artificiality of the initial transformation, deducting 3 on account of the extreme shortness and elegance of the subsequent work. Thus we obtain $20 + 15 - 3 = 32$ marks. To Mr. Michell's proof, as stated by Prof. Nanson, I should assign a slightly higher mark, say 40.

We may therefore arrange the proofs according to marks, thus:

1. Mr. Berry's second proof, - - - - - - 28
2. Mr. Michell's proof (my form), - - - - 32
3. Mr. Berry's third proof (his form), - - - 34
{ 4. Mr. Berry's first proof, - - - - - 40
{ 5. Mr. Michell's proof (his form), - - - - 40
6. Mr. Berry's third proof (ordinary form), - - 42
7. Prof. Nanson's proof, - - - - - - 108

And this fairly represents my opinion of their respective merits: possibly,[*] however, I have penalized Nos. 2 and 5 too little on the score of artificiality and 4

---

[*] This I know to be Mr. Berry's opinion. He would put 2 below 3 and 4, and 5 below 6. To compare 3 and 4 is very difficult, and they might fairly be bracketed.

Since writing this note I have recollected another proof which I had forgotten, and which is given in §173 (Ex. 1) of Mr. Bromwich's *Infinite Series*. Mr. Bromwich there too much on the score of theoretical difficulty. I conclude, however, with some confidence that Mr. Berry's second proof is distinctly the best. But whether this be so or not, one thing at any rate should be clear from this discussion: whatever method be chosen, the evaluation of the integral (1) is a problem of very considerable difficulty. This integral (and all the other standard definite integrals) lie quite outside the legitimate range of school mathematics.

G. H. HARDY.

I have been asked for my opinion of the difficulty of Prof. Dixon's proof, as compared with those of the proofs I discussed in my previous note on this subject (*Gazette*, v. 5, p. 98). The conciseness and ingenuity of the proof no one could question. Dr. Bromwich and Mr. Whipple have also sent me proofs that I had not seen or forgot to mention, and it seems appropriate that I should attempt to assign marks, according to the principles I then stated, to all three. Before I do so, however, I should like to mention two general principles suggested by Dr. Bromwich as comments on my system of marking, as I am myself of opinion that their acceptance makes my method fairer and easier to apply.

The first is, that when a proof involves two inversions resembling one another both in principle and detail, then the mark for the second should be reduced by one-half, *e.g.* from, 10 to 5.

The second rests on the following considerations.

Let us consider, for clearness, the case of an integration of a series over a finite range. Then certain integrations stand out by themselves as peculiarly easy to justify, viz. those in which the series is uniformly *and absolutely* convergent in virtue of "Weierstrass's M-Test" (Bromwich, *Infinite Series*, p. 113 and also p. 207). Such a case is that involved in

$$\int_0^\pi \frac{\cos n\theta \, d\theta}{1 - 2p\cos\theta + p^2} = \frac{1}{1-p^2} \int_0^\pi \left\{1 + 2\sum^\infty p^k \cos k\theta\right\} \cos n\theta \, d\theta = \frac{\pi p^n}{1-p^2}$$

---

proves that, provided the real parts of a and b are positive or zero,

$$\int_0^\infty (e^{-ax} - e^{-bx}) \frac{dx}{x} = \int_0^\infty \left(\frac{1}{a+y} - \frac{1}{b+y}\right) dy.$$

Putting $a = 1, b = i$, we obtain

$$\int_0^\infty (e^{-x} - \cos x) \frac{dx}{x} = 0, \quad \int_0^\infty \frac{\sin x}{x} dx = \frac{1}{2}\pi.$$

This proof, as presented by Mr. Bromwich, should be marked at about 45. It has the advantage of giving the values of several other interesting integrals as well, so that (as in the case of Prof. Nanson's proof) it is hardly fair to contrast this mark with the considerably lower marks obtained by some of the other proofs.

28. The Integral $\int_0^\infty \frac{\sin x}{x} dx$ and Further Remarks on the Integral

(where, of course, $|p| < 1$). Cases in which the series is uniformly but not absolutely convergent are intrinsically more difficult, depending as they do on one or other of the tests called by Dr. Bromwich "Abel's" and "Dirichlet's" (*l.c.* pp. 113–4), and, in the long run, on "Abel's Lemma" (p. 54). An example of this kind is the integration in what I called in my former note, "Mr. Berry's second proof."

A similar distinction is to be made in regard to proofs of the continuity of an infinite series, and, *mutatis mutandis*, proofs of the legitimacy of other types of inversions, although it cannot, of course, always be presented in so clear-cut a form.

Dr. Bromwich suggests that 15 marks at least should be assigned for any inversion of the more difficult type, and I am disposed to adopt this suggestion. The 10 or 15 marks are, of course, only intended as a rough standard by which to make a beginning, and would generally be modified in any particular case.

On looking through my former note, in the light of these remarks, I am disposed to make the following alterations in my original marks.

Proof 2.* Assign 25 instead of 20 for the inversion of the integrations, making the total 37 instead of 32. Also add 5 to the total for Proof 5, making it 45.

Proof 4. Add 5 marks in respect of the first inversion, making the total 45.

Proof 7. Deduct say 28, leaving 80.

Dr. Bromwich's proof, mentioned in the last footnote of my former note (which we may call Proof 8), I should now mark at about 50 instead of 45. In so far as it is concerned with this particular integral, it differs little from Proof 4. I proceed to consider the three new proofs.

Proof 9 (*Dr. Bromwich's second proof*). We have

$$\int_0^{\frac{1}{2}\pi} e^{-x\sin\theta} \cos(x\cos\theta) d\theta = \int_0^{\frac{1}{2}\pi} R[e^{ixe^{i\theta}}] d\theta$$

$$= \int_0^{\frac{1}{2}\pi} \left(1 - x\sin\theta - \frac{x^2}{2!}\cos 2\theta + \frac{x^3}{3!}\sin 3\theta + \cdots \right) d\theta$$

$$= \frac{1}{2}\pi - x + \frac{x^3}{3\cdot 3!} - \frac{x^5}{5\cdot 5!} + \cdots$$

$$= \frac{1}{2}\pi - \int_0^x \frac{\sin t}{t} dt.$$

---

* The numbers are those assigned to the proofs in my final list (*Gazette*, v. 5, p. 98).

But the original integral is numerically less than

$$\int_0^{\frac{1}{2}\pi} e^{-2x\theta/\pi} d\theta = \frac{\pi}{2x}(1 - e^{-x}) < \frac{\pi}{2x},$$

and the result follows on making x tend to ∞. This proof is due to Schlömilch, *Übungsbuch der höheren Analysis*, p. 173. It is to be observed that it and Proofs 2 and 5 are really variations of the same theme, and closely connected with the proof by contour integration. This being so, it is clear that the straightforward proof by contour integration should have a better mark than either; and I deduct 2 from the marks for Proofs 3 and 6, which I had perhaps marked a little severely.

As both integrations of a series in Proof 9 are of the simplest type, I give 10 marks for the first integration, 5 for the second, 5 for the summation of the trigonometrical series and 10 for the final step in the argument, making 30 in all. I add 5 more because the proof has a certain artificiality, though less than Mr. Michell's, as contrasted with such proofs as 1 and 4; the total is therefore 35.

Proof 10 (*Mr. Whipple's proof*). This depends on the transformations

$$\int_0^\infty \frac{\sin x}{x} dx = \lim_{h \to 0} h \sum_1^\infty \frac{\sin nh}{nh} = \lim_{h \to 0} \frac{1}{2}(\pi - h) = \frac{1}{2}\pi.$$

This proof has the great merit of being absolutely natural and straightforward in idea; unfortunately, however, the inversion is one of a very difficult character.* Also the assumption of a knowledge of the sum of the trigonometrical series is a rather serious matter. I cannot give less than $30 + 15 = 45$ marks. At the same time, the proof seems to me a most interesting and suggestive one.

Proof 11 (*Prof. Dixon's proof*).† This proof I find harder to mark than any. It is exceedingly concise, and it avoids all inversions of whatever kind in the most ingenious way; its difficulties are, indeed, of a kind which rather baffle my rules. Rules or no rules, I cannot regard it as on a par with Proof 1, nor, I think, as being really quite as simple as 3 or 9. At a rough guess, let us say 36. As it is absurd to try to make very fine distinctions, I alter the marks of 9 and

---

* See a paper by Dr. Bromwich and myself in the *Quarterly Journal*, vol. xxxix, p. 222; also Cesàro, *Algebraische Analysis*, p. 699.
† *Gazette*, vi, p. 225.

28. The Integral $\int_0^\infty \frac{\sin x}{x} dx$ and Further Remarks on the Integral

2 from 35 and 37 to 36 each also. My final list, then, works out as follows:

1. Proof 1 (*Mr. Berry's second*) - - - 28.
2. Proof 3 (*Mr. Berry's form of his third*) - 32.
3. ⎧ Proof 9 (*Dr. Bromwich's second*) - - - 36.
   ⎨ Proof 11 (*Prof. Dixon's*) - - - - 36.
   ⎩ Proof 2 (*My form of Mr. Michell's*) - - 36.
6. Proof 6 (*Ordinary form of Mr. Berry's third*) 40.
7. ⎧ Proof 4 (*Mr. Berry's first*) - - - - 45.
   ⎨ Proof 5 (*Mr. Michell's original*) - - - 45.
   ⎩ Proof 10 (*Mr. Whipple's*) - - - - 45.
10. Proof 8 (*Dr. Bromwich's first*) - - - 50.
11. Proof 7 (*Prof. Nanson's*) - - - - 80.

As I remarked before, Proofs 7 and 8, especially the first, are heavily penalised because they include proofs of a great deal of extraneous matter. As regards absence of serious theoretical difficulty, I think 9 is possibly the simplest of all. I regret the low position of 4 and 10; each is a good proof, proceeding to its goal by a straight and natural route, but unfortunately meeting with difficult theoretical obstacles. Finally, I may point out that Proof 1 is sometimes presented in a somewhat different form (see Bromwich, *Infinite Series*, p; 468), in which the integration of a series may be made to depend upon an "M-test," and which may possibly deserve a slightly lower mark.

IV
Tributes

29

Dr. Glaisher and the "Messenger of Mathematics"

(*Messenger of Mathematics*, 58 (1929), pp. 159–160)

J.W.L. Glaisher, founding editor of the *Messenger of Mathematics* and Fellow of the Royal Society. © London Mathematical Society

James Whitbread Lee Glaisher, who died on 7 Dec. 1928, at the age of 80, had been editor of the *Messenger of Mathematics* since the foundation of the journal in its present form in 1871. He was helped by other editors (W. Allen Whitworth, C. Taylor, W. J. Lewis, and R. Pendlebury) until 1877, after which he had sole control. He contributed to the second number of Vol. 1 in May 1871, and his last contribution appeared in Vol. 57 in July 1927. Few mathematicians can have expected that the *Messenger* would survive Glaisher's death, and in fact it dies with him. The present volume, which he had begun, is the last; and in future the *Messenger* will be absorbed in the new Oxford series of the *Quarterly Journal*, to begin in April of this year.

Various notices of Glaisher have appeared already, the most important being those by Prof. Forsyth in the *Journal of the London Math. Soc.* (Vol. 3, pp. 101–112) and by the Master of Trinity (Sir J. J. Thomson) in the *Cambridge Review*. I do not propose to repeat what has been said about him there, but only to add a few words about him as a mathematician, and in particular as an editor.

Glaisher was, I suppose, the last of the old school of mathematical editors, the men who, like Liouville, contrived to run mathematical journals practically unaided. At any rate it is safe to say that no one will again attempt to run two simultaneously, as Glaisher did. Glaisher got a little assistance from friends, but for the most part he accepted or rejected contributions on his own responsibility. The best editors have of course a wonderful power of judging the quality of a manuscript without attempting to read it; but the task becomes steadily more difficult as mathematics grows and specialises, and a man cannot always be worrying his friends for opinions. The position of an editor, without a properly constituted body of experts behind him, grows more and more thankless, and it is hardly likely that the experiment will be repeated.

At the same time a private journal, controlled autocratically by a kindly and discriminating editor, can serve many useful purposes, and particularly that of giving early encouragement to beginners, on whom societies, with councils and referees, are sometimes rather severe. As Glaisher said to me himself when I submitted my own first paper to him, he 'never discouraged a young man from rushing into print'. It is then only natural that a very large proportion of well known English mathematicians should have begun their career as authors in the *Messenger*. Baker, Barnes, Burnside, Elliott, and Forsyth, for example, all began there. It is a little more surprising to find that J. J. Thomson and Jeans both published their first mathematical papers in the *Messenger*, and that both of them started as experts in the theory of numbers.

The *Messenger* was of course always a 'minor' journal; it did not aim at the standards of the *Acta* or the *Annalen*, or even at that of the *Proceedings* of the London Mathematical Society. It occupied a comparatively humble position in the mathematical world, but a useful, individual, and honourable position, and we must all regret its extinction even if we accept it as inevitable.

I should like to end with a few words about Glaisher's position as a mathematician, because I think that he has generally been underestimated. He wrote a great deal, of very uneven quality, and he was 'old fashioned' in a sense which is most unusual now; but the best of his work is really good. This work is almost all arithmetical, but it belongs to a peculiar region of the theory of numbers; neither to the 'classical' theory on the one hand nor to the full-blooded 'analytic' theory on the other, but to the 'semi-analytic' theory of Kronecker, Liouville, Ramanujan, Mordell, or Bell, in which we apply to arithmetic not general principles of function theory but special properties of particular functions such as the elliptic modular functions.

The standard problem of this theory is that of the representation of numbers by sums of squares. Glaisher studied this problem in a series of elaborate papers in Vols. 36–39 of the *Quarterly Journal*, in which he considers not merely representations by any even number of squares from 2 to 18 but also 'classified' representations in which stated numbers of the squares are odd or even. Two of his most important results, concerning the representation of numbers $4k+3$ by 10 squares, and even numbers by 12, had been anticipated, as he points out himself, by Eisenstein and by Liouville respectively; but he proves many other very interesting theorems, and there are still unsolved problems about the new arithmetical functions which he introduced. This was probably Glaisher's most important work, but he did much more, and anyone who will take the trouble to work through the index of Dickson's *History* will probably be surprised at the number of striking theorems associated with Glaisher's name.

30
David Hilbert

(*Journal of the London Mathematical Society*, 18 (1943), pp. 191–192)*

David Hilbert, a mathematician of great range, who made contributions to invariant theory, geometry, functional analysis, and mathematical logic. The 23 Problems that he posed at the 1900 International Congress of Mathematicians in Paris were very influential in the development of 20th century mathematics, and they came to be known as the **Hilbert Problems**, bringing fame to anyone who solved one of them.

---

* Reprinted with the kind permission of the publisher, the London Mathematical Society.

From the Editors

David Hilbert died in Göttingen in 1943, at the height of World War II, and there was clearly little if any communication between the mathematical community that had remained in Germany and the mathematicians in the rest of Europe and in the United States. Still, Hardy explained, because Hilbert was such a towering figure in mathematics, his death had to be noted in Britain. Here he writes glowingly of Hilbert's place in the mathematical world and promises an expanded obituary at the end of the War. Of course, by 1945 when the War ended, Hardy was in failing health and died two years later before he could take up the task.

David Hilbert, who had been an Honorary Member of the Society since 1901, died during the summer, and the Council feel that the death of so great a mathematician should not pass, even momentarily, unnoticed; but the difficulties of obtaining any adequate account of his work and influence are at present insuperable. They have therefore decided to print this short notice and defer the publication of a more appropriate obituary until later.

Hilbert was born in 1862. He was Privatdozent in Königsberg from 1886 to 1892, when he succeeded Hurwitz as Professor Extraordinarius. He was called to Göttingen as Ordinarius in 1895, and remained there for the rest of his life. His profoundly original researches, his inspiring teaching, and his very individual personality had far-reaching effects on the development of mathematics both in his own and in other countries; it was he who did most to make Göttingen rank so long with Paris as one of the first two mathematical centres of the world.

Almost all of Hilbert's greatest contributions to constructive mathematics were made before 1911, and are recounted in Poincaré's *Rapport sur le prix Bolyai*, published in 1911 in the *Acta Mathematica*. After 1918 he was occupied almost exclusively with logic and the foundations of mathematics. It had always been his habit to concentrate on one subject at a time, to make or renew it by his discoveries, and then to leave it and forget it. He began with algebra and invariant theory (1888–92); and passed successively to algebraic theory of numbers (1892–9), foundations of geometry (1898–1903), calculus of variations and Dirichlet's principle (1899–1905), and integral equations (1901–12). There is little overlapping: thus he published nothing on number-theory after 1899, except his solution of "Waring's Problem", an isolated effort in a quite different part of the subject. There was a short period (1912–7) during which he occupied himself with the mathematical foundations of physics, but after this he became a "logician", and remained one for the rest of his active life.

There is none of these fields in which Hilbert did not make discoveries of outstanding importance, and several which he transformed entirely. In three cases he gave a systematic account of his work, in the *Bericht über die Theorie der algebraischen Zahlkörper* (1897), the *Grundlagen der Geometrie* (of which many editions and translations have been published), and the *Grundzügen eine allgemeinen Theorie der linearen Integralgleichungen* (1912). The first of these is reprinted in Vol. 1 of his *Gesammelte Abhandlungen* (1932–6). Vol. 3 of the *Abhandlungen* contains an account of Hilbert's life, and a reasoned estimate of his work, by O. Blumenthal, and this and Poincaré's *Rapport* between them give a fairly comprehensive picture of one of the greatest mathematicians of our time.

31

Edmund Landau

G. H. Hardy and H. Heilbronn

(*Journal of the London Mathematical Society* 13 (1938), pp. 302–310)*

Edmund Landau gave the first systematic presentation of analytic number theory.

Edmund Landau, an honorary member of this Society since 1924, was born in Berlin on 14 February 1877, and died there on 19 February 1938. He was the son of Professor Leopold Landau, a well-known gynaecologist. After passing

---

* Reprinted with the kind permission of the publisher, the London Mathematical Society.

through the "French Gymnasium" in Berlin, he entered the University of Berlin as a student of mathematics, and remained there, apart from two short intervals in Munich and Paris, until 1909. His favourite teacher was Frobenius, who lectured on algebra and the theory of numbers. Landau worked through these lectures very thoroughly, and used his notes of them throughout his life. He took his doctor's degree in 1899, and obtained the "venia legendi", or right to give lectures, in 1901.

In 1909 Landau succeeded Minkowski as ordinary professor in Göttingen. The University of Göttingen was then in its mathematical prime; Klein and Hilbert were Landau's colleagues, and young mathematicians came to Göttingen for inspiration from every country. After the war the University recovered its position quickly, so that Landau had always ample opportunities of training able pupils, and often had a decisive influence on their careers.

In 1933 the political situation forced him to resign his chair. He retired to Berlin, but still lectured occasionally outside Germany. In 1935 he came to Cambridge as Rouse Ball Lecturer, and gave the lectures which he developed later into a Cambridge Tract. He continued to take an active interest in mathematics, and his last lectures were in Brussels in November 1937, only a few months before his sudden death.

Landau's first and most abiding interest was the analytic theory of numbers, and in particular the theory of the distribution of primes and prime ideals. In his "doctor-dissertation" (**1**) he gave a new proof of the identity

$$(1) \qquad \sum_{1}^{\infty} \frac{\mu(n)}{n} = 0$$

(conjectured by Euler and first proved by von Mangoldt), and this was the first of a long series of papers on the zeta-function and the theory of primes.

The "prime number theorem"

$$(2) \qquad \pi(x) \sim \frac{x}{\log x}$$

was first proved by Hadamard and de la Vallée-Poussin in 1896. De la Vallée-Poussin went further, and proved that

$$(3) \qquad \pi(x) = \int_{2}^{x} \frac{dt}{\log t} + O\{xe^{-A\sqrt{(\log x)}}\} = \mathrm{li}\, x + O\{xe^{-A\sqrt{(\log x)}}\},$$

for a certain positive A. The proofs of both Hadamard and de la Vallée-Poussin depended upon Hadamard's theory of integral functions, and in particular on the fact that $\zeta(s)$, apart from a simple pole at $s = 1$, is regular all over the complex plane.

In 1903 Landau found (in **2**) a new proof of the prime number theorem which does not depend upon the general theory of Hadamard. For this proof we need know only that $\zeta(s)$ can be continued "a little way over" the line $\sigma = 1$; we do not need the functional equation, the Weierstrass product, and the other machinery used in the earlier proofs. On the other hand we do not obtain quite so precise a formula as (3).

This discovery of Landau's was very important, since it permitted a decisive step in the theory of the prime ideals of an algebraic field κ. This theory depends upon the properties of the Dedekind zeta-function

$$\zeta_\kappa(s) = \sum \frac{1}{(Na)^s}, \tag{4}$$

where a runs through the integer ideals of κ (except 0), and Na is the norm of a. It was not proved until much later (by Hecke in 1917) that $\zeta_\kappa(s)$ can be continued all over the plane, but Landau had no difficulty in showing that it has properties like those of $\zeta(s)$ used in his proof of the prime number theorem. He thus obtained the "prime ideal theorem": if $\pi_\kappa(x)$ is the number of prime ideals of κ whose norm is less than x, then

$$\pi_\kappa(x) \sim \frac{x}{\log x}. \tag{5}$$

Later*, using Hecke's discoveries, he proved the formula for $\pi_\kappa(x)$ corresponding to (3).

The logic of prime number theory has developed a good deal since 1903 and even since 1917, and Landau kept fully in touch with all these developments. Thus his paper **12** contains the shortest and most direct proof of the prime number theorem known today (a proof based on the ideas of Wiener). He was also intensely interested in the logical relations between different propositions in the theory. Thus he first proved (in **6** and **10**) that (1) and (2) are "equivalent", that each can be deduced from the other by "elementary" reasoning, although there is no "elementary" proof of either.[†] It was Landau who first enabled experts to classify the theorems of prime number theory according to their "depths".

Landau's second big discovery was in an entirely different direction. Picard's theorem states that an integral function, not a constant, assumes all values with at most one exception. Picard deduced his theorem in 1879 from the

---

* C, §20.

† The proof that (2) implies (1) is given in A, §156, but that of the converse implication is later.

properties of the "modular function", and it was not until 1896 that Borel found the first elementary proof.

In 1904 Landau, studying Borel's proof, made a most important, and then very unexpected, extension (3). If a_0 and a_1 are given, $a_1 \neq 0$, and

$$f(x) = a_0 + a_1 x + \cdots$$

is regular at the origin, then there is a number $\Omega = \Omega(a_0, a_1) > 0$, depending on a_0 and a_1 only, such that $f(x)$, if regular in the circle $|x| < \Omega$, must assume one of the values 0 and 1 somewhere in the circle. It is obvious that this theorem includes Picard's theorem.

A few weeks later Schottky developed Landau's theorem further. Suppose that $a_0 \neq 0$, $a_0 \neq 1$, and $0 < \vartheta < 1$. Then there is a number $\Phi = \Phi(a_0, \vartheta)$ with the following property: if $f(x)$ is regular, and never 0 or 1, for $|x| < 1$, and $f(0) = a_0$, then

$$|f(x)| < \Phi(a_0, \vartheta)$$

for $|x| < \vartheta$. Landau's theorem is a simple corollary*.

Schottky's theorem was imperfect in one important respect, since his function Φ was unbounded near $a_0 = 0$ and $a_0 = 1$. Landau, in **13**, removed this imperfection. Suppose that $a > 0, 0 < \vartheta < 1$. Then there is a number $\Psi = \Psi(a, \vartheta)$ with the property: if $f(x)$ is regular, and never 0 or 1, for $|x| < 1$, and $|f(0)| \leq a$, then

$$|f(x)| < \Psi(a, \vartheta)$$

for $|x| < \vartheta$. In **13** he makes an important application to the theory of $\zeta(s)$ and $\zeta_\kappa(s)$.

These theorems have inspired a great amount of later work. Carathéodory, for example, found the "best" Ω in terms of the modular function. The elementary proofs have also been transformed by the discovery of "Bloch's theorem", and are developed in this way in the second edition of Landau's *Ergebnisse* (B).

This theorem and the prime ideal theorem were probably the most striking of Landau's original discoveries. We state a few more with the minimum of comment.

(1) Every large positive integer is a sum of at most 8 positive integral cubes (**9**).

This 8 is the only number which has resisted the later analytic attacks on Waring's problem.

---

* See B (ed. ii), p. 103.

(2) Every positive definite polynomial with rational coefficients can be represented as a sum of 8 squares of polynomials with rational coefficients. In particular, every positive definite quadratic with rational coefficients is a sum of 5 squares of rational linear functions; and 5 is the best possible number (**5**).

Mordell has since proved the corresponding theorem for quadratics with *integral* coefficients.

(3) If $f(x) \sim x$ when $x \to \infty$, and $xf'(x)$ increases with x, then $f'(x) \to 1$ (**8**, 218).

This theorem (which contains the kernel of a differencing process used by de la Vallée-Poussin and others in the analytic theory of numbers) is, perhaps, the first genuine example of a "O-Tauberian" theorem.

(4) If $f(x) = a_0 + a_1 x + a_2 x^2 + \cdots$ is regular, and $|f(x)| < 1$, for $|x| < 1$, then

$$|a_0 + a_1 + \cdots + a_n| \leq 1 + \left(\frac{1}{2}\right)^2 + \left(\frac{1 \cdot 3}{2 \cdot 4}\right)^2 + \cdots + \left(\frac{1 \cdot 3 \cdots 2n-1}{2 \cdot 4 \cdots 2n}\right)^2.$$

There is equality, for every n, with an appropriate $f(x)$ depending on n (**11**).

(5) A Dirichlet's series $\sum a_n e^{-\lambda_n s}$, with non-negative coefficients, has a singularity at the real point of its line of convergence (**4**).

This had been proved before for power-series by Vivanti and Pringsheim, but their method of proof cannot be extended to the general case.

(6) If $N(\sigma_0 T)$ is the number of zeros of $\zeta(s)$ in the domain $\sigma \geq \sigma_0 > \frac{1}{2}$, $|t| \leq T$, then

$$N(\sigma_0, T) = o(T).$$

This was proved, first with O and then as stated, by Bohr and Landau in their joint papers **14** and **15**. It is known that $N(\frac{1}{2}, T)$ is of order $T \log T$, so that most of the zeros of $\zeta(s)$ lie very near $\sigma = \frac{1}{2}$. This was the first successful attempt to show that the Riemann hypothesis is at any rate "approximately" true. Carlson proved later that $o(T)$ may be replaced by $O(T^a)$, where $0 < a < 1$, and Titchmarsh and Ingham have since improved Carlson's value of a.

Landau also published a very large number of new, shorter, and simpler proofs of known theorems. We mention only his well-known proof of Weierstrass's approximation Theorem (**7**). This depends upon the singular integral

$$\int_{-\frac{1}{2}}^{\frac{1}{2}} \{1 - (u-x)^2\}^n f(u) du,$$

and was perhaps the first proof in which the approximating functions are "visibly" polynomials. It is reproduced, in a more general form, in Hobson's book (vol. ii, ed. 2, 459–461).

Landau wrote over 250 papers, but it is possible that he will be remembered first for his books, of which he wrote seven.

A. *Handbuch der Lehre von der Verteilung der Primzahlen* (Leipzig and Berlin, Teubner, 1909: 2 vols., 961 pp.).

B. *Darstellung und Begründung einiger neuerer Ergebnisse der Funktionentheorie* (Berlin, Springer, 1916; second edition, 1929: 122 pp.).

C. *Einführung in die elementare und analytische Theorie der algebraischen Zahlen und Ideale* (Leipzig and Berlin, Teubner, 1918; second edition, 1927: 147 pp.).

D. *Vorlesungen über Zahlentheorie* (Leipzig, Hirzel, 1927: 3 vols., 1009 pp.).

E. *Grundlagen der Analysis* (Leipzig, Akademische Verlagsgesellschaft, 1930: 134. pp.).

F. *Einführung in die Differentialrechnung und Integralrechnung* (Groningen, Noordhoff, 1934: 368 pp.).

G. *Über einige neuere Fortschritte der additiven Zahlentheorie* (Cambridge Tracts in Mathematics, No. 35, 1937: 94 pp.).

Of these books, E and F are elementary, and we say nothing about them, interesting and individual as they are. All the rest are works of first-rate importance and high distinction.

Landau was the complete master of a most individual style, which it is easy to caricature (as some of his pupils sometimes did in an amusing way*), but whose merits are rare indeed. It has two variations, the "old Landau style", best illustrated by the *Handbuch*, which sweeps on majestically without regard to space, and the "new Landau style" of his post-war days, in which there is an incessant striving for compression. Each of these styles is a model of its kind. There are no mistakes—for Landau took endless trouble, and was one of the most accurate thinkers of his day—no ambiguities, and no omissions; the reader has no skeletons to fill, but is given every detail of every proof. He may, indeed, sometimes wish that a little more had been left to his imagination, since half the truth is often easier to picture vividly than the whole of it, and the very completeness of Landau's presentation sometimes makes it difficult to grasp the "main idea". But Landau would not, or could not, think or write vaguely, and a reader has to read as precisely and conscientiously as Landau wrote. If he will do so, and if he will then compare Landau's discussion of a theorem

---

* For example in a mock *Festschrift* written on the occasion of his declining an invitation to leave Göttingen for another university.

with those of other writers, he will be astonished to find how often Landau has given him the shortest, the simplest, and in the long run the most illuminating proof.

The *Handbuch* was probably the most *important* book he wrote. In it the analytic theory of numbers is presented for the first time, not as a collection of a few beautiful scattered theorems, but as a systematic science. The book transformed the subject, hitherto the hunting ground of a few adventurous heroes, into one of the most fruitful fields of research of the last thirty years. Almost everything in it has been superseded, and that is the greatest tribute to the book.

Landau would not publish a second edition of the *Handbuch* (which must necessarily have been a new book), but preferred to incorporate the results of later researches in his *Vorlesungen*, which is no doubt his *greatest* book. This remarkable work is complete in itself; he does not assume (as he had done in the *Handbuch*) even a little knowledge of number-theory or algebra. It stretches from the very beginning to the limits of knowledge, in 1927, of the "additive", "analytic", and "geometric" theories. Thus part 6 (vol. i, pp. 235–269) carries the solution of Waring's problem to where it stood before Vinogradov's recent work. Part 12 (vol. iii, pp. 201–328) contains practically everything then known about "Fermat's last theorem", and the rest of the book is conceived on the same scale. In spite of this enormous programme, Landau never deviates an inch from his ideal of absolute completeness. For example, he never refers to his *Algebraische Zahlen*, but proves from the beginning everything he needs.

The richness of content of the book, and the power of condensation it shows, are astonishing. Thus the classical theorems about decompositions into two, three, and four squares are proved in twenty-eight pages (vol. i, pp. 97–125). And Landau can find room (vol. i, pp. 153–171) for four different evaluations of Gauss's sums.

The *Vorlesungen* is not only Landau's finest book but also, in spite of the great difficulty and complexity of some of the subject matter, the most agreeably written. The style here is the rather informal style of his lectures, which he was persuaded by his friends to leave unchanged.

The *Algebraische Zahlen* gives a short and self-contained account (pp. 1–54) of the theory of algebraic numbers and ideals, intended as an introduction to the proofs of the prime ideal theorem and its refinements which occupy the remainder of the book. He does not go so deeply into the algebraic theory as, for example, Hecke, being content with what is required for his applications.

The *Ergebnisse* is probably Landau's most *beautiful* book. It contains a collection of elegant, significant, and entertaining theorems of modern function theory: Hadamard's and Fabry's "gap theorems", Fatou's theorem, the most striking "Tauberian" theorems, Bloch's theorem, the Picard-Landau group of theorems, and the fundamental theorems concerning "schlicht" functions. It is one of the most attractive little volumes in recent mathematical literature, and the most effective answer to any one who suggests that Landau's mathematics was dull.

Finally, his last work, the Cambridge Tract originating from his Rouse Ball lecture, gives an account of Vinogradov's "Waring" and Schnirelmann's "Goldbach" theorems, and of a group of half solved "elementary" problems of additive number theory which open a new field of research for young and unprejudiced mathematicians. There is a review of this tract by Mr. Ingham in the current volume of the *Mathematical Gazette*, to which we should have little to add.

Landau was certainly one of the hardest workers of our times. His working day often began at 7 a.m. and continued, with short intervals, until midnight. He loved lecturing, more perhaps even than he realized himself; and a lecture from Landau was a very serious thing, since he expected his students to work in the spirit in which he worked himself, and would never tolerate the tiniest rough end or the slightest compromise with the truth. His enforced retirement must have been a terrible blow to him; it was quite pathetic to see his delight when he found himself again in front of a blackboard in Cambridge, and his sorrow when his opportunity came to an end.

No one was ever more passionately devoted to mathematics than Landau, and there was something rather surprisingly impersonal, in a man of such strong personality, in his devotion. Everybody prefers to do things himself, and Landau was no exception; but most of us are at bottom a little jealous of progress by others, while Landau seemed singularly free from such unworthy emotions. He would insist on his own rights, even a little pedantically, but he would insist in the same spirit and with the same rigour on the rights of others.

This was all part of his passion for order in the world of mathematics. He could not stand untidiness in his chosen territory, blunders, obscurity, or vagueness, unproved assertions or half substantiated claims. If X had proved something, it was up to X to print his proof, and until that happened the something was nothing to Landau. And the man who did his job incompetently, who spoilt Landau's world, received no mercy; that was the unpardonable sin in Landau's eyes, to make a mathematical mess where there had been order before.

Landau received many honours in his lifetime. He was a member of the Academies of Berlin, Göttingen, Halle, Leningrad, and Rome; but no honour seemed to please him quite so much as his election to honorary membership of this society, and he came specially from Germany to attend our sixtieth anniversary dinner. This was natural, since there was no country where his reputation stood quite so high as in England, and none where his work has borne more fruit.

References

[This list contains only papers referred to in the notice.]

1. "Neuer Beweis der Gleichung $\sum \mu(n)/n = 0$", *Inaugural-Dissertation* (Berlin, 1899).
2. "Neuer Beweis des Primzahlsatzes und Beweis des Primidealsatzes" *Math. Annalen*, 56 (1903), 645–670.
3. "Über eine Verallgemeinerung des Picardschen Satzes", *Berliner Sitzungsberichte* (1904), 1118–1133.
4. "Über einen Satz von Tschebyschef", *Math. Annalen*, 61 (1905), 527–550.
5. "Über die Darstellung definiter Funktionen durch Quadrate", *Math. Annalen*, 62 (1906), 272–285.
6. "Über den Zusammenhang einiger neuerer Sätze der analytischen Zahlentheorie", *Wiener Sitzungsberichte*, 115, 2a (1906), 589–632.
7. "Über die Approximation einer stetigen Funktion durch eine ganze rationale Funktion" *Rend. di Palermo*, 25 (1908), 337–345.
8. "Beiträge zur analytischen Zahlentheorie", *Rend. di Palermo*, 26 (1908), 169–302.
9. "Über eine Anwendung der Primzahltheorie auf das Waringsche Problem in der elementaren Zahlentheorie", *Math. Annalen*, 66 (1909), 102–105.
10. "Über die Äquivalenz zweier Hauptsätze der analytischen Zahlentheorie", *Wiener Sitzungsberichte*, 120, 2a (1911), 973–988.
11. "Abschätzung der Koeffizientensumme einer Potenzreihe", *Archiv d. Math. u. Phys*, (3), 24 (1916), 250–260.
12. "Über den Wienerschen neuen Weg zum Primzahlsatz.", *Berliner Sitzungsberichte* (1932), 514–521.

With Harald Bohr.

13. "Über das Verhalten von $\zeta(s)$ und $\zeta_\kappa(s)$ in der Nähe der Geraden $\sigma = 1$", *Göttinger Nachrichten* (1910), 303–330.
14. "Ein Satz über Dirichletsche Reihen mit Anwendungen auf die ζ-Funktion und die L-Funktionen", *Rend. di Palermo*, 37 (1914), 269–272.
15. "Sur les zéros de la fonction $\zeta(s)$ de Riemann", *Comptes rendus*, 158 (1914), 106–110.

32
Gösta Mittag-Leffler
(*Journal of the London Mathematical Society* 3 (1928), pp. 156–160)*

Gösta Mittag-Leffler, a Swedish mathematician, was the first professor of mathematics at Stockholm University. He was a member of 30 learned societies including the Royal Society of London. He was a strong advocate of women's rights, and played a key role in Marie Curie being awarded a Nobel Prize. He founded the journal *Acta Mathematica*.

---

* Reprinted with the kind permission of the publisher, the London Mathematical Society.

Magnus Gösta Mittag-Leffler, who died last year at the age of eighty-one, the oldest of our honorary members, was for many years the recognized leader of the Scandinavian school of mathematics.

Mittag-Leffler was a remarkable man in many ways. He was a mathematician of the front rank, whose contributions to analysis had become classical, and had played a great part in the inspiration of later research; he was a man of strong personality, fired by an intense devotion to his chosen study; and he had the persistence, the position, and the means to make his enthusiasms count. He thus won for himself a position unlike that of any other Scandinavian mathematician, and one which it was a little difficult for anyone who had not met him in his own country to realize. I can remember well the occasion when he lectured for the last time to a Scandinavian Congress, at Copenhagen in 1925, and the whole audience rose and stood as he entered the room. It was a reception rather astonishing at first to a visitor from a less ceremonious country; but it was an entirely spontaneous expression of the universal feeling that to him, more than to any other single man, the great advance in the status of Scandinavian mathematics during the last fifty years was due.

Mittag-Leffler had been overwhelmed with honours; he was a doctor, for example, of four English or Scottish universities; he had been a president or vice-president of endless congresses; and he was an honorary member of almost every scientific society in the world. He travelled continually, and there was no mathematician who was so familiar a figure abroad. But to the outside world he was, above everything, the editor of the *Acta Mathematica*[*][†], the famous journal which he founded, and, with the co-operation of a committee of the four Scandinavian countries[‡], edited for forty-five years.

Periodicals are the most important material facts in the mathematical world. The *Acta* took its place in the front rank of mathematical periodicals from the beginning, and its rank and standard have never declined. It has always been the most completely international of all mathematical journals. All such journals are international to some extent; *Crelle's Journal* made its reputation on the works of Abel; the leading German periodicals of the present day, the *Mathematische Annalen* and the *Mathematische Zeitschrift*, are very largely international; and we know, even from our own short experience, how essential to a new journal is some sort of international connexion. The *Acta* was always

---

[*] Honorary member of the society 1892–1927.
[†] A fuller account of Mittag-Leffler's activities in connexion with the *Acta* will be found in Nörlund's notice of him in vol. 50. Prof. Nörlund very kindly allowed me to see the proof-sheets of this notice, and I have drawn from it freely.
[‡] Including Finland.

international in a more thoroughgoing sense than this. It would hardly have been practicable, in 1881, to support a mathematical journal of the first rank on Scandinavian mathematics alone; and Mittag-Leffler understood this, and set himself from the beginning a standard which, without the co-operation of mathematicians of all countries, it would have been quite impossible to maintain.

Two great mathematicians in particular, Poincaré and Cantor, contributed freely to the early volumes. It is difficult to realize to-day how Cantor had to struggle for recognition. The mathematicians of his time, however, were singularly slow to understand the overwhelming importance of his work; and Mittag-Leffler, a younger man, was one of the first to recognize it, and to apply Cantor's ideas successfully in the region of "ordinary" analysis. In Mittag-Leffler's last published paper he recalls this with very justifiable pride*. But Mittag-Leffler was always a good judge of the quality of the work submitted to him for publication. Even in his later years, when most of the editorial work was delegated to others, he retained that curious sense which enables the great editor to feel the value of work at which he has hardly glanced; and it was this gift which enabled him to carry the heavy responsibility which he had assumed in a way which would hardly have been possible to any other mathematician. It was a heavy responsibility in more ways than one, for the finances of the *Acta*, like those of other journals, were imperilled by the war, and there was a period when Mittag-Leffler had to supplement the Government subventions heavily himself. Like other journals, the *Acta* has overcome its troubles, and Mittag-Leffler could no doubt feel, when the last three volumes were dedicated to him as a *Festschrift* on his eightieth birthday, that the future of his great foundation was secure.

Mittag-Leffler's own work is occupied almost entirely with the general theory of functions analytic in Weierstrass's sense. He was an excellent writer. His account of the foundations of the theory of elliptic functions, modelled, no doubt, on Weierstrass's lectures which he had attended in Berlin two years before, published first in Helsingfors in 1876, and translated into English in 1923 in the *Annals of Mathematics*, is a really admirable piece of exposition, and makes one regret that he never wrote a treatise on a larger scale. He occupied himself on various occasions with fundamentals in the theory of functions. For example, he gave a proof of Cauchy's theorem, in 1875, which marked at the time a very definite advance. Like Goursat after him, he divides the contour

---

* A short note on a very interesting historical article by Schönflies ["Die Krisis in Cantor's mathematischem Schaffen", *Acta Math.*, 50 (1927), 1–23]. This article reproduces a number of Cantor's letters to Mittag-Leffler.

into elementary contours by a network, and argues directly from the definition of the derivative in every mesh; but he assumes what amounts substantially to uniform differentiability, and so misses what is really characteristic in Goursat's proof. He also wrote a number of memoirs on linear differential equations.

His best known and most characteristic work, however, is concerned with the problem of the analytic representation of a one-valued function (or one-valued branch of a function) with assigned singularities, and falls into two halves associated with the periods of his life following on the years 1877 and 1900. It was in 1877 that Mittag-Leffler first published the classical "Mittag-Leffler's theorem" of the text-books, a theorem suggested, of course, by Weierstrass's factor-theorem, but in no sense a corollary of it. If $a_1, a_2, \ldots$ is any sequence of numbers such that $|a_n|$ tends to infinity with n, and

$$g_n\left(\frac{1}{z-a_n}\right) = \frac{A_{n,1}}{z-a_n} + \cdots + \frac{A_{n,v_n}}{(z-a_n)^{v_n}}$$

is an arbitrary polynomial associated with a_n, then there is a one-valued analytic function $f(z)$ which is regular in any part of the plane free from the points a_n and has at a_n a pole of the type g_n. Such a function may be defined by a series

$$f(z) = \Sigma \left\{ g_n\left(\frac{1}{z-a_n}\right) - h_n(z) \right\}$$

where h is an appropriate polynomial in z; and the most general $f(z)$ differs from this one by an arbitrary integral function. In a later memoir—his first contribution to the *Acta*, in 1884—Mittag-Leffler gave various generalizations of his theorem, of which treatises such as those of Forsyth and Osgood give a full account.

In the second series of memoirs, the first of which appeared in 1900, we find the problem of analytic continuation envisaged in an entirely modern form. Mittag-Leffler considers the general problem of continuation for an element of any analytic function regular, say, at $z = a$. He introduces for the first time the idea, now well established as of fundamental importance in all such investigations, of the "star" (*Hauptstern*) associated with a; the region obtained by removing from the plane the further segments of all radii issuing from a and passing through singular points. He solves in various manners and in explicit form the problem of representing the function by an analytical expression convergent throughout the whole interior of the star. Some of these solutions are series of polynomials in z, such as

$$\lim_{n\to\infty} \sum_{m_1=0}^{n^2} \sum_{m_2=0}^{n^4} \cdots \sum_{m_n=0}^{n^{2n}} \frac{f^{(m_1+m_2+\cdots+m_n)}(a)}{m_1! m_2! \ldots m_n!} \left(\frac{z-a}{n}\right)^{m_1+m_2+\cdots+m_n};$$

while others are limits of associated functions, such as

$$\lim_{a\to 0} \sum_{n=0}^{\infty} \frac{c_n z^n}{\Gamma(1+an)},$$

where we suppose that a is zero and that $\Sigma c_n z^n$ is the Taylor's series of $f(z)$. Representations similar to the last, but involving slightly different convergence factors, such as n^{-an}, have also been used with much effect by other writers, such as Lindelöf and Le Roy. In particular, the analytic continuation of $(1-z)^{-1}$ over the plane cut along $(1, \infty)$ is given by

$$\lim_{a\to 0} E_a(z) = \lim_{a\to 0} \sum \frac{z^n}{\Gamma(1+an)},$$

a very interesting special function of whose properties Mittag-Leffler himself and Wiman made a detailed study, which may have suggested to the latter his well known theorem concerning integral functions of order less than $\frac{1}{2}$. A glance at Bieberbach's article in the *Enzyklopädie** is enough to show the range of later investigations which have their root in Mittag-Leffler's researches.

Mittag-Leffler had two homes between which he divided his time in Sweden. His principal residence was in Djursholm, in the suburbs of Stockholm. Here was his library, the finest mathematical library in the world, perhaps, at any rate for a working mathematician, and now in the hands of the Mittag-Leffler Institut and the Swedish Academy of Sciences. All books and periodicals were there (or all, at any rate, that any sane man could want), and a colossal collection of "separata"; and if one got tired one could read the correspondence of all the mathematicians in the world, or enjoy the view of Stockholm from the roof. He had also a country house at Tällberg, some two hundred miles to the north, in the beautiful and rather primitive Dalarne country. It was a delightful experience to stay there, and there perhaps Mittag-Leffler appeared at his best, a most entertaining mixture of the great international mathematician and the rather naive country squire. He was a strong nationalist, in spite of his internationalism, as anyone who lived in so beautiful a country well might be; and he loved his house and his garden and his position as the landowner of the countryside. Even his tennis court, "the most northerly tennis court in the world"—that shapeless mass of disintegrating rubble, with holes in which the ball remained stationary where it pitched—he regarded with an entirely disarming pride. I am sure there can be no one who has ever stayed at Tällberg who is not saddened by the thought that he will never stay there again.

---

* "Neuere Untersuchungen über Funktionen von komplexen Variabeln", *Enzykl. der Math. Wiss.*, II C4.

Mittag-Leffler played a very great part in the history of the mathematics of his time. He had exceptional opportunities, and exactly the blend of qualities required to take advantage of them to the full. There have been greater mathematicians during the last fifty years, but no one who has done in his way more for mathematics.

V
Book Reviews

33

Book Review of A First Course in the Differential and Integral Calculus *and* A Treatise on the Integral Calculus, founded on the Method of Rates

(*Mathematical Gazette* 4 (1907), pp. 307–309)*

A First Course in the Differential and Integral Calculus. **By W. F. Osgood. Pp. xv, 423. New York: The Macmillan Company. 1907.**

A Treatise on the Integral Calculus, founded on the Method of Rates. **By W. Woolsey Johnson. Pp. v, 440. New York: John Wiley & Sons. 1907.**

No one who wishes to realise the enormous advance that has been made during the last ten or twenty years in the teaching of the Calculus could do better than to read and contrast these two books, published in the same country and in the same year, but differing completely in their objects and methods and representing one the old school and the other the new.

It is no disparagement of Prof. Johnson's book to say that it might have been written thirty years ago. It is a book of a frankly old-fashioned type, another *Todhunter* or *Williamson*. It may be better or worse than *Todhunter*; on the whole we prefer it to *Todhunter*: but it is with such books as *Todhunter* that we must compare it.

We must not be understood as implying that Prof. Johnson's book may not be of considerable utility. There is plenty of room for an improved *Todhunter*. It is, as Prof. Osgood himself remarks, of the utmost importance that the formal side of the Calculus should be taught thoroughly in a first course; no one can

---

* Reprinted with the kind permission of the publisher of *The Mathematical Gazette*: The Mathematical Association.

be a good mathematician who cannot differentiate

$$\arccos\left(\frac{\alpha\cos x+b}{a+b\cos x}\right),$$

or show a little ingenuity in devices of integration. And it is at any rate open to argument that the best way to impart this most necessary training is by means of a course in which the theoretical difficulties clustering around the foundations of the Calculus are deliberately or unconsciously ignored. That some of these difficulties should, at first, be passed over in silence, or almost in silence, is indeed obviously necessary. It would be ridiculous to insist that nobody should learn how to determine maxima and minima until he can prove that a continuous function attains its upper and lower limits, or to calculate areas until he can prove their existence. But there are different ways of passing over difficulties. We may simply and absolutely ignore them: that is a course for which there is often much to be said. We may point them out and avowedly pass them by; or we may expand a little about them and endeavour to make our conclusions plausible without professing to make our reasoning exact. There is only one course for which no good defence can ever be found. This course is to give what profess to be proofs and are not proofs, reasoning which is ostensibly exact, but which really misses all the essential difficulties of the problem. This was Todhunter's method, and it is one which Prof. Johnson too often adopts.

Thus, when he professes to prove that "an integral can be differentiated with respect to a quantity independent of the current variable and the limits by differentiating the expression under the integral sign," he is professing to prove what he himself later on acknowledges to be untrue by arguments entirely destitute of validity. And what is gained? Why not simply say, "it is natural to suppose, and can be proved to be in fact often true, that an integral..."? The reader would find the applications of the principle just as interesting, and the author, when confronted with the necessity of explaining why what he says he has proved is not true of the integral

$$\int_0^\infty \frac{\sin\alpha x}{x}dx,$$

would not have been obliged to invent the untrue and absurd explanation that "this results from the fact that, although α occurs in it, the expression is not really a function of α."

But let us attempt to judge the book by more appropriate standards. The first two chapters, on methods of integrations seem to us fairly good and interesting, though far from ideal. There is a lack of system about the treatment. Surely, in 125 pages, more might be said as to what are the general classes of functions whose integrals can always be found. It is not even pointed out that

$$\int R(\cos x, \sin x) dx$$

can always be evaluated by the substitution $\tan \frac{1}{2} x = t$, although this substitution is used in particular cases; and the treatment of algebraical functions is equally unsystematic.

The most interesting sections are those on approximate integration (in Ch. III.) and on applications to probability (in Ch. IV.). The chapter on definite integrals (Ch. V.) is deplorable theoretically, but contains many interesting results and examples. But we could wish that Ch. VI. had been omitted. The end of a text-book on the Integral Calculus is not the right place for an introduction to the theory of functions of a complex variable, and the introduction itself is neither sound nor particularly clear.

Prof. Osgood's book is, of course, a work of an entirely different character. His task is a much more difficult one, for he is showing the way over a road the roughness of which he knows. On the whole, his book seems to us the best elementary treatise on the Calculus in the English language; only *Lamb* and *Gibson* seriously compete with it, and of the three books the latest seems to us the most lucid, the best arranged, and (what is very important) the most *level* in its standard of interest and difficulty. Unfortunately there is hardly enough technical and analytical complication to suit a Cambridge course. If there is one chapter which we would single out for praise it is Ch. XIV., on "partial differentiation," "the total differential," "small errors," and so on; or perhaps Ch. X. on "Mechanics," the 30 odd pages of which might well supersede the whole of some well-known text-books of "Dynamics."

The book is one, we fear, likely to make considerable demands on the knowledge and discrimination of the teacher; for Prof. Osgood is a firm believer in a *gradual* introduction to fundamental difficulties, and anyone using the book as a text-book must be ready to parry the occasional awkward question which comes a little before its time. For this reason we sometimes wish for more detail in the text. It is a big jump from the rather innocent attitude towards limits adopted at the beginning to the formal definition for functions of two variables given in Ch. XIV. In the case of one variable no such definition is ever given. And we should have liked to have seen included at any rate a short and popular sketch of Dedekind's theory of number. It is not very difficult to explain in general terms, and enables a good many gaps to be filled in, notably in connection with the 'Fundamental Principle' (p. 246) that an increasing function approaches a limit or tends to infinity.

One minor point suggests itself. We do not like the notation "$\lim_{x=a}$." Messrs. Leatherm and Bromwich's "$x \to a$" is much clearer and much less ugly. And why "$\lim_{n=\infty}$"? It seems to us even more essential to use a different symbol here. For Prof. Osgood does not need to be told that nothing is ever "equal to ∞."

34
Book Review of The Psychology of Invention in the Mathematical Field

(*Mathematical Gazette* 30 (1946), pp. 111–115)*

Jacques Hadamard, French mathematician, whose most important result was his proof of the prime number theorem in 1896.

---

* Reprinted with the kind permission of the publisher of *The Mathematical Gazette*: The Mathematical Association.

The Psychology of Invention in the Mathematical Field. **By J. Hadamard. Pp. xiii, 143. 13s. 6d. 1945. (Princeton University Press; Humphrey Milford)**

This is, apart from one famous lecture of Poincaré*, the first attempt by a mathematician of the first rank to give a picture of his own modes of thought and those of other mathematicians. Professor Hadamard has earned the thanks of all of us for his courage in undertaking such a task, formidable from both its psychological and its mathematical difficulties, and there is one thing at any rate which he has certainly established, that the most important of the qualifications required is to be a good mathematician. He can sometimes quote illuminating remarks from "outsiders", philosophers, poets, or even literary critics, but nothing appears more clearly from the book than the comparative futility of the speculations of psychologists who have not been mathematicians. Galton alone, of the outside authorities quoted, shows up really well; a few pages from Poincaré or Hadamard about theorems they have found themselves have ten times the value of the theories of Gall, Möbius†, Nicolle, Paulhan, Souriau, and the rest of them. One might add that "inductive" enquiries by questionnaire, such as that conducted by *L'enseignement mathématique* in 1902, have on the whole proved equally unilluminating.

We must therefore all be grateful to Professor Hadamard for this stimulating little volume, written with all the authority of one of the greatest mathematicians of the last fifty years, and with the most charming frankness concerning his own achievements, triumphs and failures alike. Indeed it is the most personal parts of the book which seem to me the most attractive. A lesser mathematician would never have explained to us how, by some odd mischance, he failed to discover "Jensen's theorem", or how, having found the "Lorentz group" of transformations, he dismissed it as "obviously devoid of physical meaning". A lesser mathematician may perhaps also derive some comfort from the recital of such misadventures.

I must begin, however, with two criticisms which I hope may not seem ungracious. The book is too condensed—I regret myself that it is not at least fifty per cent. longer. And I regret still more that it is not written in French: after all, there would soon have been an English translation. It is true that Professor Hadamard's English is (as anyone who has talked with him, or heard him lecture, can testify) far better than that of most Frenchmen, most even of those who have lived for years in an English-speaking country. But to write for publication to so general an audience, on a subject so packed with subtleties as this, is a test

---

* Ch. 3 ('L'invention mathématique') of *Science et méthode*.
† The neurologist, not the mathematician.

34. Book Review of The Psychology of Invention

quite different from conversation, or lecturing, or the composition of a technical memoir. The result is that, while the meaning is almost always clear, and the language generally "correct", it lacks the idiomatic force of Poincaré's French, and the whole book, especially in its earlier and more definitely psychological chapters, gives a certain impression of crabbedness and congestion. Writing in French, Professor Hadamard would naturally have been more expansive.

In this respect I find a considerable difference between the earlier and later chapters. The first part of the book is dominated by discussions of the "unconscious" and its role in discovery, an obscure subject which might daunt even an experienced psychologist, and here Professor Hadamard is wisely diffident and tentative in his conclusions. The contents of the later chapters are more miscellaneous, and include much more that is both unquestionable and easy to follow.

I must begin by saying something about the earlier and more difficult chapters. The main facts enumerated in them seem beyond dispute. That unconscious activity often plays a decisive part in discovery; that periods of ineffective effort are often followed, after intervals of rest or distraction, by moments of sudden illumination; that these flashes of inspiration are explicable only as the result of activities of which the agent has been unaware—the evidence for all this seems overwhelming. Poincaré's is the classical example, but a quite ordinary mathematician can recognise similar experiences. How these unconscious activities are related to those of a more normal kind, to fully conscious work or reflection on the fringe of consciousness, how they function and what is the proper language in which to describe them, are terribly difficult questions. But all of us can remember problems, even on the examination level, whose solution has come to us suddenly when we have all but forgotten them, and it is hard to believe that the rest given by forgetfulness is a sufficient explanation.

Poincaré's experiences (in the discovery of the "theta-fuchsian" functions and the transformations which govern them) were exceptional in some ways, but most of us can remember experiences similar in kind; we have only to recall the best work we have done as the result of prolonged and strenuous effort. The typical course of events is something as follows. There is a first stage of fully deliberate activity, possibly with some, but certainly with unsatisfactory, results—Poincaré's were entirely negative. Then a rest, complete or partial, compulsory or deliberate, the result of other occupation or diversion to different problems, followed by a moment of sudden illumination. Then a second period of conscious effort, this time successful, in which the broad outlines of the solution become clear. Then, very likely after long delay, the final stage of what Hadamard calls "precision", in which the results are "written up" and put in order, a tiresome and subsidiary but essential process. These four periods seem the minimum, but of course there may be more; Poincaré's experiences were a good deal more elaborate, and he had at least two moments of unexpected inspiration.

The mystery lies entirely in the early stages, and first in the initial stage of "preparation", which Hadamard considers in detail in Ch. 4. It is plain that during this stage, however futile it may have seemed, we have done something essential; we have shaken up our ideas in a way which somehow makes later illumination possible. To "discover" is to combine ideas fruitfully, and we have at any rate formed a mass of combinations. These have seemed useless, and most of them have been really uninteresting; but the process of forming them has been less unfruitful than it has seemed, since it has set in motion the unconscious machinery which would never have operated without such an initial disturbance. It is a most difficult process to describe, and neither Poincaré nor Hadamard does more than suggest vague images; but the vaguest image of a mathematician may be more profitable than the theory of a psychologist who has never made a mathematical discovery in his life, and Poincaré has one of which Hadamard seems to approve. He compares the ideas which are the future elements of our combinations to "the hooked atoms of Epicurus". While our mind was inactive, these atoms were motionless; they were "hooked to the wall", but our efforts set them in motion. "After this shaking up imposed upon them by our will, they do not return to their primitive rest, but continue their dance freely", and have impacts with one another resulting in fresh combinations. In the new combinations thus formed, indirect results of the original conscious work, lie the possibilities of apparently spontaneous inspiration.

So far this is plausible enough, but we have still to *select* from among all these combinations: "invention is discernment, choice", and where and how is this choice made? This is the most puzzling question, and I cannot feel that either Poincaré or Hadamard points at all clearly to any satisfactory reply. It seems plain that our unconscious activities must have included some process of selection, since most of our unconscious combinations never rise to our consciousness at all. Poincaré and Hadamard find the solution in the unconscious working of our aesthetic sense. "The privileged unconscious phenomena, those capable of becoming conscious, are those which affect our emotional sensibility most profoundly.... In what region of the mind does this sorting take place? Surely not in consciousness, which, among all possible combinations, knows only the right ones.... To the unconscious belongs the most delicate and essential task, that of selecting those which satisfy our sense of beauty and are therefore most likely to be fruitful...." It may be so, though I cannot say that I find it very convincing; but I am no psychologist, and my distaste for all forms of mysticism may be prejudicing me unduly.

It is admitted that there are more commonplace explanations of some at any rate of the phenomena. Our mind, when we return to the problem, is freshened by its rest; we have escaped from "interferences which block progress during the stage of preparation"; we have "got rid of false leads and hampering assumptions, and can approach the problem with a more open mind"; and it is quite natural that we should be more successful. But I am forced to agree with Hadamard that "while such explanations can be admitted in some cases, in others they are contradicted by the facts" : they do not account for experiences like that of Poincaré when he boarded the omnibus at Coutances. We seem driven to admit that "the unconscious is not merely automatic, it has tact and delicacy", even that "it knows better how to divine than the conscious self, since it succeeds where that has failed". But I do not *like* this kind of language, and if, with Poincaré, we begin asking "is not therefore the subliminal self superior to the conscious self?", then I have an uncomfortable feeling that we are rather near talking nonsense (and I gather from a remark on p. 42 that Hadamard does not altogether disagree with me).

It is something of a relief to pass to the later chapters, which are full of interesting and less controversial matter. In particular Ch. 6 ("Discovery as a synthesis") is a long and important one. It contains a detailed discussion, with many examples, of the use of *signs*, both by mathematicians generally and by Hadamard himself; and here I find his conclusions, though they seem to be well substantiated, rather astonishing. He may be justified in his contempt for Max Müller's view that thought is conducted almost entirely by *words*, but he goes to the other extreme; it seems that his mind, like Galton's, is of just the opposite type. "I insist that words are entirely absent from my mind when I really think" (and here "words" are to be understood in a very wide sense, including, for example, algebraical symbols). "I use them only when dealing with easy calculations; whenever the matter becomes difficult, they become too heavy a baggage for me. I use concrete representations, but of a quite different nature.... Words remain absolutely absent from my mind until I come to the moment of written or oral communication." It is an extreme view, stated with surprising emphasis, but it is supported by definite instances; Hadamard explains, for example, his "picture" of the proof of the infinitude of the primes. And, although I should have expected his case to be exceptional, he has found, as a result of quite extensive enquiries, that most mathematicians agree with him. There were only two among those whom he consulted who confessed to belonging to the "verbal" type, Birkhoff, who "visualized algebraical symbols", and Pólya, who seems dominated by words even more than I am myself. For my own part I must confess that I think almost entirely in words and formulae, written if possible, or visualised as printed on paper, and that for this reason

I find thought almost impossible if my hands are cold and I cannot write in comfort. If Hadamard thinks with his legs ("except in the night when I cannot sleep, I never find anything except by pacing up and down the room"), then I think with my fingers. It may be a humiliating confession, but surely I have one illustrious companion in sin—did not Euler say, somewhere, that he thought with his pen?

Another arresting discussion is that, in Ch. 7, of "different types of mathematical minds", with special reference to the familiar contrast between the "intuitive" and "logical" types. The subject is of course one about which a great deal of nonsense has been talked. Even in 1893 we find Klein declaring that "a strong native space intuition seems to be an attribute of the Teutonic race, while the critical and purely logical sense is more developed in the Latin and Hebrew races"—and we have all heard this doctrine later in much cruder forms. It is pleasant to find that Hadamard preserves his sense of humour, and quotes French dicta of equal absurdity.

Hadamard recalls Poincaré's comparison of two pairs of famous mathematicians. Poincaré, like Hadamard, rises above the nationalistic prejudice which stakes out a claim to the gifts which sound most impressive, and chooses two pairs of mathematicians of the same nationality, Bertrand (perhaps a little overshadowed in such company) and Hermite, Riemann and Weierstrass. And here Hadamard reveals a very interesting difference of opinion. Bertrand, he agrees, "had visibly a concrete and spatial view of every question", but it seems to him absurd to class Hermite as a "logician", "Nothing can appear to me more directly contrary to the truth.... Methods always seemed to be born in his mind in some quite mysterious way.... I can hardly imagine a more perfect type of an intuitive mind than Hermite's." He agrees that Hermite "was not used to thinking in a concrete way" and "had a positive hatred for geometry", and concludes that the association, so often suggested, of physics and geometry with intuition, analysis with logic, may sometimes have very little foundation. I think that one might also quote Ramanujan, an "intuitive" mathematician, surely, if ever there was one, for the same purpose.

About Riemann and Weierstrass, Hadamard agrees better with Poincaré; and no doubt this antithesis is generally accepted and corresponds broadly to the facts. Yet even here one must be cautious about accepting such distinctions too readily. I do not know whether Hadamard has studied Siegel's analysis of Riemann's *Nachlass* on the Zeta-function. Riemann, we were told, was the outstanding example of a mathematician dominated by broad and general ideas, and here he reveals himself as a formalist quite of Ramanujan's type. It is enough to shake anybody's confidence in these facile generalisations.

Here Hadamard suggests two criteria of his own, based on his earlier discussions of the "unconscious" (while warning us that we must not expect to find them always concordant). First, we might reasonably describe a mind as "intuitive" if its original combinations of ideas are formed in a comparatively deep layer of the unconscious, as "logical" if this layer is comparatively superficial. Secondly we might (reverting to Poincaré's image of the "hooked atoms") define the intuitive mind as one in which the initial disturbances of the atoms are notably random, and scattered, the logical mind as one in which, even from the beginning, they follow comparatively narrow and convergent paths. The criteria seem plausible enough, but it is significant that, in the particular case which Hadamard considers most closely, they lead to contradictory results. Galois, he finds, was a highly "intuitive" mathematician according to the first criterion, a highly "logical" one according to the second.

I have no space to refer to more of Hadamard's many arresting discussions—the whole book is packed with provocative matter. But I cannot refrain from one final expression of satisfaction. In his last chapter he gives a short discussion of the motives which inspire research, and I am naturally delighted to find that his views agree substantially with my own. Indeed he states the case more strongly than I should have dared to do myself. That little first-rate mathematics is done with a view to immediate application; that "it seldom happens that mathematical researches are undertaken directly in view of a given practical aim"; that "practical questions are most often solved by means of existing theories", and that the applications, however important, are usually remote in time; that, in short, research is normally inspired by "the common motive of all scientific work, the desire to know and to understand"—all this is surely (whatever the ardent young politicians of the Archimedeans may say) common ground among ninety per cent of mathematicians, and the experiences of an Hadamard or a Hilbert merely confirm those of their humbler colleagues.

35

Book Review of Differential and Integral Calculus

(*Mathematical Gazette* 7 (1914), p. 337)*

From the Editors

This is a mixed review of a now probably long-forgotten book on calculus. What makes it unusual is the surprising closing paragraph where Hardy scolds the author for including suggestions on the French language, in particular a suggestion for pronouncing the name of Cauchy. We feel that the accent on the first syllable which seems to offend Hardy is very common in the United States today. Hardy's calling attention to this prompted the editors of his Collected Papers *to scold Hardy in turn for being so "severe," particularly when, as they point out, a common pronunciation of the name had been "Corky" at Cambridge in the previous century—continuing a long tradition of the English imposing their own pronunciations on foreign words, particularly when dealing with names borrowed from the French.*

Differential and Integral Calculus. **By Lorrain S. Hulburt. Pp. xviii, 481. 1912. (Longmans, Green & Co.)**

This book has many good points. It is readable and fairly accurate, and the examples are simple and interesting. The geometrical parts are the best, but even in the analytical parts a good deal is done correctly which is bungled hopelessly in many books with wide circulations: I may instance the treatment

---

* Reprinted with the kind permission of the publisher of *The Mathematical Gazette*: The Mathematical Association.

of differentials and the differentiation of x^n. The author does not pretend to be rigorous in his treatment of fundamentals, and the compromise which he attempts to set up between rigour and simplicity is often a very reasonable one. Sometimes he is less successful; and I append a few criticisms of particular passages, which might be useful if the book should reach a second edition.

P. 5. $3 - (2 - 1) = 7 - (3 + 2)$ is certainly not an 'identity,' for it contains no variables. The author's example contradicts his own definition.

Pp. 10 *et seq.* The author seems to suggest that a discontinuity of a function is necessarily accompanied by a failure in its definition.

P. 68. The 'definition' of an 'increasing function' is not a definition at all, but a mere tautology.

Pp. 88 *et seq.* The treatment of the exponential limit is bad. The author *does* profess to prove that

$$\lim \left(1 + \frac{1}{n}\right)^n = e,$$

when n is restricted to be a positive integer, but his proof is fallacious. He then tacitly assumes that his proof applies to the case in which $n \to -\infty$. He recognises that he has not proved everything, but not that he has proved nothing.

Pp. 240 *et seq.* All the discussion of the definite integral is also bad. It would have been much better to give no proof at all. The proof of the existence of a definite integral is substantially simply a proof that a certain type of area exists. If the latter proposition, which involves all the difficulties of the former, is to be assumed, nothing can be gained by not assuming the former as well. The most that can be done profitably is to give the obvious geometrical reasons for supposing the two problems to be identical.

Pp. 388 *et seq.* To define $e^{i\theta}$ as meaning $\cos\theta + i\sin\theta$, and then to 'deduce' De Moivre's theorem by assuming that $(e^{i\theta})^n = e^{ni\theta}$, is palpably absurd.

Pp. 410 *et seq.* The use of the term 'consecutive' in the treatment of envelopes is unfortunate.

I may add that I cannot regard an elementary mathematical book as a very suitable place for instruction in the rudiments of French, and that in any case I have grave doubts whether "Kō'sheé" is a very accurate phonetic rendering of the name of the great mathematician.

36
Abridged Review of An Introduction to the Study of Integral Equations

(*Mathematical Gazette*, 5 (1910), pp. 208–209)*

***An Introduction to the Study of Integral Equations.* M. Bôcher. (Cambridge Tracts in Mathematics and Mathematical Physics, No. 10.)**

Prof. Bôcher's tract is rather different in character from any of its predecessors in the Cambridge Series. Most of these have dealt with subjects selected from "classical" theories, with the general outlines of which every mathematician is familiar. The general theory of Integral Equations, on the other hand, is a product of the last ten years; it is still possible to be a mathematician of the highest reputation and to know next to nothing about it.

It has hitherto been very difficult for anyone who has not kept pace with the development of the theory from the beginning to make good his deficiencies afterwards. The subject is intrinsically difficult, and has been approached from different points of view, which have at first sight but little in common. The literature is scattered, and new contributions appear almost every month. It has been a severe handicap to anyone who wished to "get up" the subject that there has been no connected account of those parts of the theory which may be regarded as tolerably complete. At the same time, so much still remains confessedly incomplete that a treatise on the subject would probably be superseded almost as soon as it was written, and it is not surprising that no one has been willing to write one. The publication of a "Tract" is an ideal compromise, and Prof. Bôcher's should be one of the most valuable of the series.

The most striking characteristic of Prof. Bôcher's exposition is the closeness with which he follows the historical order of the development of the theory. Here, we think, he has probably been wise, though the method is one that has its disadvantages as well as its advantages. The advantages are obvious; the

---

* Reprinted with the kind permission of the publisher of *The Mathematical Gazette*: The Mathematical Association.

historical order is always the most interesting and the easiest to follow, and the reader is better able to preserve his sense of proportion when different sides of the theory are presented to him in turn. The chief disadvantage is that, when the author's space is as strictly limited as it is here, and when so much of it is devoted to the work of the pioneers, the more systematic modern theories are apt to receive less than their due. And in this tract we certainly wish that less space had been given to Volterra and more to Hilbert and his followers, and in particular that room had been found for some general account of the connection between Integral Equations and Differential Equations, and of the application of this theory to expansions such as Fourier's—to us at any rate the most remarkable and interesting of all its applications.

Still, it would be unreasonable to expect everything in 70 pages, and the wonder is that Prof. Bôcher has been able to give us so much as he has without compressing his argument beyond the limits of intelligibility. He is, indeed, considering the amount of information he contrives to give us in so short a space, extraordinarily lucid and readable throughout, and almost succeeds in making a difficult subject seem easy; and it is in many ways an excellent thing that he should have given us so full an account of the work of Abel and the other precursors of the theory—work which those who are familiar only with the writings of the German school might be in danger of forgetting.

37

Review of Higher Algebra

(*Mathematical Gazette*, 7 (1913), pp. 21–24)

Higher Algebra. **By Charles Davison, Sc.D. 6s. 1912. (Cambridge University Press.)**

The publication of this book by the Cambridge Press can only be attributed to reprehensible carelessness on the part of its expert advisers.

A reviewer who receives a book on "Higher Algebra" naturally turns first to the chapters on limits, convergence of series, and the exponential, logarithmic, and binomial series. Dr. Davison's treatment of all these subjects can only be described as hopelessly uneducated. He shows no kind of conception of the logical relations between different parts of the theory, and his definitions and proofs are not only extraordinarily slovenly, but are full of the grossest blunders. It would be waste of time to justify these remarks by a large number of criticisms of detail, but I may give a few illustrations from Chapter V. Dr. Davison defines a *series* in §64, and a convergent series in §65, both wrongly. As he never defines a *limit* at all, and postpones to §180 any sort of explanation of what he means by a limit, this is only natural. The most important theorems in the chapter are contained in §§68, 73, 75, 78, and 79. In these sections Dr. Davison merely repeats the traditional blunders of English text-books of twenty years ago. In the first three cases, for example, he assumes what he professes to prove.

Nor can I say honestly that I think that Dr. Davison is very much happier when he gets away from the difficulties of limits and convergence. He is completely mistaken, for instance, in supposing that, in §28, he has established the possibility of expressing a given rational function as a sum of partial fractions. Chapter IV, on "Complex Quantity," is a morass of confusion. I am quite unable to disentangle what the author regards as definition and what as proof. The proof (§164) that the arithmetic mean is greater than the geometric is unsound, as has been pointed out by a previous writer in the *Gazette*. Finally, in §170, Dr. Davison proves that "every rational integral equation of the nth degree has

n, and only n, roots" without a word of explanation that he is assuming the existence of at least one root.

I have selected these examples more or less at random. The book is, in my opinion, a thoroughly bad one, which ought never to have been published. The fact that it appears under the *aegis* of a University Press leads me to think that I should say so with more emphasis than I should have otherwise considered necessary.

38
Review of Leçons de Mathématiques Générales

(*Mathematical Gazette*, 7 (1914), p. 338)

Leçons de Mathématiques Générales. Par L. Zoretti. Pp. xvi + 753. 1914. (Gauthier-Villars.)

This is a very remarkable book, to which I should like to call the attention of schoolmasters and of all whose business it is to teach mathematics to intelligent pupils who are not mathematical specialists.

I should hardly have thought it possible to cover so much ground in a single volume. M. Zoretti assumes no knowledge in his reader beyond that of a little geometry, algebra, and trigonometry; and he carries him far beyond the limits of Part I. of the Mathematical Tripos. There are chapters, for example, on elliptic functions, Fourier's Series, partial differential equations, vectors in space, and elementary differential geometry of two and three dimensions. And in spite of the extraordinary variety of the subjects with which he deals, M. Zoretti always seems to be treating them at his leisure and to have plenty of space to spare.

Naturally M. Zoretti does not profess to give a full and rigorous treatment of fundamentals. He does not prove, for example, that a series of positive terms must converge or diverge to infinity, or establish the existence of implicit functions or of the definite integral. But he is always perfectly clear and consistent. "Ma méthode est ... admettre franchement tout ce qui présente des difficultés sérieuses, ou simplement de trop longs calculs; donner, quand c'est possible sans longueurs, soit la démonstration, soit au moins ses grandes lignes."

Herein lies the great difference between such a book as this and the majority of the English books on "Practical Mathematics" which appear in such profusion nowadays. M. Zoretti is "practical" enough. He has a most admirable chapter on methods of numerical calculation, calculating machines, and so forth, a chapter which, ignorant as I am in such matters, I read with the greatest interest: and in general he has the possibility of practical application always before him. But with him to be "practical" does not mean, as too often with English writers, merely to be ignorant of all that makes mathematics a science worth learning. For M. Zoretti is a real mathematician who has done distinguished work in the theory of functions, and knows perfectly well where the real difficulties of analysis lie.

A Last Word

Hardy in 1920, at the height of his powers, shifted from Cambridge to Oxford and when he arrived there he gave an inaugural lecture entitled "Some Famous Problems of the Theory of Numbers and in Particular Waring's Problem." We did not include this in our collection of Hardy's works, but we include here part of that text. In it, Hardy reflects on a life spent in mathematics and, in so doing, properly gets the last word.

"There are subjects in which only what is trivial is easily and generally comprehensible. Pure mathematics, I am afraid, is one of them; indeed it is more: it is perhaps the one subject in the world of which it is true, not only that it is genuinely difficult to understand, not only that no one is ashamed of inability to understand it, but even that most men are more ready to exaggerate than to dissemble their lack of understanding.... For my own part, I trust that I am not lacking in respect either for my subject or myself. But, if I am asked to explain how, and why, the solution of the problems which occupy the best energies of my life is of importance in the general life of the community, I must decline the unequal contest: I have not the effrontery to develop a thesis so palpably untrue. I must leave it to the engineers and the chemists to expound, with justly prophetic fervor, the benefits conferred on civilization by gas engines, oil, and explosives. If I could attain every scientific ambition of my life, the frontiers of the Empire would not be advanced.... It is possible that the life of a mathematician is one which no perfectly reasonable man would elect to live. There are, however, one or two reflections from which I have sometimes found it possible to extract a certain amount of comfort. In the first place the study of mathematics is, if an unprofitable, a perfectly harmless and innocent occupation, and we have learnt that it is something to be able to say that at any rate we do no harm. Secondly, the scale of the universe is large, and, if we are wasting our time, the waste of the lives of a few university dons is no such overwhelming catastrophe. Thirdly what we do may be small, but it has a certain character of permanence; and to have produced anything of the slightest permanent interest, whether it be a copy of verses or a geometrical theorem, is to have done something utterly beyond the powers of the vast majority of men."

Sources

1. Albers, Donald J., "Freeman Dyson: Mathematician, Physicist, and Writer." *College Math. J.* 25:1 (1994), 2–21.
2. _____ "John Todd—Numerical Mathematics Pioneer," *College Math. J.* 38:1 (2007), 2–23.
3. _____ (with G. L. Alexanderson) *Fascinating Mathematical People*, Princeton University Press, 2011.
4. _____ (with G. L. Alexanderson) *Mathematical People/Profiles and Interviews*, Birkhäuser Boston, 1985; second edition, A K Peters, 2008.
5. Alexanderson, Gerald L., and Dale H. Mugler, *Lion Hunting & Other Mathematical Pursuits/A Collection of Mathematics, Verse, and Stories, by Ralph P. Boas, Jr.*, Mathematical Association of America, Washington, DC, 1995.
6. _____ *The Random Walks of George Pólya*, Mathematical Association of America, Washington, DC, 2000.
7. Berndt, Bruce C., and Robert A. Rankin, *Ramanujan: Letters and Commentary* (History of Mathematics, vol. 9), American Mathematical Society-London Mathematical Society, Providence, RI, 1995.
8. Birkhoff, Garrett, "The Rise of Modern Algebra—1936–1950," *Men and Institutions in American Mathematics* (D. Tarwater, ed.), Texas Tech University, Lubbock, TX, 13 (1976), 65–85.
9. Bohr, Harald, *Collected Works*, Dansk Mat. Forening, 1953, vol. 1, pp. xxvii–xxviii.
10. Bollobás, Béla, ed. *Littlewood's Miscellany*, Cambridge University Press, 1986.
11. Burkhill, Harry, "G. H. Hardy (1877–1947)," *Math. Spectrum* 28 (1995–1996), 25–31.
12. Conway, Flo, and Jim Siegelman, *Dark Hero of the Information Age/In Search of Norbert Wiener/The Father of Cybernetics*, Basic Books, New York, 2005.
13. Corry, Leo, and Norbert Schappacher, "Zionist Internationalism through Number Theory." Edmund Landau at the Opening of the Hebrew University in 1925," *Science in Context* 23:4 (1925), 427–471.

14. Crilly, Tony, *Arthur Cayley: Mathematician Laureate of the Victorians*, Johns Hopkins University Press, Baltimore, 2006.
15. Curbera, Guillermo P., *Mathematicians of the World, Unite!*, A K Peters, Wellesley, MA, 2009.
16. Dauben, J. W., "Mathematicians and World War I: the International Diplomacy of G. H. Hardy and Gösta Mittag-Leffler as Reflected in their Personal Correspondence," *Hist. Math.* 7:3 (1980), 261–288.
17. Dawkins, Richard, *An Appetite for Wonder*, Harper Collins, New York, 2013.
18. Derbyshire, John, *Prime Obsession/Bernhard Riemann and the Greatest Unsolved Problem in Mathematics*, Joseph Henry Press (National Academies Press), Washington, DC, 2003.
19. Eves, Howard, *Mathematical Circles, Adieu*, Prindle, Weber & Schmidt, Boston, 1977.
20. _____ *Mathematical Circles Revisited*, Prindle, Weber & Schmidt, Boston, 1971.
21. _____ *Mathematical Circles Squared*, Prindle, Weber & Schmidt, Boston, 1972.
22. Fauvel, John, Raymond Flood, and Robin Wilson, *Oxford Figures/800 Years of the Mathematical Sciences*, Oxford University Press, 2000.
23. Flood, Raymond, Adrian C. Rice, and Robin Wilson, *Mathematics in Victorian Britain*, Oxford University Press, 2011.
24. Grattan-Guinness, I., "The Interest of G. H. Hardy, F.R.S., in the philosophy and the history of mathematics," *Notes Rec. R. Soc. Lond.* 55:3 (2001), 411–424.
25. _____ "Russell and G. H. Hardy: A Study of Their Relationship," *J. of the Bertrand Russell Archives* (New Series), 11:2 (Winter 1991), 165–179.
26. Gratzer, Walter, *Eurekas and Euphorisms, Oxford Book of Scientific Anecdotes*, Oxford University Press, 2002.
27. Greene, (Henry) Graham, "An Austere Art," (Review of *A Mathematician's Apology*), *The Spectator* 165 (December 19, 1940), 682.
28. Halmos, Paul R., *I Want To Be a Mathematician/An Automathography*, Springer, New York, 1985; second printing, MAA, 2005.
29. Hardy, Godfrey Harold (with S. Ramanujan) "Asymptotic formulæ in combinatory analysis," *Proc, London Math. Soc.*, (2) 17 (1918), 75–115.
30. _____ *Bertrand Russell and Trinity/A College Controversy on the Last War*, Cambridge University Press, 1942.
31. _____ "Camille Jordan 1838–1922," *Proc. Royal Soc.* (A) 104 (1922), xxiii–xxvi.
32. _____ "The Case against the Mathematical Tripos," *Math. Gazette* 13 (1926), 61–71.

33. _____ (with P. V. S. Aiyar and B. M. Wilson, editors) *Collected Papers of Srinivasa Ramanujan*, Cambridge University Press, 1927.
34. _____ *A Course of Pure Mathematics*, Cambridge University Press, 1908.
35. _____ "David Hilbert," *J. London Math. Soc* 18 (1943), 191–192.
36. _____ *Divergent Series*, Clarendon Press (Oxford University Press), 1949.
37. _____ (with H. Heilbronn), "Edmund Landau," *J. London Math. Soc.* 13 (1938), 302–310.
38. _____ "Ernest William Hobson," *J. London Math. Soc.* 9 (1934), 225–237.
39. _____ "A formula for the prime factors of any number," *Messenger of Math.* 35 (1906), 145–146.
40. _____ (with W. W. Rogosinski), *Fourier Series* (Cambridge Tract), Cambridge University Press, 1944.
41. _____ (with M. Riesz), *The General Theory of Dirichlet's Series* (Cambridge Tract), Cambridge University Press, 1915.
42. _____ "Dr. Glaisher and the 'Messenger of Mathematics,'" *Messenger of Math.* 58 (1929), 159–160. (Excerpt published in *Amer. Math. Monthly* 90:3 (1983), 211.)
43. _____ "Gösta Mittag-Leffler," *J. London Math. Soc.* 3 (1928), 156–160.
44. _____ "The Indian Mathematician Ramanujan," *Amer. Math. Monthly* 44 (1937), 137–155. (Reprinted in *Ramanujan/Twelve Lectures on Subjects Suggested by His Life and Work*, Cambridge University Press, 1940, pp. 1–21.)
45. _____ (with J. E. Littlewood and G. Pólya), *Inequalities*, Cambridge University Press, 1934.
46. _____ "The integral $\int_0^\infty \frac{\sin x}{x} dx$," *Math. Gazette* 5 (1909), 98–103; 8 (1916), 301–303.
47. _____ *The Integration of Functions of a Single Variable* (Cambridge Tract), Cambridge University Press, 1905.
48. _____ "An Introduction to the Theory of Numbers," *Bull. Amer. Math. Soc.* 35 (1929), 773–818.
49. _____ (with E. M. Wright), *An Introduction to the Theory of Numbers*, Clarendon Press (Oxford University Press), 1938.
50. _____ "The *J*-Type and the *S*-Type among Mathematicians," *Nature* 7 (1934), 134, 250.
51. _____ "Mathematical Analysis," *Nature* 109 (1922), 437.
52. _____ "Mathematical Proof," *Mind* 38 (1929), 1–25.
53. _____ "A Mathematical Theorem about Golf," *Math. Gazette* 29 (1945), 226–227.
54. _____ *A Mathematician's Apology*, Cambridge University Press, 1940.
55. _____ "Mathematics," *The Oxford Magazine* 48 (1930), 819–821.
56. _____ "Mathematics in War-time," *Eureka* 1:3 (1940), 5–8.

57. _____ "Mendelian proportions in a mixed population," *Science* (N.S.) 28 (1908), 49–50.
58. _____ "Mr. Russell as a religious teacher," *Cambridge Magazine* (1917), 119–135.
59. _____ (with J. E. Littlewood) "A new solution of Waring's problem," *Quart. J. Math.* 48 (1920), 272–293.
60. _____ "A note on two inequalities," *J. London Math. Soc.* 11 (1936), 167–170.
61. _____ "On the representation of numbers as the sum of any number of squares, and in particular of five," *Trans. Amer. Math. Soc.* 21 (1920), 255–284, *Trans. Amer. Math. Soc.* 29 (1924), 845.
62. _____ *Orders of Infinity: The 'Infinitärcalcul' of Paul Du Bois-Reymond*, (Cambridge Tract), Cambridge University Press, 1910.
63. _____ "Prime Numbers," *British Association Report* 10 (1915), 350–354.
64. _____ "Prof. H. L. Lebesgue For. Mem. R. S.," *Nature* 152 (1943), 685.
65. _____ "Question 13979," *Educational Times*, 71 (1899), 61–62.
66. _____ *Ramanujan: Twelve Lectures on Subjects Suggested by His Life and Work*, Cambridge University Press, 1940.
67. _____ "S. Ramanujan F.R.S.," *Nature* 105 (June 17, 1920), 494–495.
68. _____ Review of *The Calculus for Beginners*, by W. M. Baker. *Math. Gazette* 7 (1913), 21.
69. _____ Review of *Die complexen Veränderlichen und ihre Funktionen*, by G. Kowalewski. *Math. Gazette* 6 (1912), 345–346.
70. _____ Review of *Differential and Integral Calculus*, by L. Hulburt. *Math. Gazette* 7 (1914), 337.
71. _____ Review of *Einleitung in die Funktionentheorie* (I), by O. Stolz & J. A. Gmeiner. *Math. Gazette* 3 (1905), 184–86.
72. _____ Review of *Einleitung in die Funktionentheorie*, by O. Stolz & J. A. Gmeiner. *Math. Gazette* 3 (1905), 304.
73. _____ Review of *Elements of the Differential and Integral Calculus*, by W. A. Granville. *Math. Gazette* 7 (1913), 24.
74. _____ Review of *Estudio elemental de la prolongación analitica*, by P. Peñalver y Bachiller. *Math. Gazette* 6 (1912), 346.
75. _____ Review of *A First Course in the Differential and Integral Calculus* by W. F. Osgood; *A Treatise on the Integral Calculus, founded on the Method of Rates*, by W. Woolsey Johnson. *Math. Gazette* 4 (1907), 307–309.
76. _____ Review of *Leçons sur les fonctions discontinues*, by René Baire. *Math. Gazette* 3 (105), 231–234.
77. _____ Review of *Leçons sur les principes de l'Analyse*, by R. d'Adhemar. Tome I. *Math. Gazette* 6 (1912), 346–347.

78. _____ Review of *Mysticism in Modern Mathematics*, by Hastings Berkeley. *Cambridge Review* 31 (1910), lit. suppl., xiii–xiv.
79. _____ Review *of Principia Mathematica*, vol. 1, by A. N. Whitehead and B. Russell. *Times Literary Supplement* 504 (1911), 321–322.
80. _____ Review of *The Principles of Mathematics*, by Bertrand Russell. *Times Literary Supplement* 88 (1903), 263.
81. _____ Review of *The Psychology of Invention in the Mathematical Field*, by J. Hadamard. *Math. Gazette* 30 (1946), 111–115.
82. _____ Review of *A Source Book in Mathematics*, by David Eugene Smith. *Nature* 126 (1930), 197–198.
83. _____ Review of *Les Spectres Numériques*, by M. Petrovitch. *Math. Gazette* 10 (1920), 77.
84. _____ Review of *Theoretische Arithmetik* (II), by O. Stolz & J. A. Gmeiner. *Math Gazette.* 2 (1903), 312–313.
85. _____ Review of *The Thirteen Books of Euclid's Elements*, by T. L. Heath. *Times Literary Supplement* 409 (1909), 420–421.
86. _____ Review of *What Is Mathematics?*, by R. Courant and H. Robbins. *Nature* 150 (1942), 673–675.
87. _____ "Some famous problems in the theory of numbers and in particular Waring's problem," Inaugural Lecture, Oxford, 1920.
88. _____ "Some notes on certain theorems in higher trigonometry," *Math. Gazette* 3 (1906), 284–288.
89. _____ "The Theory of Numbers," *British Association Report* 90 (1922), 16–24; reprinted in *Nature* 110 (1922), 381–385.
90. _____ "What Is Geometry?" *Math. Gazette* 12 (1925), 309–316.
91. Harman, Peter, and Simon Mitton, eds., *Cambridge Scientific Minds*, Cambridge University Press, 2002.
92. Hauptman, Ira, *Partition*, 2003. (Excerpt at http://www.youtube.com/watch?v=rjlsysAEnZM)
93. Hobson, E. W., and A. E. H. Love, editors, *Proceedings of the Fifth International Congress of Mathematicians*, Cambridge University Press, 1913.
94. Hodges, Andrew, *Alan Turing: The Enigma* (Centenary Edition), Princeton University Press, 2012.
95. Hoffman, Paul, *The Man Who Loved Only Numbers*, Hyperion, New York, 1998.
96. James, Ioan, *Remarkable Mathematicians from Euler to von Neumann*, Cambridge University Press/Mathematical Association of America, Cambridge, 2002.
97. Kanigel, Robert, *The Man Who Knew Infinity/The Life of the Genius Ramanujan*, Scribner's, New York, 1991.

98. Körner, T.W., "Foreword" to *A Course of Pure Mathematics*, by G. H. Hardy, Cambridge University Press, 1908.
99. Leavitt, David, *The Indian Clerk*, Bloomsbury Books, London, 2007.
100. _____ *The Man Who Knew Too Much*, W. W. Norton, New York, 2006.
101. Littlewood, J. E., *Mathematical Miscellany*, Methuen & Co., York, 1953.
102. _____ "Preface," in *Divergent Series*, by G. H. Hardy, Oxford University Press, 1949, p. vii.
103. Littlewood, J. E., with G. Pólya, L. J. Mordell, E. C. Titchmarsh, H. Davenport, and N. Wiener, "Two Statements Concerning the Article on G. H. Hardy," *Bull. Amer. Math. Soc.* 55 (1949), 1082.
104. Livio, Mario, *Is God a Mathematician?*, Simon & Schuster, New York, 2009.
105. Mordell, L. J., "Hardy's *A Mathematician's Apology*," *Amer. Math. Monthly* 77 (1970), 831–836.
106. Nadis, Steve, and Shing-Tun Yao, *A History in Sum/100 Years of Mathematics at Harvard 1825–1975*, Harvard University Press, 2013.
107. Newman, M. H. A., et al., "Godfrey Harold Hardy (1877–1947)," *Math. Gazette* 32 (1948), 49–51.
108. Pólya, George (with Gerald L. Alexanderson), *The Pólya Picture Album/Encounters of a Mathematician*, Birkhäuser-Boston & Basel, 1987.
109. _____ "Some Mathematicians I Have Known," *Amer. Math. Monthly* 76 (1969), 746–753.
110. Reid, Constance, *Hilbert*, Springer, New York, 1970.
111. _____ *The Search for E. T. Bell/Also Known as John Taine*, Mathematical Association of America, Washington, DC, 1993.
112. Rice, Adrian C., and Robin J. Wilson, "The Rise of British Analysis in the Early 20th Century: the Role of G. H. Hardy and the London Mathematical Society," *Hist. Math.* 30 (2003), 173–194.
113. Roberts, Siobhan, *King of Infinite Space/Donald Coxeter, the Man Who Saved Geometry*, Walker, Toronto, 2006.
114. Royden, Halsey, "A History of Mathematics at Stanford," in *A Century of Mathematics in America, Part II*, Peter Duren, ed., American Mathematical Society, Providence, RI, 1989, pp. 237–277.
115. Segal, Sanford L, *Mathematicians under the Nazis*, Princeton University Press, 2003.
116. Segel, Joel, ed., *Recountings: Conversations with MIT Mathematicians*, A K Peters, Wellesley, MA, 2009.
117. Snow, C. P., "Foreword," *A Mathematician's Apology*, by G. H. Hardy, Cambridge University Press, 1940, pp. 9–58.
118. _____ "The Mathematician on Cricket," *Saturday Book* (8th year), Hutchinson, London, 1948, pp. 65–73.
119. _____ *The Two Cultures*, Cambridge University Press, 1993.

120. _____ "The Two Cultures and the Scientific Revolution," The Rede Lecture, 1959.
121. Tattersall, James, and Shawnee McMurran, "Dame Mary L. Cartwright," *College Math. J.* 32:4 (2001), 242–250. (Reprinted in Albers and Alexanderson, *Fascinating Mathematical People*, Princeton University Press, 2011.)
122. *The Tercentenary of Harvard College/A Chronicle of the Tercentenary Year, 1935–1936*, Harvard University Press, 1937.
123. Titchmarsh, E. C., "Obituary: Godfrey Harold Hardy," *J. London Math. Soc.* 25 (1950), 81–101.
124. _____ "Obituary," *Notices of the Fellows of the Royal Society* 6 (1949), 447–458.
125. _____ "The Riemann Zeta Function, and Lattice Point Problems," *J. London Math. Soc.* 25 (1950), 125–128.
126. Wiener, Norbert, *Ex-Prodigy*, Simon & Schuster, New York, 1953.
127. _____ "Godfrey Harold Hardy, 1877–1947," *Bull. Amer. Math. Soc.* 55 (1949), 72–77.
128. Woolf, Leonard, *Sowing: An Autobiography of the Years 1880–1904*, Hogarth Press, London, 1960.
129. Young, Laurence Chisholm, *Mathematicians and Their Times/History of Mathematics and Mathematics of History*, North-Holland, Amsterdam, 1981.

Acknowledgments

We are grateful to all the individuals and organizations named below for their generous help in the creation of this book.

For permission to reproduce images in their control, we thank the following by name accompanied by the page number (s) on which the image appears. For copyrighted material, the Source number (s) follows the name. Our failure to obtain a necessary permission for the use of any copyrighted material included in this edition is inadvertent and will be corrected in future printings following notification in writing to the Publisher of such omission accompanied by appropriate documentation.

IMAGES

Jennifer Andrews (née Titchmarsh) (pp. 30, 219)
Cambridge University Press (pp. 123, 124)
Curtis Brown Group Ltd. London (pp. 269, 272)
Harold Boas (pp. 32, 141)
Freeman Dyson (p. 109)

The London Mathematical Society (pp. 219, 325)
The MIT Museum (p. 29)
St. Catherine's School (pp. 116, 117)
The Master and Fellows of Trinity College, Cambridge (pp. 9, 10, 12, 14, 25, 39, 72, 103)

The Mathematical Institute of the University of Oxford (*G H Hardy's Oxford years* posters section)
The Warden and Scholars of Winchester College (pp. 6, 7)
The Donald J. Albers Collection (pp. 34, 143, 152, 154, 155)
The Gerald L. Alexanderson Collection (pp. 17, 18, 21, 26, 28, 49, 52, 329, 333, 343)
The George Pólya Collection (pp. 15, 17, 19, 22, 23, 31, 33, 107, 159,165, 355)

Access to materials from the archives of St. Catherine's School, Bramley, Surrey, was kindly given by the Headmistress, and research undertaken by Richard and Rosemary Christophers, the School Archivists, located several photos of Gertrude Hardy as well as details of her career.

Thanks to Sandy Paul of the Wren Library of Trinity College, Cambridge, who provided extraordinary service in finding photos of Hardy and Littlewood.

We are grateful to Miss Suzann Foster, Archivist of Winchester College, for locating photos of Hardy as a schoolboy.

Robin Wilson was of enormous help in securing permission from the Mathematical Institute of the University of Oxford to reproduce the *G H Hardy's Oxford Years* posters. The posters were conceived by him with the assistance of Raymond Flood and Dyrol Lumbard. He also contributed an article about J.E.Littlewood, which was abridged for this book.

June Barrow-Green of the Open University is the Librarian of The London Mathematical Society. She was instrumental in securing permission to reproduce photos.

Thanks to Edmund Robertson of the University of St. Andrews who guided us to an early photo of Titchmarsh (p. 138).

TEXT MATERIALS

Organization Name (Source Number)

The American Mathematical Society (7, 48, 103, 114)
Cambridge University Press (30, 66, 96, 98, 117)
Eureka/Cambridge University Mathematical Society (56)
Historia Mathematica (16, 112)
Hutchinson/Curtis Brown Group/London (118)

Hyperion/Hachette Book Group (95)
Journal of the Bertrand Russell Archives (25)
Robert Kanigel (97)
The Mathematical Gazette/The Mathematical Assn. (32, 46, 53, 68, 69, 70, 71, 72, 73, 74, 75, 76, 77, 88, 90, 107)
The London Mathematical Society (7, 35, 37, 43, 60, 123, 125)

The Mathematical Association of America (1, 2, 5, 6, 44, 96, 105, 109, 121)
Mathematical Spectrum/Applied Probability Trust (11)
Messenger of Mathematics (42)

Mind/Oxford University Press (52)
Oxford Magazine (55)

Oxford University Press (23, 102)
Princeton University Press, (115)
Quarterly Journal of Mathematics/Oxford Univ. Press (59)
Royal Society London (24)
Simon & Schuster (104, 126)
Walker Books/Bloomsbury Pub. Inc. (113)

Thanks to Freeman Dyson who gave permission to reproduce letters to C. P. Snow and G. H. Hardy.

Special thanks to Robert Kanigel for granting permission to reprint the "Epilogue" from his magnificent book *The Man Who Knew Infinity*, which is the basis for the forthcoming movie of the same title.

John Stillwell wrote the article "Cricket for the Rest of Us" especially for this volume.

This book is a joint publication of Cambridge University Press (CUP) and the Mathematical Association of America (MAA), and has had the benefit of support from the MAA and CUP. Carol Baxter and Beverly Ruedi of MAA Press did a great job with production. We are indebted to CUP Editors Roger Astley in the Cambridge office, who smoothed the way for us in England and Kaitlin Leach in the New York office, who gathered important reviews of the manuscript and coordinated with the MAA in Washington, DC.

Leonard Klosinski provided outstanding photographic assistance in the reproduction of many images.

Mary Jackson and Mary Long gave invaluable help with the arduous tasks of obtaining permissions and in preparing the index.

Geri Albers and Penny Dunham provided substantial and continuing support throughout the three-year development of this book.

Index

Abel, Niels Henrik, 4, 10, 11, 54, 71, 98, 104, 144, 147, 270, 288, 319, 344, 366
Abel's functional equation, 63
Abel's lemma, 319
Acta Mathematica, 75, 82, 156, 298, 326, 330, 344, 345, 346
Aiyar, Ramaswami, 53
Aiyar, Seshu, 48, 51, 53
Albers, Donald J., 143, 145
Alexander, James Waddell, 154
algebraic numbers, 172–175
algebraic theory of numbers, 330
American Association for the Advancement of Science, 165, 199
American Mathematical Society, 165
analytic functions, 50, 295, 303, 307
analytic theory of numbers, 16, 58, 63–64, 65, 68, 160, 295, 303, 334, 337, 339
Annals of Mathematics, 345
Apollonius, 237
Apostles, The, 11
Archimedeans, 287, 361
Archimedes, 4, 28, 254, 275, 292
Argand's diagram, 308
Aronszajn, Nachman, 111
asymptotic formulae, 295–299
Athens, University of, 79
Atiyah, Sir Michael Francis, 11, 149
Austen, Jane, 21
axiom of choice, 18

Babylonian mathematics, 4
Bach, Johann Sebastian, 102, 111
Bailey, Wilfrid Norman, 57, 61, 70, 83

Baire, René, 131
Baker, H. F., 150–151, 326
Barnes, Ernest William, 103, 114, 326
Beethoven, Ludwig van, 26, 102
Bell, Eric Temple, 109, 326
Berlin, University of, 78, 334
Bernoulli numbers, 59
Berry, A., 313–315, 316, 317, 318, 319, 321
Bertrand, Joseph Louis François, 360
Bertrand Russell & Trinity/A college controversy of the last war, 27, 134–135
Besicovitch, A. S., 89, 141, 147
Bessel functions, 223
Bieberbach, Ludwig, 28, 78, 347
Binet, Jacques, 70
Birkhoff, Garrett, 133, 142
Birkhoff, George David, 152–153, 359
Birmingham, University of, 79
Blackwell, David, 142, 143
Bloch's theorem, 336, 340
Bloomsbury Group, 44
Blumenthal, Otto, 331
Boas, Ralph Philip, Jr., 141, 142, 143–144, 145–146
Bôcher, Maxime, 365–366
Bohr, Harald, 8–9, 16–17, 22–23, 77, 104, 157, 160, 201, 220, 341
Bohr, Niels Henrik, 8, 22
Bollobás, Béla, 139, 140, 149, 150, 156
Bonaparte, Napoleon, 26
Bondi, Christine, 111
Borel, Emile, 128, 336
Bosanquet, Lancelot Stephen, 44, 97
Bowden, Leon, 15

Bradman, Sir Donald (Australian cricketer), 275, 281
"Bradman class," 281, 282
Briggs, Henry, 11
British Association, 89, 97, 199, 207
British Science Association, 199, 207
Bromwich, Thomas John l'Anson, 48–49, 307, 313, 317, 318, 319–320, 321, 354
Browning, Robert, 24
Burgess, Guy Francis, 11
Burkill, H., 138–139
Burnside, William Snow, 326
Bush, Vannevar, 29

calculus of variations, 330
California Institute of Technology (Caltech), 11, 76, 89, 137, 146
Cambridge, University of, 291, 292
　Christ's College, 268, 274
　Girton College, 30–31, 32, 34
　Gonville and Caius College, 37, 49
　King's College, 118
　St. John's College, 293
　Trinity College, 6, 8, 9, 15, 20, 21, 27, 35, 38, 44, 54, 73, 74, 77, 80, 82, 90, 98, 103–104, 105, 106, 109–110, 111, 112, 114, 118, 144, 151, 153, 256, 257, 270, 274, 293, 326
Cantor, Georg, 134, 238, 345
Cape Town University, 102
Carlson, Fritz David, 230, 337
Carr, George Shoobridge, 49–50, 61–62
Cartwright, Dame Mary, 30–32, 35, 76, 81, 83, 95–96, 106, 149, 150
Case of the Philosopher's Ring, The, 3
Cassels, John William Scott, 11
Cauchy, Augustin-Louis, Baron, 28, 167, 363
　Cauchy's inequality, 230
　Cauchy's theorem, 58, 151, 205, 297, 303, 312, 314, 345
Cayley, Sir Arthur, 11, 50, 70, 102, 105, 152, 250, 254, 296
Cesàro, Ernesto, 320
Chandrasekhar, S., 84
Chapman, Sydney, 147, 149

Chauvenet Prize, 78, 165
Chebyshev [Tchebychef], Pafnutiĭ L'vovich, 68, 186, 187, 202–203
Chicago, University of, 76
Christ, Jesus, 26, 27, 151
Christ's College, Cambridge, 268, 274
Chrystal, George, 307
Coates, John Henry, 11
Cocks, Clifford, 148
Collins, Randall, 3
combinatory analysis, 303
common cartesian geometry, 240, 241, 242
congruence properties of partitions, 62, 82
Conrad, Joseph, 139
continued fractions, 62
"Conversation Class" (see also: Hardy-Littlewood lectures), 32, 34, 105, 141, 142
Copley medal, 81, 106
Coulson, Charles Alfred, 93, 149
Courant, Richard, 35, 130, 157
Cours d'Analyse, 93, 112, 128, 256
Course of Pure Mathematics, A, 123, 126, 145
Coxeter, Harold Scott Macdonald, 153, 154
Cramér, Harald, 24
Cramér, Marta, 24
Cranleigh School, 5–6
Crelle's Journal, or *Journal für die reine und angewandte Mathematik*, 344
cricket, 5, 25, 27, 30, 76–77, 79–80, 97–98, 105, 112, 116, 121, 133, 137, 138, 148, 149, 154, 267–276, 277–283
Curbera, Guillermo P., 157
Curie, Marie, 343
"curious inequality of Carlson, a," 230

Darboux, Jean Gaston, 237
Darwin, Charles Robert, 5
Daubin, Joseph W., 156
Davenport, Harold, 113, 147, 157, 159
David (slayer of Goliath), 26, 27
Davison, Charles, 367

Dawkins, Richard, 140
Day at the Oval, A, 275–276
Dedekind, J. W. Richard, 353
  Dedekind zeta-function, 335
deMoivre, Abraham,
  deMoivre's theorem, 305, 306
De Morgan, Augustus, 220
  De Morgan Medal, 89, 93, 105–106, 220
Derbyshire, John, 18
Dickson, Leonard Eugene, 60, 70, 114, 166, 172, 185
Dictionary of National Biography, 166
Dienes, Paul, 96
differential geometry, 148
Diophantine approximation, 8, 298
Diophantine equations, 58–59
Diophantus, 176
Dirac, Paul Adrien Maurice, 77, 288
Dirichlet, Peter Gustav Lejeune, 68, 127, 172, 191, 214, 319
  Dirichlet series, 161, 204, 297, 337
  Dirichlet's principle, 330
Disraeli, Benjamin, 26, 140
distribution of primes, 65, 104, 295
divergent series, 16, 40, 106, 144
Divergent Series, 106, 144
Dixon, A. L., 93–94, 97, 318, 320, 321, 326
Donaldson, Simon Kirwan, 149
Dougall-Ramanujan identity, 61
Drobot, Vladimir, 15
Duillier, Fatio de, 11
Durrell, Gerald, 119
Dyson, Freeman John, 80, 109–114, 123, 128, 145

Eddington, Sir Arthur Stanley, 64, 84, 238, 288
Edinburgh, University of, 79
Einstein, Albert, 26, 78, 84, 89, 97, 137, 140, 148, 209, 238, 239, 241, 264, 288, 293
Eisenstein, F. Gotthold M., 327
elasticity, 93
Elements, 35–36, 182, 183

Elgar, Sir Edward William, 1st Baronet, 250
Elliott, Edwin Bailey, 93–95
elliptic functions, 50, 57–58, 60, 63–64, 70, 93, 295, 298, 369
Ergodic theorem, 152
Esson, William, 90, 149
Euclid, 35–36, 134, 182, 183, 184, 185, 191, 196, 200, 201, 209, 212, 238, 239, 242
Euler, Leonhard, 4, 18, 49, 62, 70, 127, 147, 186, 200, 201, 211, 212, 228, 229, 230, 288, 295, 296, 334
  Eulerian second integral, 40
  Euler's constant, 173
  Euler's equation, 58, 59, 62, 70
  Euler's formula (for the zeta-function), 201, 204
  Euler's identity ($e^{i\pi}$), 305
  Euler's identity (for the sum of reciprocal powers), 67, 186, 201, 204
Eureka, 287
Eves, Howard, 146, 153

Fabry's gap theorem, 339
Farey, John, 166–168
  Farey dissection, 303
  Farey series, 166–168
Fejér, Lipót, 18–19
Fender, Percy G. H., 26–27, 140, 270
Fenner's (cricket ground at Cambridge), 269, 270–271, 272, 273, 274
Fermat, Pierre de, 4, 127, 147, 176, 180, 288
  Fermat's theorem, 172, 178, 181, 183, 194, 200, 339
Ferrar, William Leonard, 96, 97
Fields Medal, 4, 157
Finnegan's Wake, 274
Fisher, H. A. L., 92
Flood, Raymond, 87, 149
football (American), 99, 121, 132, 133, 153
Ford, Henry, 26, 27
Forster, E[dward] M[organ], 11
Forsyth, Andrew Russell, 11, 314, 326

Fourier, Jean-Baptiste-Joseph, 148, 366
  Fourier kernels, 63
  Fourier series, 8, 16, 106, 148, 160, 369
Fourier Series, 106
Frobenius, Ferdinand Georg, 334
functional analysis, 329

Gaitskell, Hugh, 92
Galois, Évariste, 128, 361
Galsworthy, John, 139
Galton, Sir Francis, 359
gamma function, 40
Gauss, Carl Friedrich, 4, 28, 65–66, 68, 127, 147, 148, 180, 186, 200, 202, 203, 211, 254, 275, 288, 289, 339
  Gaussian numbers, 176, 180, 182
genetics, 20, 225, 231–234, 285
Gibbs, Josiah Willard, 165
Girton College, Cambridge, 30–31, 32, 34
Gladstone, William Ewart, 137
Glaisher, James Whitbread Lee, 128–129, 295, 325–327
Gödel, Kurt, 160
Goldbach conjecture, 8, 16, 69, 158, 188, 304
golf, 285–286
Gonseth, Ferdinand, 22
Gonville and Caius College, Cambridge, 37, 49
Göttingen, 330
Göttingen, University of, 291, 330, 334, 338
Goursat, Édouard Jean-Baptiste, 314, 345
Granville, William Anthony, 130
Grattan-Guinness, Ivor, 27, 160
Gratzer, Walter, 150, 160
Greats (Oxford), 250, 253, 266
Greene, [Henry] Graham, 5, 145
Greenhill, Alfred George, 59, 70
Grundlagen der Mathematik, 243

Hadamard, Jacques Salomon, 8, 65, 128, 140, 189, 203, 205, 220, 334, 335, 355–361
  Hadamard's gap theorem, 339

Haemolytic disease, 20
Haldane, John Burdon Sanderson, 48, 78, 92, 289
Halley, Edmond, 11
Halmos, Paul Richard, 155, 156
Hamburg, University of, 292
Hamilton, Sir William Rowan, 140
Hamlet, 21
Handbuch der Lehre von der Verteilung der Primzahlen, 200, 338, 339
Hardy, Gertrude Edith, 5, 34, 35, 79, 80, 115–120, 146, 276
  and cricket, 115, 116
  art, 115, 116
  caring for G. H. Hardy, 118–119
  cricket mishap, 115
  French, 115
  glass eye, 117
  "Hardicanute," 117
  Latin, 116
  mathematics, 116
  teaching, 116–119
  tennis, 116
Hardy, G[odfrey] H[arold],
  "Analytical Geometry of the Plane" lectures, 94
  and anti-Semitism, 27–28, 78
  and atheism, 20
  and cricket, 25, 26, 27, 148–149, 150
  and God, 21–22, 79
  and lists, 24, 26–27
  and religion, 21
  competition against God, 23
  devotion to mathematics, 259
  feud with God, 97
  heart attack, 80
  loyalty to Cambridge, 148–149
  loyalty to Oxford, 148–149
  mutton, 6, 13, 139
  on his life in mathematics, 371–372
  research, 95, 98
  squash, 80
  suicide attempt, 80
  teaching, 33–34, 94, 110–112
Hardy-Littlewood Seminar (see also Conversation Classes), 5, 31–33

Hardy-Weinberg law, 20, 225, 231–234, 285
Harvard University, 12, 47, 69, 77, 79, 84, 133, 141, 291, 292
  Tercentenary Days, 47, 84
Hasse, Helmut, 151, 157
Hauptman, Ira, 3
Hausdorff, Felix, 121
Heath, Sir Thomas, 35–36
Hecke, Erich, 335
Heilbronn, Hans Arnold, 333
Heine, Heinrich, 26, 27
Heisenberg, Werner Karl, 84
Herman, R. A., 103, 256
Hermite, Charles, 360
Heywood, H. Byron, 251, 264
Higman, Graham, 149
Hilbert, David, 24, 37, 129, 134, 159, 160, 193, 243, 293, 302, 303, 329–331, 334, 361, 366
  Hilbert's problems, 329
Hill, A. V., 77
Hilton, Peter J., 142
History of the Theory of Numbers, 166, 167, 185
Hobbes, Thomas, 27
Hobbs, Sir John Berry (cricketer), 78, 151, 272, 282
"Hobbs class," 275, 281, 282
Hobson, Ernest William, 11, 77, 98, 105, 132, 145, 174, 309, 312, 337
Housman, A[lfred] E[dward], 23–24, 129, 135
Hulburt, Lorrain S., 363
Hurwitz, Adolf, 330
Huxley, Aldous Leonard, 111
Huxley, Thomas, 5
hydrostatics, 94
hyperbolic functions, 8
hypergeometric functions, 57

ideal numbers, 196
Indian Clerk, The, 3
inequalities, 16, 89, 225
Inequalities, 145
Ingham, Albert Edward, 64, 71, 337, 340

Institute for Advanced Study, 109
integral equations, 16, 330, 365
International Congresses of Mathematicians, 156, 157, 165, 329
  Bologna (1928), 157
  Oslo (1936), 157
  Paris (1900), 329
  Stockholm (1924) (rescheduled), 156
  Strasbourg (1920), 126, 156
  Toronto (1924), 157
  Zürich (1932), 157, 165
Introduction to the Theory of Infinite Series, An, 318
Introduction to the Theory of Numbers, An, 105, 126
Invariant Society, Oxford, 98
invariant theory, 93, 329, 330
"iterated prisoner's dilemma," 110

Jacobi, Carl Gustav Jacob, 57, 60, 62
James, Henry, 271
James, Ioan M., 146, 150, 158
Jeans, James Hopwood, 326
Johnson, W. Woolsey, 351, 352
Jordan, Camille, 93, 112, 128, 130, 256
Journal of the London Mathematical Society, 16, 96, 219, 225, 326
Jowett, Benjamin, 291, 292
Jung, Carl, 84

Kanigel, Robert, 5–6, 7, 11, 25
Keynes, John Maynard, 11, 13
King's College Chapel, Cambridge, 118
Klein, Felix, 334, 360
Königsberg, University of, 330
Körner, Thomas William, 123, 150
Kô′ sheé, 364
Kowalewski, Gerhard, 131
Kronecker, Leopold, 326
Kummer, Ernst Eduard, 194, 196

Lagrange, Joseph-Louis, 301
  Lagrange's theorem, 301
Laguerre, Edmond Nicolas, 62
  Laguerre's formula, 70

Landau, Edmund Georg Herman, 57, 64, 68, 70–71, 75, 77, 95, 113, 114, 156, 160, 161, 166, 187, 196, 200, 220, 333–341
Landowska, Wanda, 111
Laplace, Pierre-Simon, Marquis de, 57
Leavitt, David, 3, 158
Lebesgue, Henri Léon, 128, 129
   Lebesgue integral, 90, 151
Lefschetz, Solomon, 157
Legendre, Adrien-Marie, 68–70, 186, 202, 203
   Legendre's theorem, 68
Lenin, Vladimir, 78, 151
LeRoy, Edouard, 347
Levinson, Norman, 28–29
Lighthill, [Michael] James, 111–112, 113
Lindelöf, Ernst Leonhard, 347
   Lindelöf's hypothesis, 222, 223
Lindemann, [Carl Lewis] Ferdinand, 174
Linfoot, Edward Hubert, 94, 95, 97
Liouville, Joseph, 174, 175, 326
Littlewood, John Edensor, 7–8, 10, 14, 15, 16, 24, 30, 31–33, 37, 60, 66, 68, 70, 78, 81, 82, 87, 88, 90, 91, 95–96, 98, 101–102, 105–106, 110, 114, 115, 121, 126, 139, 140, 141, 142, 144, 145, 146, 147, 150, 151, 155, 159, 214, 220, 221, 222, 230, 288, 298, 299, 301
Liverpool University, 105
Livio, Mario, 148
Lloyd George, David, 1st Earl, 13
Lobachevskii, Nikolai Ivanovich, 239
London Mathematical Society, 54, 81, 89, 93, 96, 105, 106, 145, 146, 149, 150, 199, 219, 220, 333, 340
London, University of, 115
Longford, Lord (Frank, 7th Earl of Longford), 92
Lord's Cricket Ground, 273, 275
Lorentz group, 356
Love, Augustus, 93, 94, 95, 97, 256
Lucas, Édouard, 200

Macauley, F. S., 102
MacMahon, Percy Alexander, 91, 296

Magdalen College, Oxford, 102, 139, 293
Man Who Knew Infinity, The (film), 3
Manchester University, 79, 104
Mangoldt, Hans Karl Friedrich von, 334
Massachusetts Institute of Technology (MIT), 29, 158
Mathematical Association, 89, 99, 105, 235, 247, 250
Mathematical Association of America, 165, 235, 250
Mathematical Gazette, The, 235
Mathematical Institute (at Oxford), 89, 94, 96, 146, 291, 294
mathematical logic, 329
Mathematician's Apology, A, 3, 4, 5, 13, 20, 37, 81, 88, 106, 110, 112, 123, 134, 145, 146, 147, 148, 158, 267, 268, 271, 275, 287
Mathematician's Miscellany, A, 106, 139, 140, 149, 150, 156
Mathematische Annalen, 344
Mathematische Zeitschrift, 344
Maxwell, James Clerk, 11, 102, 288
   Maxwell's equations, 257
McMurran, Shawnee, 149
Mellin, Hjalmer, 63
Mendelian characters, 20
"Mendelian Proportions in a Mixed Population," 225
Mercer, James, 150
Mersenne, Marin, 212–213
   Mersenne numbers, 212
Messenger of Mathematics, 235, 325, 326
Michelangelo Buonarroti, 26, 27
Michell, John Henry, 316–317, 320, 321
Mill, John Stuart, 237
Milne, Edward Arthur, 93, 96, 97, 149
Milton, John, 24
Minkowski, Hermann, 209, 334
Mittag-Leffler, Gösta, 37, 75, 82, 156, 343–348
   membership in national academies, 340
Mock-Theta functions, 54
Moore, Eliakim Hastings, 8

Mordell, Louis Joel, 11, 48, 58, 77, 96, 146–147, 148, 159, 194, 326, 337
Morrell, Lady Ottoline, 43–44
Mozart, Wolfgang Amadeus, 102
Mugler, Dale H., 142, 144, 146
Mussolini, Benito, 79, 139

Namakkal (goddess), 50, 51
Nandy, Ashis, 83
Nanson, Edward J., 305, 312, 315–316, 317, 321
Netto, Eugen, 296
Neugebauer, Otto, 157
Neville, Eric Harold, 54
New College, Oxford, 6, 21, 73, 74, 77, 90, 92, 94, 95, 98, 99, 105, 146, 149, 150
Newman, Maxwell H. A., 160–161
Newton, Sir Isaac, 5, 6, 11, 19, 20, 21, 44, 250, 254, 275
Nobel Prize, 5, 25, 84, 343
non-Euclidean geometry, 94, 244
Nörlund, Niels Erik, 344
Northwestern University, 142
Norway, King of, 78
"Note on Two Inequalities, A," 225
nuclear engineering, 110

Offord, Cyril, 106
Ohio State University, The, 76
(*On the*) *Origin of Species*, 5
orders of infinity, 40
Osgood, William Fogg, 351, 353, 354
Oslo, University of, 78
Oxford, University of, 291–294
 Exeter College, 293
 Lincoln College, 293
 Magdalen College, 102, 139, 293
 New College, 6, 21, 73, 74, 77, 90, 92, 94, 95, 98, 99, 105, 146, 149, 150
 Oriel College, 293
 Pembroke College, 293
 Trinity College, 293
 University College, 293
 Wadham College, 293
 Worcester College, 293

Oxford School of Mathematics, 292–294
Oxford University Press, 225

Paley, Raymond Edward Alan Christopher, 106
Paris, University of, 291, 330
Partition, 3
partitions of integers, 16, 57, 295, 296, 298
Penrose, Sir Roger, 93, 149
Piaget, Jean, 84
Picard, C. Émile, 130, 335, 336, 340
Plato, 26
Poincaré, Henri, 129, 140, 330, 331, 345, 356–361
Poinsot, Louis, 140
Poisson, Siméon, 140
Poisson's summation formula, 63
Pólya, George, 13, 15–16, 18, 19, 21, 22, 38, 96, 107, 140, 142, 145, 148, 149, 151, 159, 170, 184, 230
Pólya, Stella, 21, 22, 24, 35
Poncelet, Jean Victor, 237
Powers, R. E., 213
prime ideals, 335
prime number theorem, 65–66, 68, 69, 189–190, 200–201, 202, 203, 204, 205, 335, 355
Princeton University, 11, 76, 77, 89, 121, 132, 146, 153, 154, 291, 292
Principia Mathematica, 27, 35, 133
Principles of Mathematics, The, 27
Pringsheim, Alfred, 337
Projective Geometry, 243
Proust, Marcel, 270, 274
Putnam (William Lowell) Mathematical Competition, 249
Pythagoras, 134, 173, 174, 292
 Pythagorean triangles, 175

quadratic forms, 176
quadratic reciprocity, law of, 59
quantum electrodynamics, 110
Quarterly Journal of Mathematics, 89, 96, 326
Quillen, Daniel Gray, 149

Radó, Tibor, 70
Ramanujan, Srinivasa, 24, 36–37, 39,
   40–41, 44, 47–65, 68–71, 73–85,
   76–85, 90, 96, 111, 115, 121,
   127–128, 146, 295, 303, 326, 360
   and Carr's *Synopsis*, 49–50
   Collected Papers, 48, 54, 55, 271
   death, 73, 74–75, 76
   Erode (birthplace), 48
   Government College of Kumbakonam,
   41, 50, 53, 54
   High School of Kumbakonam, 41, 48,
   50, 53, 54
   Kumbakonam, 48, 53, 73
   Madras, 58, 73, 83, 96
   Madras, University of, 53, 70
   marriage, 53
   Namakkal (goddess), 50, 51
   Pachaiyappa's College, Madras, 53
   Port Trust of Madras, 40, 41, 53
   religious views, 51–52
   Subramanyan scholarship, 53
Rao, Ramachaundra, 48, 51, 53
Reid, Constance, 137, 159
Remak, Robert, 303
Rh-blood-groups, 20
Rice, Adrian C., 145, 149
Riemann, [Georg Friedrich] Bernhard, 4,
   17, 18, 127, 147, 148, 203, 220, 222,
   360
   Riemann hypothesis, 16–17, 18, 22,
   29, 70–71, 79, 104, 139, 158, 203,
   221, 337
   Riemann zeta-function, 8, 17, 62–63,
   69, 201, 203, 204, 219–224,
   297–298, 334, 360
Riesz, Marcel, 77, 79, 161
Robbins, Herbert, 35
Roberts, Siobhan, 153
Rockefeller Fellowship, 151
Rogers, Claude Ambroise, 113
Rogers, L[eonard] J[ames], 55, 57, 62, 71
   Rogers-Ramanujan's identities, 61–62
Rogosinski, Werner Wolfgang, 106, 142
Romer, Sir Robert (Lord Justice), 262
Rouse Ball, W[alter] W[illiam], 93, 105,
   213, 254

Rouse Ball Chair at Cambridge, 105,
   334, 340
Rouse Ball Chair at Oxford, 93
Royal Society of London, 5, 11, 38, 54,
   79, 81, 82, 104, 106, 114, 154, 219,
   269, 343
Royden, Halsey Lawrence, 151
Russell, Bertrand Arthur William, 5, 11,
   13, 18, 21, 27, 36, 43–45, 74, 79, 90,
   257, 290, 293
   Principia Mathematica, 27, 35, 133
   Russell's Paradox, 134
Ruth, George Herman "Babe", 76, 133

Sadlierian Chair of Pure Mathematics,
   Cambridge, 11, 77, 80, 98, 105, 106,
   268
St. Catherine's School, 115, 116, 117,
   118, 119
St. John's College, 293
St. Paul's School, 102
Savilian Chair in Geometry, Oxford, 11,
   30, 38, 73, 74, 87, 88, 89, 90–91, 92,
   126, 149, 150, 219
Schnirelmann's Goldbach theorems, 340
Schottky, Friedrich Hermann, 336
Schrödinger, Erwin, 293
Segal, Sanford L., 157
Selberg, Atle, 4, 221
Shah, N. M., 304
Shakespeare, William, 21, 23, 24, 38
Shelley, Percy Bysshe, 24
Siegel, Carl Ludwig,
Skewes, Stanley, 66, 71
   Skewe's number, 66, 71
Smith, David Eugene, 35
Smith, Henry, 127, 207, 291, 292
Smith's Prize, 8, 104, 114, 255
Snow, C[harles] P[ercy], 1st Baron, 3, 8,
   12–13, 74, 77, 79, 80–87, 88,
   109–113, 151, 267, 277, 281, 282
Spring, Sir Francis, 53
Stanley, G. K. (Gertrude), 10, 35, 70
Stockholm University, 156, 161, 343
Stokes, George Gabriel, 102
Strachey, [Giles] Lytton, 11, 152
Stridsberg, Erik, 303

Index 393

Swedish Academy of Sciences, 347
Sylvester, James Joseph, 11, 87, 90, 95, 102, 106, 127, 296
  Sylvester Medal, 79, 106
  Sylvester's theorem, 296
Szegő, Gábor, 170, 184

Tagore, Sir Rabindranath, 24–25
Tamarkin, Jacob David, 157
Tannery, J., 296
Tattersall, James, 149
Tauberian theorems, 153, 160, 299, 337, 340
Taussky-Todd, Olga, 34
Taylor series, 347
tennis, 76, 80, 105, 133, 138
Tennyson, Alfred Lord, 11, 24
Theodorus, 173
Thomson, J[oseph] J[ohn], 73, 326
Thomson, William (Lord Kelvin), 102
Times of London, The, 138
Times crossword, *The*, 119
Titchmarsh, Edward Charles, 11, 13, 16–17, 20, 21, 27, 30, 63, 92, 95–98, 105, 137, 138, 139, 142, 143, 146, 149, 159, 219, 220, 337
transcendental numbers, 172–175
trigonometrical series, 298
Trinity College, Cambridge, 6, 8, 9, 15, 20, 21, 27, 35, 38, 44, 54, 73, 74, 77, 80, 82, 90, 98, 103–104, 105, 106, 109–110, 111, 112, 114, 118, 144, 151, 153, 256, 257, 270, 274, 293, 326
Tripos, Mathematical, The, 7–8, 50, 82, 89, 102, 103, 133, 147, 151, 249–266, 269, 287, 292, 293, 369
Tschebycheff (see under: Chebyshev)
Turing, Alan Mathison, 154

Vallée-Poussin, Charles Jean de la, 65–66, 189, 203, 205, 220, 334, 337
Vaughn-Williams, Ralph, 250
Veblen, Oswald, 76, 89, 96, 133, 154, 157, 209, 243

Vinogradov, I. M., 339, 340
Volterra, Vito, 8, 366
von Staudt, Karl George Christian, 59
von Staudt's theorem (von Staudt-Clausen theorem), 59, 70
Vorlesungen über Zahlentheorie, 166, 338, 339

Wallis, John, 11
Walter, Sir Gilbert, 53
Ward, Morgan, 193
Waring, Edward, 302
  Waring's problem, 8, 10, 16, 91, 113–114, 159, 301–304, 330, 336, 339, 340, 371
Watson, George Neville, 48, 54–55, 58, 62–63, 69–70, 90, 96, 102
Waynflete Chair of Pure Mathematics, Oxford, 93
Weierstrass, Karl Theodor Wilhelm, 345, 360
  Weierstrass product, 335
  Weierstrass's factor theorem, 346
  Weierstrass's M-test, 318
Western A. E., 181, 211, 215
Westfield College, 10
Weyl, [Klaus Hugo] Hermann, 22, 223
What Is Mathematics?, 35, 130
Whipple, F. J. W., 318, 320, 321
Whitehead, Alfred North, 11, 27, 36, 104, 133, 245
  Principia Mathematica, 27, 35, 133
Whitehead, Henry, 149
Whittaker, Edmund Taylor, 8, 49
Wiener, Norbert, 20, 28–29, 79, 81, 121, 139, 144, 150–151, 158, 159–160
Wilcox, Ella Wheeler, 24
Wilson, B. M., 304
Wilson, John,
  Wilson's theorem, 172, 176, 177, 183, 200
Wilson, Robin J., 87, 101, 104, 145, 149, 155
Wiman, Anders, 347
Winchester College, 6, 7, 15, 74, 80, 112, 115, 128, 270
Wittgenstein, Ludwig, 11, 44, 129

Woolf, Leonard, 38
Wrangler, 8, 102, 128, 249, 251, 258
Wrangler, Senior, 8, 10, 103, 151, 257, 262
Wright, E[dward] M[aitland], 70, 71, 95, 106, 126

Yao, Shing Tung, 153
Young, Grace Chisholm, 74, 148
Young, John Wesley, 243
Young, Lawrence Chisholm, 74, 77, 148, 152, 158
Young, William Henry, 74, 90–91, 145, 148,

Zentralblatt für Mathematik, 157
zeta-function (See Riemann zeta-function)
Zoretti, M. Ludivic, 369–370
Zygmund, Antoni, 96

About the Editors

Don Albers served as Director of Publications and Associate Executive Director of the Mathematical Association of America (MAA) from 1991–2006 and Editorial Director of MAA Books from 2006–2012. He was Editor of *The College Mathematics Journal* and Founding Editor of *Math Horizons*. He has written or co-authored six books, including *Mathematical People* and *Fascinating Mathematical People*—both with G. L. Alexanderson. Prior to his association with the MAA, he was Professor of Mathematics and Special Assistant to the President of Menlo College.

Gerald L. Alexanderson is the Valeriote Professor of Science at Santa Clara University, where he has taught mathematics since developing an interest in problems under the guidance of George Pólya and enthusiasm for analytic number theory from a course with Gabor Szegö at Stanford. Long active in the Mathematical Association of America he has served as Secretary and as President, as well as Editor of *Mathematics Magazine*. He is author of many articles and reviews, and has co-authored or co-edited textbooks, problem collections, books of interviews, as well as biographies and books on history. This book is probably number 18 if my count is correct.

William Dunham served as Koehler Professor of Mathematics at Muhlenberg College (*emeritus*, 2014) and taught mathematics history as a visiting professor at Harvard, Princeton, Penn, and Cornell. In 2009, he was a visiting scholar at the University of Cambridge, and the following year he recorded the DVD course "Great Thinkers, Great Theorems" for The Teaching Company. He is the author of four books: *Journey Through Genius* (Wiley, 1990), *The Mathematical Universe* (Wiley, 1995), *Euler: The Master of Us All* (MAA, 1999), and *The Calculus Gallery* (Princeton, 2005). Dunham is currently a Research Associate in Mathematics at Bryn Mawr College.

Printed in the United States
By Bookmasters